Microwave Radar

Imaging and Advanced Concepts

For a listing of recent titles in the *Artech House Radar Library*,
turn to the back of this book.

Microwave Radar

Imaging and Advanced Concepts

Roger J. Sullivan

Artech House
Boston • London
www.artechhouse.com

Library of Congress Cataloging-in-Publication Data
Sullivan, Roger J.
Microwave radar : imaging and advanced concepts / Roger J. Sullivan.
p. cm. — (Artech House radar library)
Includes bibliographical references and index.
ISBN 0-89006-341-9 (alk. paper)
1. Radar. 2. Microwaves. 3. Synthetic aperture radar. I. Title. II. Series.

TK6575 .S82 2000 00-029989
621.3848—dc21 CIP

British Library Cataloguing in Publication Data
Sullivan, Roger J.
Microwave radar : imaging and advanced concepts. — (Artech House radar library)
1. Radar 2. Microwave imaging
I. Title
621.3'848

ISBN 0-89006-341-9

Cover design by Igor Valdman

685 Canton Street
Norwood, MA 02062

International Standard Book Number: 0-89006-341-9
Library of Congress Catalog Card Number: 00-029989

10 9 8 7 6 5 4 3 2 1

To Susan, Andrew, Barbara, Cathy
Philip, John, Betsy, and Betty

In Memory of Barbara, Jack, and Jean

Contents

Foreword

During my 38 years with the Department of Defense (DoD), it was my privilege to be involved one way or another with the development of many air-to-ground intelligence, surveillance, reconnaissance, and targeting radar systems, including the Pave Mover ground moving target indicator (GMTI)/synthetic aperture radar (SAR) system; the stealthy Tacit Blue radar system; the helicopter-borne standoff target acquisition system (SOTAS) GMTI; the operational E-8C Joint Surveillance Target Attack Radar System (Joint STARS), which grew out of Pave Mover; the U-2 advanced SAR system (ASARS) and its current improved version; the high-altitude endurance Global Hawk and DarkStar unmanned aerial vehicle (UAV) radar sensors; and other classified systems. I continue to be impressed at the tremendous growth in the development of radar technology and system capability through outstanding individual contributions to system analyses, electronic and antenna hardware design, digital signal processing, and data analysis.

A few unique people like Dr. Roger Sullivan have the talent to explain and teach the math, physics, and engineering on which these complex radar systems are built. In my last job as Deputy for Technology of the Defense Airborne Reconnaissance Office (DARO), I commissioned Dr. Sullivan to teach a course in advanced radar to interested people on the DARO and National Reconnaissance Office (NRO) staffs. It became apparent that he had unique teaching skills in providing information in a clear and understandable fashion. His course included introductory material as well as systematic, quantitative descriptions of key aspects of modern microwave radar, including detailed treatments of SAR and inverse SAR (ISAR) imaging, airborne-moving-target indication (AMTI), GMTI, and other related radar systems. I take great pride that I encouraged him to put his material into a book, to serve both as a text for the new student as well as a refresher for the seasoned radar engineer. I believe the profession will be greatly served by this book.

In 1997–1998, I called on Dr. Sullivan at the Institute for Defense Analyses (IDA), along with a number of leading radar experts in the Air Force, the Defense Advanced Research Projects Agency (DARPA), the Office of the Secretary of Defense (OSD), the Massachusetts Institute of Technology (MIT) Lincoln Laboratory, MITRE Corporation, and the industry, to assist in what was called the "Advanced Radar Study" that I was privileged to lead at the request of the Office of the Under Secretary of Defense (Acquisition and Technology) and the U.S. Congress. We reviewed classified and unclassified advanced radar technology and reached the conclusion that we are standing on the verge of new exciting advances in radar, even more revolutionary than what we have seen over the past 25 years. The significant advances in future radar are based on recent breakthroughs in active electronically scanned array (AESA) microwave radar hardware and advanced digital processing and have occurred from investments by DoD and the industry in programs like the F-22, F/A-18 E/F, Joint Strike Fighter (JSF), ground-based radar (GBR), and space-based radar programs. Such advances have enabled a new generation of economical, modular/scaleable multimode radars. AESA radars can provide agile beams of simultaneous/interleaved pulse-to-pulse AMTI, GMTI, SAR imaging, and target-recognition waveforms. The radars provide very high radiated power at high efficiency, low loss with manufacturing economy, high reliability, and graceful degradation.

The major technology contributions are:

- Low-cost, high-efficiency microwave monolithic integrated circuit (MMIC) transmit/receive (T/R) modules;
- Planar multilayer manifolds that distribute both dc power, digital controls, and RF signals;
- Microelectro mechanical (MEM) switches that provide high-reliability polarization switching and digital-time-delay modules at each radiating element;
- Frequency hop/chirp exciters/receivers with capability for very wide bandwidth and large tuning range;
- Embedded low-cost digital commercial signal processors, providing 50–100 gigaoperations per second per cubic foot, with the promise of teraoperations per second per cubic foot in the next five years with attendant high-speed memory and interconnection fabric;

- Space time adaptive processing (STAP) for cancellation of moving ground clutter as well as adaptive spatial nulling of jammer and interference signals;
- Algorithms that use the high update rates and accuracy to provide automatic moving-target tracking of a large number (thousands) of moving targets;
- Algorithms for automatic surface and airborne moving target recognition using high-range resolution (HRR) profiles and SAR/ISAR imaging.

It is exciting to see the potential of advances that will revolutionize air and ground battle information. Providing high-resolution and high-update MTI information turns the slowly updated moving "blobs" of today's MTI radars into recognized targets that are tracked over wide areas and provides revolutionary, continuous, real-time information about the nature and dynamics of ground vehicle movements, such as:

- Automatic vehicle count through an area, sorted by direction, vehicle source/destination detection, convoy recognition, and track history of vehicle movements;
- Ultra-high-resolution SAR spot images with two to three times the resolution of current fielded systems;
- Interleaved, simultaneous GMTI, SAR, and target-recognition modes that allow the use of cued high-resolution spot imaging rather than low-resolution wide-area imaging, which is difficult to analyze.

Dr. Sullivan's book, *Microwave Radar: Imaging And Advanced Concepts,* is an outstanding contribution to both radar professionals and students in understanding the principles of the coming revolutionary advances in radar.

John N. Entzminger, Jr.

Preface

This book begins with radar fundamentals, discusses imaging radar, including synthetic aperture radar (SAR) in some detail, and also covers other important topics such as space-time adaptive processing (STAP), air-to-air radar, moving target indication (MTI), bistatic and low-probability of intercept (LPI) radar, and such applications as weather radar and ground-penetrating radar (GPR). It is divided into four major sections:

- Part I: Radar Fundamentals
- Part II: Imaging Radar
- Part III: Pulse-Doppler and MTI Radar
- Part IV: Special Radar Topics (STAP, Bistatic and LPI Radar, Weather Radar, and Ground-Penetrating Radar)

Part I begins with the definitions of frequency and wavelength, and proceeds through: the basic *radar range equation*; radar *antennas* and electronics; propagation through atmosphere, fog, and rain; *radar cross section* of targets; external noise; *detection* of targets in noise and clutter; the *radar ambiguity function*; *pulse-compression*; accuracy of radar measurements; and *monopulse.*

Part II applies these principles to the case of a rotating target, or a fixed scene observed by a moving radar (*synthetic aperture radar—SAR*), to develop the concept of *imaging radar.* The fixed-radar-rotating-target case is typically called *inverse SAR* (*ISAR*). However, this term is misleading, since no mathematical inverse is involved, and ISAR is actually a more fundamental concept than SAR. The discussion begins with ISAR and then proceeds to SAR. Part II also covers various aspects of SAR imagery, including *superresolution, sidelobe-reduction,* and *automatic target-recognition* (*ATR*).

In Part III the discussion shifts to *pulse-doppler radars,* which are designed to detect and analyze moving targets. Part III includes a discussion of phase noise, the use of several *pulse-repetition frequencies* (*PRFs*) with the *Chinese Remainder Theorem* to resolve ambiguities in range and velocity, and the use of airborne radar to detect moving targets in the air or on the ground.

Finally, Part IV covers several special topics of interest in modern radar analysis: *STAP, bistatic radar, LPI radar, weather radar,* and *ground-penetrating radar.*

I hope that student, teacher, and researcher find the book helpful.

Roger J. Sullivan

Acknowledgements

In the preparation of this book, many people provided valuable and essential contributions; without these, the result would not have been possible. First, I express my deep appreciation to the Institute for Defense Analyses (IDA) and its staff for their constant encouragement and support, not only for my writing the book but also for my developing and teaching a radar course that served as its basis.

Special thanks to five individuals for their excellent and extensive accomplishments: Mary Smith and Joanne Aponick, for administrative support during the preparation of the course and the book, respectively; Gary Franklin, for preparing almost all the original figures; Mike Tuley, for graciously reviewing the entire manuscript and making many essential suggestions for improvement; and John Entzminger, both for his initial encouragement and for contributing the Foreword. Furthermore I thank the anonymous Artech House reviewer for suggesting a number of valuable improvements.

I am grateful to Gene Goldstein (Raytheon), Troy Schilling (IDA), and Steve Wilson (University of Virginia) for each carefully reviewing several chapters. For many other important contributions to the book and to the predecessor course, I also thank the following (at IDA unless otherwise noted): Parney Albright, Dale Ausherman (Veridian ERIM International—Veridian EI), Liz Ayers, Gerald Benitz (MIT Lincoln Laboratory—MITLL), Tom Blair, Norman Butman, Bryant Centofanti (Northrop-Grumman Corporation—NGC), Libba Colby, Pat Coleman (MRI, Inc.), Amy Cranford, Shirley Crowell, Amnon Dalcher, Paulette Davis (Whitman, Requardt, and Associates), Stuart DeGraaf (NGC), Dennis Deriggi, Anthea DeVaughan, Ben Edwards, Chuck Everett, Jim Fienup (Veridian EI), John Frasier, Barry Fridling (Joint Theater Air and Missile Defense Organization), Ron Goodman (Veridian EI), Ruth Greenstein,

Dianna Gregory, Don Grissom, Bob Guarino (NGC), David Hart, Kent Haspert, Jim Heagy, Art House, Bill Jeffrey (Defense Advanced Research Projects Agency), Ken Koester (NGC), Kevin Leahy (NGC), Dick Legault, Pat Lequar, Les Novak (MITLL), Ron Majewski (Veridian EI), Phil Major, Keith Meador, Tom Milani, Jeff Nicoll, Sam Park, Richard Perry (Mitre), David Potasznik, Giovanna Prestigiacomo, Jim Ralston, Ken Ratkiewicz, Mike Rigdon, Deanna Saunders, Dan Sheen (Lockheed-Martin), Jim Silk, Maile Smith, Larry Stucki, Mark Stuff (Veridian EI), Dorothy Taylor (MRI, Inc.), Susan Taylor, Jim Ward (MITLL), Lynn Welch, and all those who have participated in the IDA radar course. I also express appreciation to Ron Easley, Andy Schultheis, and the staff of System Planning Corporation, and to Jack Walker and the staff of Veridian ERIM-International, from whom I have learned so much. To these, to the many others who helped and encouraged me, and especially to my wonderful, patient wife Susan, I express my sincere appreciation.

Part I:
Radar Fundamentals

1

Introduction to Radar

1.1 Definition of Radar

Radar is defined as "a device for transmitting electromagnetic (EM) signals and receiving echoes from objects of interest (*targets*) within its volume of coverage" [1]. *Radar* was originally an acronym for *radio detection and ranging.*

Many existing useful systems are described by that same definition, except that they utilize sound waves rather than EM waves. Such systems are typically called *sonar* (or, in some applications, *ultrasound*). By that definition, bats navigate using sonar rather than radar. Sonar principles are very similar to those of radar; however, sonar is beyond the scope of this book.

Unless otherwise specified, this book assumes that EM energy is transmitted from the radar hardware to an antenna, radiated from the antenna, scattered from one or more external objects, collected back at the transmitting antenna, and received by the radar hardware. Such a radar is called *monostatic,* that is, the transmitting and receiving locations are the same. Occasionally, reference is made to *bistatic* radars, in which the transmitting and receiving locations are different, or to *multistatic* radars, which involve transmission from one or more locations and reception at one or more locations. Radar is discussed in a number of books [2–16].

1.2 Brief History

The history of radar, especially the early history, is a fascinating subject in itself. Excellent treatises have been prepared by Budieri [17], Burns [18], Swords [19], and others. Briefly, the history of radar is as follows. In 1886, Heinrich

Hertz confirmed radio wave propagation. The first "radar" was patented in 1904 by Huelsmeyer, of Dusseldorf, Germany. Huelsmeyer called his device "Hertzian-wave projecting and receiving apparatus adapted to indicate or give warning of the presence of a metallic body, such as a ship or a train, in the line of projection of such waves" [18, 20]. Taylor and Young accomplished systematic ship detection at the Naval Research Laboratory, Washington, D.C., in 1922, and Hyland performed the first aircraft detection in 1930. In 1941, a U.S. Army radar detected Japanese planes approaching Pearl Harbor, but the supervisor in charge decided that the signals were spurious. The rapid development of the Chain Home radar system in Britain was essential to Britain's successful defense against air attacks by Germany during World War II. Following the war, radar burgeoned to produce the many types in use today [14].

1.3 Electromagnetic Waves

As James Clerk Maxwell showed in 1865, the fundamental equations of electricity and magnetism (Maxwell's equations) predict the existence of EM waves consisting of fluctuating electric and magnetic fields and propagating with a speed that can be calculated from fundamental electromagnetic properties of free space:

$$\begin{aligned} \epsilon_0 &= \text{permittivity of free space} = 8.85 \cdot 10^{-12}\ \text{kg}^{-1}\ \text{m}^{-3}\ \text{s}^2\ \text{coul}^2 \\ \mu_0 &= \text{permeability of free space} = 4\pi \cdot 10^{-7}\ \text{kg m coul}^{-2} \\ c &= \frac{1}{\sqrt{\epsilon_0 \mu_0}} = 2.998 \cdot 10^8\ \text{m/s} \end{aligned} \tag{1.1}$$

The calculated speed matches the observed speed of light, thus demonstrating that light is an EM wave. Hertz demonstrated the existence of nonvisible EM waves, called radio waves. We now know that EM waves form a continuous spectrum, including radio, infrared, visible, ultraviolet, X rays, and gamma rays. Usually, the term *radar* refers to a system that utilizes radio waves.

This section is a very brief summary of EM waves. A full treatment is available in many texts, for example, Jackson [21] and Stratton [22]. EM theory can be expressed equivalently in several different systems of units; this book uses the rationalized meter-kilogram-second (MKS) system [22, pp. 16–23].

Figure 1.1(a) illustrates a simple electromagnetic wave in a vacuum, as seen at an instant of time. The electric field (**E**) and the magnetic field (**H**) are each sinusoidal in space. (Boldface type indicates vectors.) The two fields

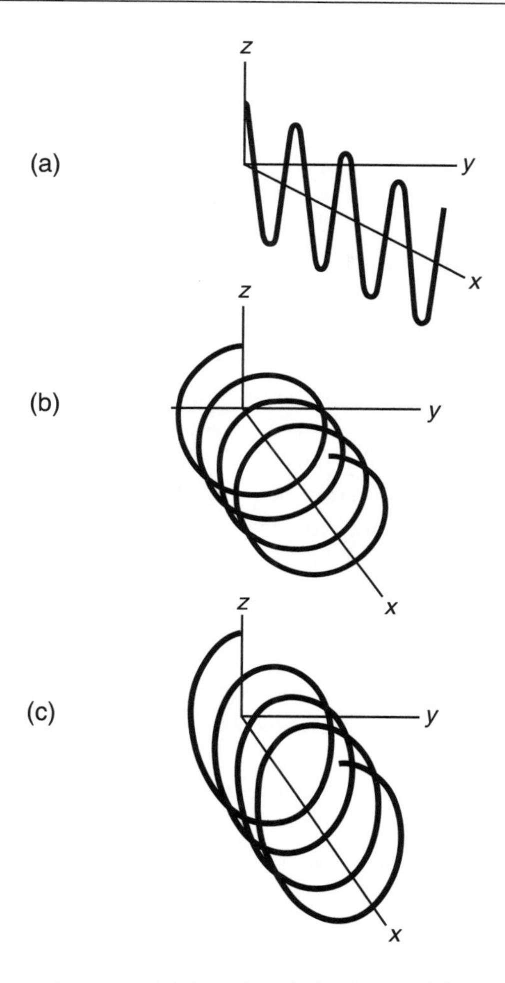

Figure 1.1 Electromagnetic waves: (a) linearly polarized wave; (b) circularly polarized wave; (c) elliptically polarized wave.

are in phase with one another, perpendicular to each other and to the direction of propagation. The number of wave crests per second passing a given point is the frequency, f, measured in cycles per second (hertz). In radar, the frequency is usually called the *carrier frequency*. The distance between adjacent crests is the wavelength, $\lambda = c/f$. The propagation direction is characterized by a vector **k**, known as the *wave number*, where $|\mathbf{k}| = 2\pi/\lambda = 2\pi f/c$, and $\hat{\mathbf{k}} = \hat{\mathbf{E}} \times \hat{\mathbf{H}}$ (^ indicates a unit vector). It can be shown [22] that the instantaneous transmitted flux density (w/m^2) is given by $|\mathbf{S}| = |\mathbf{E} \times \mathbf{H}| = c\epsilon_0 E^2 = c\mu_0 H^2$. **S** is known as the Poynting vector.

E is measured in volts per meter and S is measured in watts per meter-squared; thus, the constant $c\epsilon_0$ has the units of 1/ohms, or siemens:

$$c\epsilon_0 = \sqrt{\frac{\epsilon_0}{\mu_0}} = \frac{1}{377\Omega} \tag{1.2}$$

The reciprocal of $c\epsilon_0$ is sometimes called the *impedance of free space.*

For the case shown in Figure 1.1(a), the direction of **E**, denoted by $\hat{\mathbf{E}}$, is constant in time (except for a ± sign). This direction, $\hat{\mathbf{E}}$, determines the polarization of the wave; if it is constant, the wave is said to be linearly polarized. If a gravitational field is present and the wave is moving approximately perpendicular to it, linearly polarized waves are traditionally characterized as either horizontally or vertically polarized. More generally, in free space **E** (or **H**) may be pointed in any direction perpendicular to the propagation direction, $\hat{\mathbf{k}}$, and it is described by its components in two orthogonal directions perpendicular to $\hat{\mathbf{k}}$. EM waves need not be linearly polarized; in general, the **E** vector may describe an elliptical helix around the direction of propagation, forming an elliptically (or, in a special case, circularly) polarized wave, as shown in Figure 1.1(b) and Figure 1.1(c).

A linearly polarized EM plane wave (i.e., **E** is constant over a plane perpendicular to **k**) can be represented by

$$\mathbf{E}(\mathbf{r}, t) = \mathbf{E}_0 \cos(\omega t - \mathbf{k} \cdot \mathbf{r} + \phi_0) \tag{1.3}$$

where **r** is the three-dimensional spatial coordinate; ω is the angular frequency, measured in radians per second ($\omega = 2\pi f$); t is time; and ϕ_0 is the phase offset (often set to zero). Then $k = \omega/c$ and $c = \lambda f = \omega/k$. The *period* of the wave is $T_p = 1/f = 2\pi/\omega$.

Because the average of the square of a sinusoid is one-half its amplitude, the average flux density (watts per meter-squared) of a plane wave is

$$|S|_{avg} = \frac{c\epsilon_0}{2} E_0^2 \tag{1.4}$$

It is useful to consider the time dependence by representing the cosine as the real part of a complex quantity. In this case, $\mathbf{E}(\mathbf{r}, t)$ becomes

$$\mathbf{E}(\mathbf{r}, t) = Re[\mathbf{E}_0 e^{j(\omega t - \mathbf{k} \cdot \mathbf{r} + \phi_0)}] \tag{1.5}$$

The magnitude at $\mathbf{r} = 0$ is $E(0, t) = E_0 \, \mathrm{Re}[\exp j(\omega t + \phi_0)] = E_0 \cos(\omega t + \phi_0)$. Here, j is the square root of -1. $E_0 \cos\phi$ can be represented as the real part of a rotating phasor in the complex plane, as shown in Figure 1.2, which is known as the Argand diagram. The argument of the cosine is the phase, and ω is the rate of change of the phase:

$$\omega = d\phi / dt \tag{1.6}$$

Loci of constant phase are called *wave fronts*; their orthogonals are called *rays*.

The speed of light (or any EM wave) in vacuum is c = 299,792,458 m/s. That expression is exact and forms the definition of the meter [23]. The value of c is very close to 300,000 km/s; in fact, c = 300,000 km/sec · (0.999308). In almost all cases, engineers use the approximate value, although for certain precise work, the former value is necessary. It is useful to remember that

$$\begin{aligned} c &= 300{,}000 \text{ km/s;} \\ &= 300 \text{ km/ms;} \\ &= 300 \text{ m/}\mu\text{s;} \\ &= 30 \text{ cm/ns } (= 0.984 \text{ ft/ns}). \end{aligned}$$

Table 1.1 is a brief summary of frequency-wavelength combinations of interest to the radar community. Table 1.2 summarizes the nomenclature typically used to describe the various radar "bands." Except for the abbreviations for high frequency (HF), very high frequency (VHF), and ultrahigh frequency (UHF), the letters are deliberately meaningless, having been chosen during World War II to provide unclassified designators for the then-classified frequency intervals.

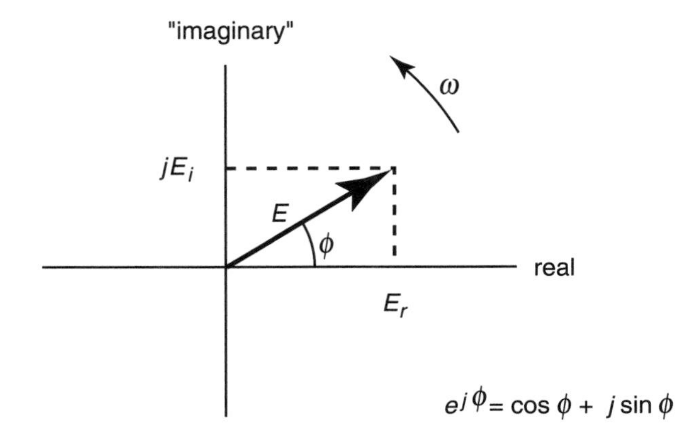

Figure 1.2 The Argand diagram.

Table 1.1
Frequency-Wavelength Combinations

Frequency (f)	Wavelength (λ)
1 MHz	300 m
10 MHz	30 m
100 MHz	3 m
1 GHz	30 cm
10 GHz	3 cm
100 GHz	3 mm

Table 1.2
Radar Bands

Band	Frequency (GHz)
HF	0.003–0.03
VHF	0.03–0.3
UHF (P)	0.3–1
L	1–2
S	2–4
C	4–8
X	8–12
Ku	12–18
K	18–27
Ka	27–40 (usually ~35)
V	40–75
W	75–110 (usually ~95)

1.4 Pulses

All radar technology considered in this book involves microwave radar ($f \sim$ 0.1 to 100 GHz, VHF through W band) with a waveform consisting of discrete pulses, that is, short intervals of transmitted energy followed by longer intervals of no transmission. Not all radars are pulsed; some are continuous wave (CW) [14].

Figure 1.3 illustrates a typical series of pulses. The simple approximation is made that the transmitted power instantaneously rises from zero to a finite value and later instantaneously falls to zero again. Actual pulses must have a finite rise time and fall time, but for many applications the simpler assumption is justified. The transmission time of a single pulse is called its pulse width, typically expressed in nanoseconds or microseconds. In Figure 1.3, it is assumed

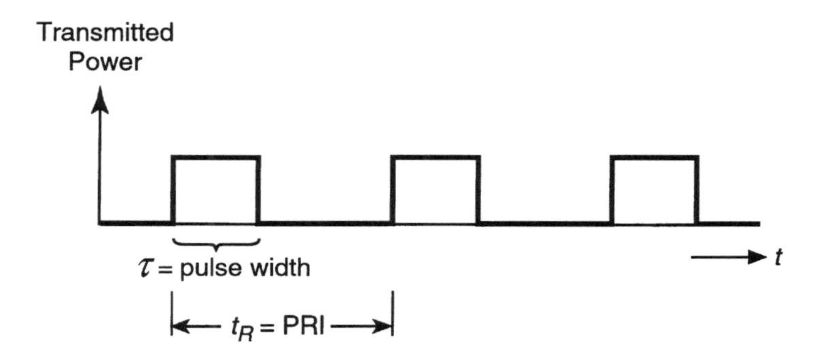

Figure 1.3 Radar pulses.

that all the pulses have the same pulse width, which is usually the case in radars. The time between pulses is the pulse repetition interval (PRI), which is usually (but not always) constant over the time necessary to transmit many pulses. If the PRI, denoted as t_R, is constant over such an interval, the reciprocal of the PRI is known as the pulse-repetition frequency (PRF). Radar PRFs are typically 0.1 to 300 kHz. This book denotes the PRF as f_R and the pulse width as τ. Their product is the fraction of time that the radar is transmitting (or that the waveform contains significant energy) and is known as the duty cycle, f_D:

$$f_D = f_R \tau \tag{1.7}$$

The duty cycle of a radar is typically expressed as a percentage; for pulsed radars, it usually is between 1% and 40%.

The peak power, P_{peak}, is the average power transmitted during a pulse. The average power, P_{avg}, is the transmitted power averaged over an extended series of pulses. For linear polarization, the power actually fluctuates sinusoidally during a pulse with a period = $T_p/2$. For that case, the peak power is not the absolute maximum power transmitted at any instant; rather, it is one-half that value, because the average value of cosine-squared is 1/2.

As an example, a radar designer might choose the following values:

$$\begin{aligned}
\tau &= 100\ \mu\text{s} \\
t_R &= 1\ \text{ms} \\
f_R &= 1\ \text{kHz} \\
f_D &= 10\% \\
P_{\text{peak}} &= 1{,}000\text{W} \\
P_{\text{avg}} &= 100\text{W}
\end{aligned}$$

1.5 Decibels

The ratio of two power levels is frequently expressed in decibels (dB), defined as

$$dB = 10\log_{10}(P_1/P_2) \quad (1.8)$$

For example, if the power ratio is 0.01, we usually say it is –20 dB. The decibel is, as the name implies, one-tenth of a *bel*, a unit named after Alexander Graham Bell, inventor of the telephone. The power ratio in bels would be $\log(P_1/P_2)$.

Table 1.3 summarizes the decibel values for power ratios from 1 to 10. It is interesting to note that many of the values are quite close to integer or half-integer values of decibels. The reader is well advised to memorize the approximate values.

1.6 Antennas

This section presents only enough introductory material to make possible a discussion of the radar equation. Sections 2.2 and 2.3 contain a more detailed discussion of antennas.

An antenna is defined as "that part of a transmitting or receiving system which is designed to radiate or to receive electromagnetic waves" [1]. It is a device through which EM waves transition (1) from being within the radar

Table 1.3
Decibel Values for Power Ratios from 1 to 10

Power Ratio	Decibel Value	Approximate Decibel Value
1	0.000	0
2	3.010	3
3	4.771	5
4	6.020	6
5	6.990	7
6	7.782	8
7	8.451	8.5
8	9.031	9
9	9.542	9.5
10	10.000	10

hardware to being in free space (or air or other medium separate from the radar) or (2) the reverse, that is, from being in free space to being within the radar. Many antennas take the form of a plate or a dish, characterized by a diameter D. For such an antenna, Section 2.2 shows that the pattern of the emitted radiation, as a function of angle, consists of a high-intensity region called the mainlobe, accompanied by adjacent regions of lesser intensity called sidelobes. That typical pattern is illustrated schematically in Figure 1.4. The direction of the mainlobe is called the *boresight* direction. In Figure 1.4, the radial distance from the pattern contour to the center of the antenna is proportional to the intensity of the radiation in watts per steradian. Between the lobes are *nulls*, which are characterized by intensities considerably less than those at the peaks of the lobes. Theoretically, the intensity at a null is zero; in practice, it is some small value.

If the antenna is planar and square with side D, if $\lambda << D$, and if immediately beyond the antenna surface the pattern of radiation intensity (the *aperture. illumination function*) is uniform, then in the plane perpendicular to the face and parallel to one edge, the angle from peak to first null of the mainlobe is λ / D, and the half-power beamwidth (the angle between the two points on the mainlobe where the power is 1/2, or −3 dB, of the power on boresight) is 0.886 λ / D. If the shape is circular with diameter D and the aperture illumination function is uniform, the peak-to-first-null beamwidth is 1.22 λ / D (an equation also seen in the theory of optical systems with circular apertures).

This section uses the following notation, generally based on Chapter 1 of [24]. Section 2.3.2 contains more details.

P_{trans} = transmitted power (watts) accepted by the antenna from the radar hardware (may be either peak or average power, depending on the context).

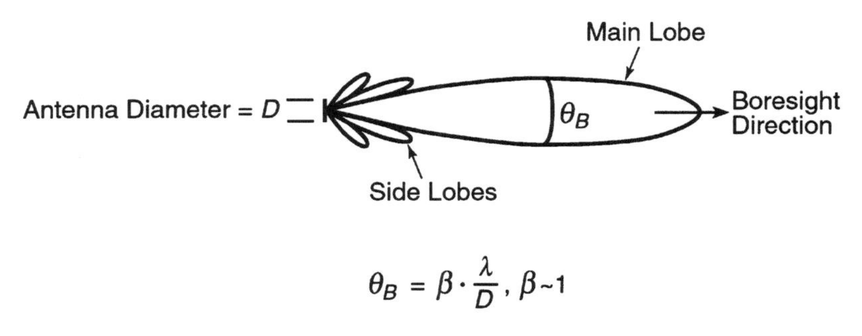

Figure 1.4 Typical antenna pattern.

P_r = power radiated by the antenna into the external environment (watts).

η_r = radiation efficiency = P_r/P_{trans}.

$\Phi(\theta, \phi)$ = radiation intensity (watts/steradian) (θ, ϕ correspond to a spherical coordinate system describing the external environment); $\Phi_{\text{average}} = P_r/4\pi$.

$D(\theta, \phi)$ = directivity of the antenna = $\Phi(\theta, \phi)/\Phi_{\text{average}} = 4\pi\Phi(\theta, \phi)/P_r$.

$G(\theta, \phi)$ = gain of the antenna = $\eta_r D(\theta, \phi)$

$S(\theta, \phi)$ = flux density (watts/meter^2) = magnitude of the Poynting vector [22, p. 133] = $\Phi(\theta, \phi)/R^2$, where R is the distance from the antenna (assumed to be many times the antenna diameter).

$A_e(\theta, \phi)$ = effective area of the antenna.

η = aperture efficiency = $A_e/A \leq 1$, where A_e is measured normal to the aperture and A is the physical area of the aperture.

P_{recd} = power received by the radar hardware after scattering from some object(s) in the external environment and collection by the antenna (may be either peak or average power, depending on the context).

If a radar with transmitted power P_{trans} were to emit its radiation uniformly over all 4π steradians of solid angle, then the power per steradian would be $P_{\text{trans}}\,\eta/4\pi$. Such an antenna (not easily physically realizable) is called an isotropic antenna. For real antennas, the peak power emitted at boresight ($\theta = \phi = 0$) is greater than the corresponding value for an isotropic antenna; the ratio is the directivity. Thus,

$$\Phi(\theta, \phi) = \frac{P_r D(\theta, \phi)}{4\pi} = \frac{P_{\text{trans}} G(\theta, \phi)}{4\pi} \tag{1.9}$$

Whenever microwave radiation passes through any passive electronic component (including passive antennas), it experiences a loss (L) in intensity due to heat dissipation, impedance mismatch, and so on. L is usually expressed as a number greater than 1 by which the input power is divided to obtain the output power. Thus,

$$\frac{G(\theta, \phi)}{D(\theta, \phi)} = \frac{P_r}{P_{\text{trans}}} = \frac{1}{L_{\text{antenna}}} \tag{1.10}$$

It is shown in Section 2.2 that

$$G_0 = G(0, 0) = \frac{4\pi A_e(0, 0)}{\lambda^2} = \frac{4\pi A\eta}{\lambda^2} \tag{1.11}$$

As long as the wavelength is much less than the aperture diameter, this relation is true independent of the aperture shape. The factor η is a product of several factors: $\eta = \eta_r \eta_i \eta_1 \eta_2 \ldots$ [24, Chap. 1]. The factor η_i is the aperture illumination efficiency, the ratio of the actual directivity to the theoretically maximum directivity. The other factors include any other effects that may reduce antenna gain, such as blockage of the aperture by the feed (Section 2.3.2).

The power per steradian transmitted in any direction, θ, ϕ can be expressed as

$$\Phi(\theta, \phi) = \frac{P_{\text{trans}} G(\theta, \phi)}{4\pi} = \frac{P_{\text{trans}} G_0 |f(\theta, \phi)|^2}{4\pi} \tag{1.12}$$

where $f(\theta, \phi)$ is the normalized complex (including phase) radiation pattern pertaining to the transmitted electric field; its magnitude is squared to apply to power. The maximum value of $|f(\theta, \phi)|$ is unity.

Many ways exist for expressing $G(\theta, \phi)$. Two that are popular in radar engineering are decibels relative to the mainlobe and decibels relative to the pattern of an isotropic antenna. The latter method is termed *dB isotropic* (dBi). For example, if the gain is 35 dB and a sidelobe gain is −45 dB relative to the mainlobe, then the sidelobe gain is −10 dBi.

If the transmission is into free space and the energy is allowed to expand freely without obstructions, then at a distance R from the radar many times the antenna diameter (the far field), the flux density (W/m^2) is

$$S(\theta, \phi) = \frac{\Phi(\theta, \phi)}{R^2} = \frac{P_{\text{trans}} G(\theta, \phi)}{4\pi R^2} \tag{1.13}$$

1.7 Radar Cross-Section (RCS)

When an incident flux density S_{incident} strikes an external object and is scattered back to the radar, the object is referred to as the target. Although that term presupposes that the radar energy is deliberately aimed at the object, the term *target* is also used when unexpected objects are detected.

When an EM wave strikes a target, the power scattered per steradian is

$$\Phi_{\text{scattered}} = S_{\text{incident}} \cdot \frac{\sigma}{4\pi} \tag{1.14}$$

The resulting flux density at a range R is given by

$$S_{\text{scattered}} = S_{\text{incident}} \cdot \frac{\sigma}{4\pi R^2} \tag{1.15}$$

where σ is defined as the radar cross section (RCS) of the target. The RCS is "four times the ratio of the power per unit solid angle scattered in a specified direction to the power per unit area in a plane wave incident on the scatterer from a specified direction" [1]. Unless otherwise specified, RCS is assumed to be the value measured when the input radiation is CW; that is, the target has reached equilibrium with the incident radiation, and transients have damped out. RCS is also assumed to be independent of S_{incident} (usually an excellent approximation). Therefore, RCS depends only on the characteristics of the target, not on those of the radar (except its frequency and polarization) or the distance to it. RCS evidently has the dimensions of area and is typically expressed in either square meters or "decibels relative to a square meter" (dBm^2, sometimes denoted dBsm). For example, an RCS of 1,000 m^2 can also be expressed as 30 dBm^2. Also, unless otherwise specified, σ is assumed to be the monostatic RCS corresponding to the backscatter direction, that is, the direction of the radar; more generally, the bistatic RCS corresponds to scattering in any other direction.

If a target scatters all the incident radiation (no absorption or transmission) and scatters it isotropically, that is, into all 4π steradians with equal intensity, then the RCS is the projected area perpendicular to the direction of the incident radiation. Most targets are not of that simple type (although Section 3.1 shows that a perfectly conducting sphere is). For a real target, the RCS can be greater or less than the projected area. We say that a target is "high RCS" if it reflects back to the radar considerably more energy than an isotropic-RCS target with the same projected area. An example is a flat plate perpendicular to the incident radiation. Similarly, a "low-RCS" target reflects considerably less energy than a corresponding isotropic-RCS target, because it reflects energy preferentially away from the backscatter direction, absorbs the energy, or both. In recent years, considerable effort has been expended in attempting to make some military aircraft as low RCS as possible to avoid detection by enemy radar; that technology is known as *stealth*. An interesting discussion of the history of stealth is given in [25].

Because the losses and gains often are not precisely known, we frequently measure the RCS of a target, σ_t(meas), by comparing its apparent RCS (σ_a) with the apparent RCS of a well-understood target (σ_{ca}), the correct RCS of which (σ_c) is known accurately. Then

$$\sigma_t(\text{meas}) = \sigma_{ta} \cdot \frac{\sigma_c}{\sigma_{ca}} \tag{1.16}$$

This procedure is known as calibration of the measured RCS. (The term *calibration* is also used to describe many other types of measurements that relate actual radar performance to absolute standards.)

1.8 Reception of Scattered Energy

EM radiation scattered from a target at range R is collected and measured at the radar. The received radiation is sometimes called an echo. The received power is the returned flux density $S(\text{W/m}^2)$ at the radar multiplied by the effective antenna area A_e, which is less than the actual area. Because $A_e = A\eta$ for a transmitted power P_{trans}, the received power is

$$P_{\text{recd}} = \left(\frac{P_{\text{trans}}G}{4\pi R^2}\right)\left(\frac{\sigma}{4\pi R^2}\right)\frac{A_e}{L} \tag{1.17}$$

where G replaces $G(\theta, \phi)$. We have also included the effect of various losses L (in the electronics, the atmosphere, etc.) that prevent the actual measured received power from being as high as a no-loss version of (1.17) would imply. Utilizing

$$A_e = \frac{G\lambda^2}{4\pi} \tag{1.18}$$

we have

$$P_{\text{recd}} = \frac{P_{\text{trans}}G^2\lambda^2\sigma}{(4\pi)^3R^4L} = \frac{P_{\text{trans}}A^2\eta^2\sigma}{4\pi R^4\lambda^2L} \tag{1.19}$$

which is the expression for the received power (peak or average, depending on context). We also define

$$\frac{P_{\text{recd}}}{P_{\text{trans}}} \equiv \Gamma \equiv \frac{G^2 \lambda^2 \sigma}{(4\pi)^3 R^4 L} = \frac{A^2 \eta^2 \sigma}{4\pi R^4 \lambda^2 L} \tag{1.20}$$

As explained in more detail in Chapter 4, a radar often observes a target for a dwell time t_{dwell}, and then a decision is made concerning the presence or absence of a target of interest. It is also common for a radar to process returned energy from multiple dwells and then to combine the results. An important parameter is the received energy, E_r. For $t_{\text{dwell}} >> \tau$, $E_r = P_{\text{avg}} t_{\text{dwell}} \Gamma$. If the dwell consists of only a single pulse, then $E_r = P_{\text{peak}} \tau \Gamma$.

1.9 Noise

The ability to detect a target in noise is determined by the signal-to-noise ratio (SNR), the ratio of the received energy E_r to the thermal noise power density (power per unit frequency) in the receiver. In any electronic system, noise is always present, due to the random thermal motion of electrons [13, 26, 27]. The noise power density is given by Blake [13; 28, Chap. 2] as

$$\rho_{\text{noise}} = kT_s \tag{1.21}$$

where k is Boltzmann's constant (1.38×10^{-23} J/K) and T_s is the system temperature. (For the moment, assume that this is the actual radar temperature multiplied by a real-world correction factor greater than unity; Section 2.2.11 discusses the concept more fully.) The unit of power density is watts per hertz, which is equivalent to joules.

1.10 Signal-to-Noise Ratio

The SNR is given by Blake [28, Chap. 2] as

$$SNR = \frac{E_r}{kT_s C_B} \tag{1.22}$$

C_B is a number known as the *bandwidth correction factor* or the *filter mismatch factor*. It represents the fact that the received signal must pass through a filter (actually a set of filters: analog filters in the hardware, digital filters in the signal-processing software, or both) before a final decision is made about the presence or absence of a target. The perfect filter is the matched filter, such that $C_B = 1$. Actual filters are, to some extent, nonoptimum, with C_B greater—often much greater—than unity.

1.11 The Radar Equation

We are now ready to state the radar equation, also known as the radar range equation.

Single pulse:

$$SNR \equiv \frac{E_r}{kT_s C_B} = \frac{P_{\text{peak}} G^2 \lambda^2 \sigma \tau}{(4\pi)^3 R^4 k T_s C_B L} = \frac{P_{\text{peak}} A^2 \eta^2 \sigma \tau}{4\pi \lambda^2 R^4 k T_s C_B L} \tag{1.23}$$

Multiple pulses:

$$SNR \equiv \frac{E_r}{kT_s C_B} = \frac{P_{\text{avg}} G^2 \lambda^2 \sigma t_{\text{dwell}}}{(4\pi)^3 R^4 k T_s C_B L} = \frac{P_{\text{avg}} A^2 \eta^2 \sigma t_{\text{dwell}}}{4\pi \lambda^2 R^4 k T_s C_B L} \tag{1.24}$$

The radar equation (or a version thereof) is considered the fundamental equation of radar.

The individual factors in the radar equation are often measured in decibels and, because they are logarithmic, added (rather than multiplied) to obtain the SNR. Table 1.4 illustrates an estimate of the single-pulse SNR of an S-band radar designed to detect small objects at a distance of 20 km.

Chapter 4 shows that, although the single-pulse SNR in Table 1.4 is too low to permit reliable target detection, SNR can be greatly increased if many pulses are used.

Each factor in the radar equation deserves discussion in significantly more detail. Chapters 2 through 4 are arranged so as to accomplish that, as summarized in Figure 1.5.

1.12 Maximum Unambiguous Range

When a radar emits a pulse, energy travels to the target at range R, is scattered, and returns to the radar. The time required for that to occur is roughly $2R/c$. In many (not all) cases, it is desirable that the radar be range unambiguous. To ensure that the observed return is from the most recently transmitted pulse, the radar must wait until the return from pulse n is received before pulse

Table 1.4
Estimate of SNR of an S-band Radar

Numerator	
$P_{peak} = 10^5$ W	50 dB W
$G^2 = 2$ (40 dB)	80 dB
$\lambda^2 = (0.1\ \text{m})^2$	–20 dBm^2
$\sigma = 0.01\ \text{m}^2$	–20.0 dBm^2
$\tau = 100$ ns	–70.0 dB s
Denominator	
$(4\pi)^3 = 1{,}984$	–(33.0 dB)
$R^4 = (20{,}000\ \text{m})^4$	–(172 dBm^4)
$kT_s = (1.38 \times 10^{-23}\ \text{J K}^{-1})$ (580 K)	–(–201.0 dB J)
$C_B = 1$	0 dB
$L = 7$ dB (estimated)	–(7.0 dB)
SNR	9 dB

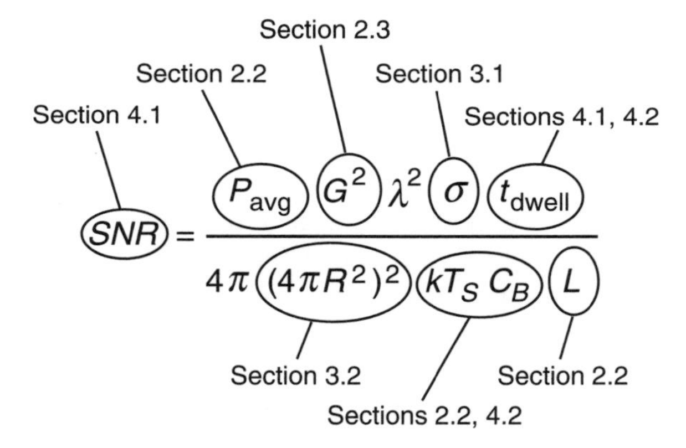

Figure 1.5 Sections in this book explaining the radar equation.

$n + 1$ is transmitted. Of course, that assumes that the user knows, at least approximately, the range to the target of interest. More precisely, $2R/c$ is the time required for the leading edge of the pulse to travel to the target and back. Because most radars cannot simultaneously transmit and receive, the radar must wait for the trailing edge of the returned pulse to be received before it can transmit again (Figure 1.6). Assuming a point target, the required minimum PRI for range-unambiguous operation is

$$\text{PRI(min)} = \frac{2R}{c} + \tau \tag{1.25}$$

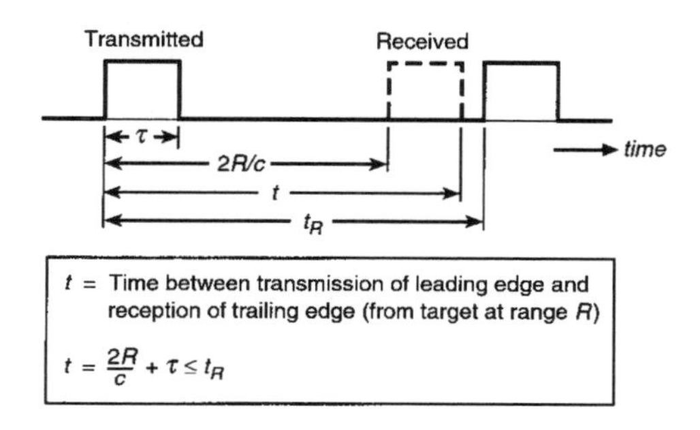

Figure 1.6 Maximum unambiguous range.

For many cases of practical interest, τ is small compared with $2R/c$. In this case, PRI(min) ~ $2R/c$; the maximum PRF for unambiguous range is

$$\mathrm{PRF(max)} \sim \frac{c}{2R} \tag{1.26}$$

and the maximum unambiguous range is

$$R_u \sim \frac{c}{2f_R} \tag{1.27}$$

Table 1.5 lists some values of R_u for several values of f_R.

If a radar is near the ground and observing a target situated near the ground, then even if the PRF is chosen to be below PRF(max) relative to the target of interest, pulse n may travel beyond the target, continuing to produce echoes, and some of those echoes may return to the radar after pulse $n + 1$ has been transmitted. The strength of that type of echo frequently is very low, due to the factor of $1/R^4$ in the radar equation. Nevertheless, the radar designer must consider that type of range ambiguity.

An interesting example of range ambiguity was related by Kettelle [29]. The Ballistic Missile Early Warning System (BMEWS), a system of large radars,

Table 1.5
Values of R_u for f_R

f_R (kHz)	R_u (km)
1	150
10	15
100	1.5

was built in the 1950s to warn the United States of an attack by Soviet Union intercontinental ballistic missiles (ICBMs). The unambiguous range of the radars was 3,000 miles, considered sufficient to measure the range of an ICBM. On the first day of operation, when the moon rose, it scattered the radar energy. The range of the moon was 275,000 miles; however, because of range ambiguity, the echo appeared to be coming from 2,000 miles away. Fortunately, this occurred the week that Soviet Premier Nikita Khrushchev "was at the United Nations in New York, pounding his shoe on the table, and the SAC (Strategic Air Command) general in charge used that fact to reinforce his belief that the alarm was false."

1.13 Coherent Radar

A radar waveform is said to be coherent if a sinusoidal signal is generated by a source within the radar (called a stable local oscillator, or STALO; see Section 2.2) that maintains a constant frequency over many pulses. The phases of returned pulses then can be compared with the STALO phase to determine relative phases between pulses. Ideally, a pulse returning from a point target at range R has its phase delayed (relative to the STALO) by

$$\Delta\phi = -2\pi \cdot \frac{2R}{\lambda} = -\frac{4\pi R}{\lambda} \tag{1.28}$$

Most modern radars are coherent; older radars were not. Coherent operation generally is necessary for the precise measurement of target velocity, as described in the following sections.

1.14 The Doppler Effect

Consider a stationary radar observing a moving point target that has a line-of-sight (LOS) velocity, v. (The total target velocity also may have a cross-track component, which is not relevant here.) A positive v indicates that the target is moving away from the radar. The radar emits a pulse consisting of N cycles during a time τ beginning at $\tau = 0$; thus, $f = N/\tau$. If the target is at range R when the leading edge of the pulse strikes it, then the leading edge echo arrives back at the radar at $t_1 = 2R/c$. The trailing-edge echo arrives at the radar at

$$t_2 = \tau + \frac{2}{c}(R + v\tau) \tag{1.29}$$

The received frequency is, therefore,

$$f_{\text{recd}} = \frac{N}{(t_2 - t_1)} = \frac{N}{\left[\tau\left(1 + \frac{2v}{c}\right)\right]} \tag{1.30}$$

Any macroscopic target must be moving at nonrelativistic speeds; thus, $v << c$, and to an excellent approximation

$$f_{\text{recd}} \cong \frac{N}{\tau}\left(1 - \frac{2v}{c}\right) = f\left(1 - \frac{2v}{c}\right) \equiv f + f_{\text{doppler}} \tag{1.31}$$

Thus, a moving target causes the returned pulse to exhibit a frequency shift:

$$f_{\text{d}} \cong f_{\text{doppler}} = -\frac{2vf}{c} = -\frac{2v}{\lambda} \tag{1.32}$$

The factor of 2 is due to the two-way nature of monostatic radar. If the target is spontaneously emitting the observed radiation (a one-way situation), the shift is $-vf/c = -v/\lambda$. The frequency shift is known as the doppler shift after its discoverer, Christian Johann Doppler (1803–1853).

1.15 Doppler Frequency as Measured by Radar

Most modern radars can precisely measure the relative phases of returned pulses; however, they typically do not directly measure the doppler frequency shift of an individual pulse. When used in a radar context, *doppler frequency* means the "rate of change of relative phase between pulses due to LOS target velocity." That is equivalent to the change in carrier frequency, as we shall now see.

A pulse emitted at time t_1, incident on a moving target that is at R when pulse 1 strikes it, produces an echo with phase change $\Delta\phi_1 = -4\pi R/\lambda$ relative to the STALO signal. The phase change of the echo from a second pulse emitted at $t_2 = t_1 + \delta t$ is

$$\Delta\phi_2 = -\frac{4\pi}{\lambda}(R + v\delta t) \tag{1.33}$$

The relative phase between these two pulses is

$$\delta\phi = \Delta\phi_2 - \Delta\phi_1 = -\frac{4\pi v \delta t}{\lambda} \tag{1.34}$$

As a series of the pulses is collected, the rate of change of phases from pulse to pulse will be

$$\frac{\delta\phi}{\delta t} = -\frac{4\pi v}{\lambda} = 2\pi \cdot f(\text{doppler})$$
$$f(\text{doppler}) = -\frac{2v}{\lambda} \tag{1.35}$$

Therefore, the doppler frequency of a moving target can be (and routinely is) measured by observing the rate of change of the phases of the pulses scattered from the target.

1.16 Maximum Unambiguous Doppler Velocity

If the relative phase of two successive pulses received from a moving target is $\delta\phi = 2\pi n$, with n a positive or negative integer not equal to zero, then the relative phase of the pulses is indistinguishable from $\delta\phi = 0$ and hence ambiguous. The corresponding ambiguous doppler velocities are often called *blind speeds*. We have

$$\delta\phi = 2\pi n = -\frac{4\pi v \delta t}{\lambda} \tag{1.36}$$

where δt = PRI. For any PRI, there is a sequence of blind speeds

$$v_{\text{blind}} = \frac{n\lambda}{2\delta t} = n\frac{f_R \lambda}{2}, \; n = 0, \pm 1, \pm 2, \ldots \tag{1.37}$$

The maximum unambiguous velocity interval Δv_{LOS} is $\Delta v_u = f_R \lambda / 2$.

Although for a given PRF the maximum unambiguous range (R_u) is independent of carrier frequency, maximum unambiguous Δv_{LOS} (Δv_u) is not. The lower the carrier frequency, the higher Δv_u becomes. Table 1.6 lists some typical values of Δv_u (m/sec).

For example, if a stationary X-band radar is designed to measure unambiguously the speed of an oncoming aircraft, the top speed of which is Mach 1, its PRF must be greater than 22 kHz.

Table 1.6
Typical Values of Δv_u

f_R (kHz)	L Band, 1 GHz (m/s)	S Band, 3 GHz (m/s)	X Band, 10 GHz (m/s)
0.2	30	10	3
1.0	150	50	15
10.0	1,500	500	150
100.0	15,000	5,000	1,500

Note: Mach 1 = 330 m/s at sea level; 1 m/s = 1.94 knots $\cong$ 2 knots.

1.17 High, Medium, and Low PRF

As PRF increases, Δv_u increases but R_u decreases. Table 1.7 lists values of f_R, R_u, and Δv_u at X band.

Consider a stationary X-band radar observing a target aircraft at $R = 150$ km, with a v_{LOS} known to be between 0 and 300 m/sec. For unambiguous range measurement, f_R must be less than 1 kHz, which is 20-fold ambiguous in velocity. On the other hand, for unambiguous Δv measurement, f_R must be greater than 20 kHz, which is 20-fold ambiguous in range. Although such ambiguities can be alleviated to some extent via switching between several PRFs (see Section 10.5), designers of radars for observing high-velocity targets (notably military air-to-air radars) must constantly consider that tradeoff. The situation is exacerbated when a radar on a high-speed aircraft is observing another aircraft flying directly toward it, in which case the relative v_{LOS} is approximately twice the v_{LOS} that would occur if the radar were stationary.

The following terminology has been adopted for air-to-air radar to describe those PRFs (note that the actual PRF values depend on range, velocity, and frequency band).

- *Low PRF* is unambiguous range, ambiguous velocity (typically < 4 kHz).

Table 1.7
Values of f_R, R_u, and Δv_u at X Band

f_R (kHz)	R_u (km)	Δv_u (m/s)
1	150	15
10	15	150
100	1.5	1,500 (Mach 4.5)

- *High PRF* is unambiguous velocity, ambiguous range (typically > 100 kHz).
- *Medium PRF* is ambiguous range and velocity (typically 4–100 kHz).

A modern military air-to-air radar uses a number of modes, including high-, medium-, and low-PRF modes. The pilot selects the mode most appropriate for the particular situation.

1.18 Range Gates

By a monochromatic transmitted pulse of width τ, we mean a pulse formed by multiplying a pure continuous tone by a rectangular function. Assume it to have negligibly short rise and fall times. Its leading edge is transmitted at $t = 0$, and its trailing edge is transmitted at $t = \tau$. Its echo is sampled at time T.

For range-unambiguous point targets (from Figure 1.7), the greatest-range observable target is at $R_{\text{greatest}} = cT/2$; a target at a greater range cannot be observed because its echo will not start to arrive until after T. Similarly, the least-range observable target is at $R_{\text{least}} = c(T - \tau)/2$, because pulses returned from targets at lesser ranges will already have ceased to be present at the sampling time T. Thus, if a point target is observed, it must lie in a range interval $R_{\text{greatest}} - R_{\text{least}} = c\tau/2$. That interval is known as a range gate. Radars frequently collect samples from many range gates via multiple time samples.

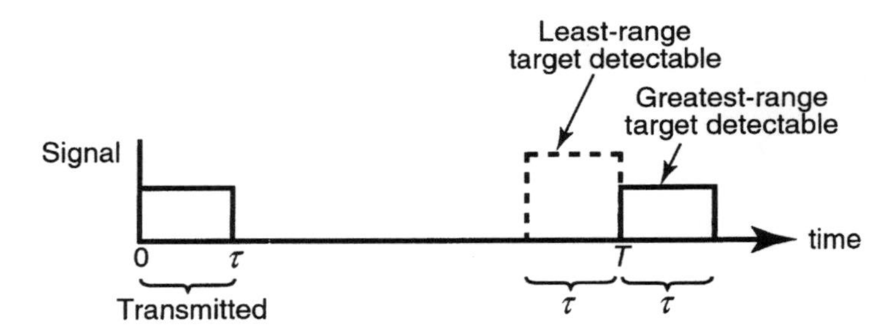

- Greatest-range target detectable: $cT = 2R_g$, $R_g = cT/2$
  Least-range target detectable: $c(T-\tau) = 2R_\ell$, $R_\ell = \frac{c}{2}(T-\tau)$
  $R_g - R_\ell = \frac{c\tau}{2}$

Figure 1.7 Range gates.

1.19 Radar Miscellany

1.19.1 Radar Displays

Radar engineers use the following terminology for various displays of radar outputs.

- *A-display* is echo intensity versus time.
- *B-display* is echo range versus azimuth
- *C-display* is echo azimuth versus elevation
- *Plan-position indicator* (PPI) is the 360-degree polar display of echo range versus azimuth.

Terms for a number of other displays are also used; a full list is given in [14, Sec. 9.5].

1.19.2 U.S. Department of Defense Radar Nomenclature

The U.S. Department of Defense designates radars according to the Joint Electronic Type Designation System (JETDS), formerly known as the Joint Army-Navy (AN) Nomenclature System, for nomenclature of military equipment. U.S. military electronic equipment is typically designated as AN/abc-n, where *a* represents the type of installation, *b* the type of equipment (radars are *P*, since *R* is used for "radio"), and *c* the purpose. For example, AN/APS-137 is an airborne search radar on a Navy P3 aircraft, AN/TPQ-37 is an Army ground-based Firefinder radar used to detect artillery shells, and AN/SPY-1 is the Aegis shipboard surveillance radar on a Navy cruiser. The full nomenclature system is summarized in [28, p. 1–19].

1.19.3 Radar Types

Some common radar types include search radars, which attempt to find targets that may or may not be present; tracking radars, which follow one or more particular targets; fire-control radars, which are operated in conjunction with a weapon to try to destroy an enemy target; and instrumentation radars, which are used to measure properties of targets under the control of the user. A two-dimensional search is performed in one angular dimension (usually azimuth), the two dimensions being angle and range; a three-dimensional search is performed in both azimuth and elevation. Johnston [30] compiled an international radar directory that at one time included 9,800 radars made by 400 manufacturers in 40 countries, most of which have since become obsolete. In 1998, Johnston produced a directory that, among other things, describes specific existing radar types as follows: 109 surface search, 64 two-dimensional air search, 38 three-dimensional mechanically scanned air search, 36 three-dimensional phased arrays (see Section 2.4), 31 multifunction, 26 weapon system, 22 air traffic control, 20 fire control, 18 police, and 18 instrumentation.

1.19.4 Radar Safety

According to the U.S. Air Force, for frequencies between 10 MHz and 300 GHz, personnel should not be continuously exposed to microwave radiation of intensity greater than 100 W/m^2. Further details are provided in [6, Sec. 14-5], which also points out that, on the boresight axis of an antenna with a uniformly illuminated circular aperture of diameter D, the maximum flux density occurs at a distance of approximately $D^2/(5\lambda)$.

References

[1] Kurpis, G. P., and C. J. Booth (eds.), *The New IEEE Standard Dictionary of Electrical and Electronics Terms*, 5th ed., New York: Institute of Electrical and Electronics Engineers, 1993.

[2] Kingsley, S., and S. Quegan, *Understanding Radar Systems*, Mendham, NJ: SciTech, 1999.

[3] Mahafza, B. R., *Introduction to Radar Analysis*, New York: CRC Press, 1998.

[4] Peebles, P. Z., *Radar Principles*, New York: Wiley, 1998.

[5] Raemer, H. R., *Radar Systems Principles*, New York: CRC Press, 1997.

[6] Edde, B., *Radar: Principles, Technology, Applications*, Upper Saddle River, NJ: Prentice Hall, 1993.

[7] Nathanson, F., *Radar Design Principles*,. 2nd ed., New York: McGraw-Hill, 1991.

[8] Toomay, J. C., *Radar Principles for the Non-Specialist*, 2nd ed., New York: Van Nostrand Reinhold, 1989.

[9] Barton, D., *Modern Radar System Analysis*, Norwood, MA: Artech House, 1988.

[10] Levanon, N., *Radar Principles*, New York: Wiley, 1988.

[11] Brookner, E. (ed.), *Aspects of Modern Radar*, Norwood, MA: Artech House, 1988.

[12] Eaves, J. L., and E. K. Reedy (eds.), *Principles of Modern Radar*, New York: Van Nostrand Reinhold, 1987.

[13] Blake, L., *Radar Range-Performance Analysis*, Norwood, MA: Artech House, 1986.

[14] Skolnik, M., *Introduction to Radar*, New York: McGraw-Hill, 1980.

[15] Brookner, E. (ed.), *Radar Technology*, Norwood, MA: Artech House, 1977.

[16] Berkowitz, R. S. (ed.), *Modern Radar*, New York, Wiley, 1965.

[17] Budieri, Robert, *The Invention That Changed the World*, New York: Simon & Schuster, 1996.

[18] Burns, Russell (ed.), *Radar Development to 1945*, London: Peregrinus, 1986.

[19] Swords, S. S., *Technical History of the Beginnings of Radar*, London: Peregrinus, 1986.

[20] Huelsmeyer, Christian, "Hertzian-wave projecting and receiving apparatus adapted to indicate or give warning of the presence of a metallic body, such as a ship or a train, in the line of projection of such waves," U.K. Patent No. 13,170, September 22, 1904.

[21] Jackson, John D., *Classical Electrodynamics*, 2nd ed., New York: Wiley, 1975.

[22] Stratton, J. A., *Electromagnetic Theory*, New York: McGraw-Hill, 1941.

[23] Anderson, H. L. (ed.), *A Physicist's Desk Reference*, New York: American Institute of Physics, 1989.

[24] Johnson, R. C., and H. Jasik (eds.), *Antenna Engineering Handbook*, 3rd ed., New York: McGraw-Hill, 1993. See also Zissis, G. J. (ed.), "Sources of Radiation," Volume 1 of Accetta, J., and D. L. Shumaker, *The Infrared and Electro-Optical Systems Handbook*, Bellingham, WA: SPIE Optical Engineering Press, 1993.

[25] Rich, B., *Skunk Works*, Boston: Little, Brown, 1994.

[26] Johnson, J. B., "Thermal Agitation of Electricity in Conductors," *Phys. Rev.*, 32, 97–109 (July 1928).

[27] Mumford, W. W., and E. H. Scheibe, *Noise Performance Factors in Communication Systems*, Dedham, MA: Horizon House, 1968.

[28] Skolnik, M. I., *Radar Handbook*, 2nd ed., New York: McGraw-Hill, 1990.

[29] Kettelle, J. D., "In Memoriam: Daniel H. Wagner 1925–1997," *Operations Research/Military Science (OR/MS) Today*, Oct. 1997, p. 64.

[30] Johnston, S. J., "The International Radar Directory: Who Makes What and Where," *Proc. 1998 IEEE Radar Conf.*, 1998, pp. 80–85.

Problems

Problem 1.1

Consider a radar with a pulsed waveform (sinusoidally varying electric field) with a peak power of 6 kW. What is the absolute maximum power emitted? Does it depend on frequency? Polarization?

Problem 1.2

Consider a flat-plate antenna (η = 0.5) of area 1 m^2 operating at mid-C band.

a. Calculate its gain.
b. Calculate its peak RCS in the midband of L, S, C, X, and Ku. What is the equation relating its gain to its RCS, as a function of λ?

Problem 1.3

A mid-C band radar has the following:

$$P_{avg} = 1 \text{ kW}$$
$$G = 30 \text{ dB}$$
$$t_{dwell} = 30 \text{ ms}$$
$$T_s = 580\text{K}$$
$$L = 6 \text{ dB}$$

Calculate the maximum range at which it can detect a target of $\sigma = -20$ dBm^2. Assume the SNR required for detection is 17 dB and $C_B = 1$.

Problem 1.4

A pulsed mid-X band radar has a maximum unambiguous velocity interval $\Delta v_u = 600$ m/s. (Assume $\tau << t_R$.)

a. What is its maximum unambiguous range R_u?
b. Suppose its frequency is changed to mid-Ku band and other parameters remain the same. Calculate Δv_u and R_u.

Problem 1.5

Calculate range gate width ΔR in both meters and feet for $\tau = 0.01$, 0.1, 1, and 10 μsec. What is the approximate relationship between τ (nsec) and ΔR (feet)?

Problem 1.6

A lossless radar with gain G and wavelength λ is located at a height H above an infinite, flat, planar, perfectly conducting surface, with boresight directed perpendicular to the surface. Compute Γ.

2

Radar Systems

This introductory discussion of radar is divided into three broad topics:

- How radar waves are generated and transmitted;
- How radar waves interact with external objects and return to the radar;
- How (fundamentally) the returned signals are processed to yield interesting information.

Those topics are discussed in Chapters 2, 3, and 4, respectively. This chapter focuses on generating radar waves, sending them outside the radar hardware through the antenna, and receiving and digitizing the returned echoes.

2.1 Fourier Transforms

This book frequently refers to *Fourier transforms* (FTs). Details of FTs can be found in many references. An extremely abbreviated summary (from Brigham [1]) is as follows. Consider a voltage signal that is a continuous function of time; $s(t)$ is the signal as expressed in the time domain. Joseph Fourier (1768–1830) showed that we can also consider such a signal as the sum of a (finite or infinite) number of signals, each at a different frequency with a particular amplitude. Thus we can express the same signal in the frequency domain as $S(f)$. If $s(t)$ is a uniform tone at frequency f_0, $S(f)$ is zero everywhere except at $\pm f_0$, at which it is a "spike;" that is known as a delta function (see Section 4.2.1 for a discussion of positive and negative frequencies). If $s(t)$ is band limited, that is, contains frequencies only within $\pm B/2$ of $\pm f_0$, then $S(f)$ is nonzero only at $\pm(f_0 \pm B/2)$, and we call B the signal bandwidth.

$s(t)$ and $S(f)$ can be determined from each other via

$$S(f) = \int_{-\infty}^{\infty} s(t)e^{-j2\pi ft}dt \tag{2.1}$$

$$s(t) = \int_{-\infty}^{\infty} S(f)e^{j2\pi ft}df \tag{2.2}$$

$s(t)$ usually is real, and $S(f)$ typically is complex.

The FT relationship is often denoted as

$$s(t) \Leftrightarrow S(f) \tag{2.3}$$

If two signals are multiplied in one domain, then their expression in the other domain is a *convolution*, denoted by the * symbol:

$$\begin{aligned} s_1(t) * s_2(t) &\equiv \int_{-\infty}^{\infty} s_1(\tau)s_2(t-\tau)d\tau \\ s_1(t) \cdot s_2(t) &\Leftrightarrow S_1(f) * S_2(f) \\ s_1(t) * s_2(t) &\Leftrightarrow S_1(f) \cdot S_2(f) \end{aligned} \tag{2.4}$$

If a signal $s(t)$ has frequency components only between $\pm f_1$, and if $s(t)$ is sampled at intervals Δt, producing

$$s_n(t) = s(t - n\Delta t),\ n = 0, \pm 1, \pm 2, \ldots \tag{2.5}$$

then the full signal $s(t)$ may be recovered from the sampled signal $s_n(t)$ as long as the sampling time Δt does not exceed $1/(2f_1)$, that is, if the sampling frequency is not less than $2f_1$ [1, Sec. 5.3]. For a particular signal, the sampling frequency necessary to recover the full signal is called the *Nyquist frequency.* According to [1, p. 83], "the bandwidth of a signal is the width of the positive frequency band where the amplitude is nonzero"; thus, for the example just discussed, $B = f_1$, $\Delta t \leq 1/(2B)$.

2.2 Radar Hardware*

Figure 2.1 is a simplified block diagram of a typical coherent radar. A brief description follows. (Edde [2], Scheer [3], Skolnik [4], and many other references provide a much more complete discussion.) The block diagram illustrates a master-oscillator power amplifier (MOPA) type of radar. Other types are also used, some of which are referred to in subsequent sections.

2.2.1 Oscillators

The *reference oscillator* (RO) is an extremely stable oscillator that provides the basic reference frequency for the radar. ROs usually operate at a frequency between 10 and 100 MHz. A common type uses a piezoelectric crystal. Like the drummer in a jazz band, the RO provides the basic "beat" to which all the other components keep time. The *stable local oscillator* (STALO) is driven by the RO (for maximum stability). It oscillates at a frequency of $f_{LO} = f_{RF} - f_{IF}$, where f_{RF} is the carrier frequency (Section 1.3) and f_{IF} is the *intermediate frequency* (IF). The *coherent oscillator* (COHO), also driven by the RO, oscillates at the IF. f_{IF} is usually (but not always) less than f_{RF}. The RO, STALO, and COHO are referred to as the oscillator group.

Many multimode radars require that pulses have time-varying frequencies or that successive pulses have different frequencies. In such cases, the oscillator group becomes much more extensive. A common design involves an RO driving a step-recovery diode (SRD) or a series of phase-locked loops (PLLs) to generate a series of harmonic frequencies, which then can be multiplied and otherwise combined to produce the desired set of frequencies.

Ideally, an oscillator with frequency f_{RF} would produce a single-frequency tone. In reality, any oscillator produces a signal that includes a noise component with essentially random phase—the phase noise. In frequency space, the FT of the ideal signal would be a spike, or delta function, at f_{RF}, whereas the FT of a real oscillator output exhibits a peak at f_{RF} with "skirts" that fall away on either side. Furthermore, because of nonidealities in the oscillator or associated circuitry, additional spurious peaks, or spurs, may occur at frequencies other than f_{RF} [3, 5] (also see Section 10.2).

2.2.2 Waveform Generation

A control computer or resident software typically provide waveform information about what signals the radar is to transmit, including the frequency, pulse width, PRF, start and stop times, pulse characteristics, and any other relevant

*This section is based largely on material prepared by Dr. James Ralston of IDA. It uses a formalism that considers all frequencies to be positive.

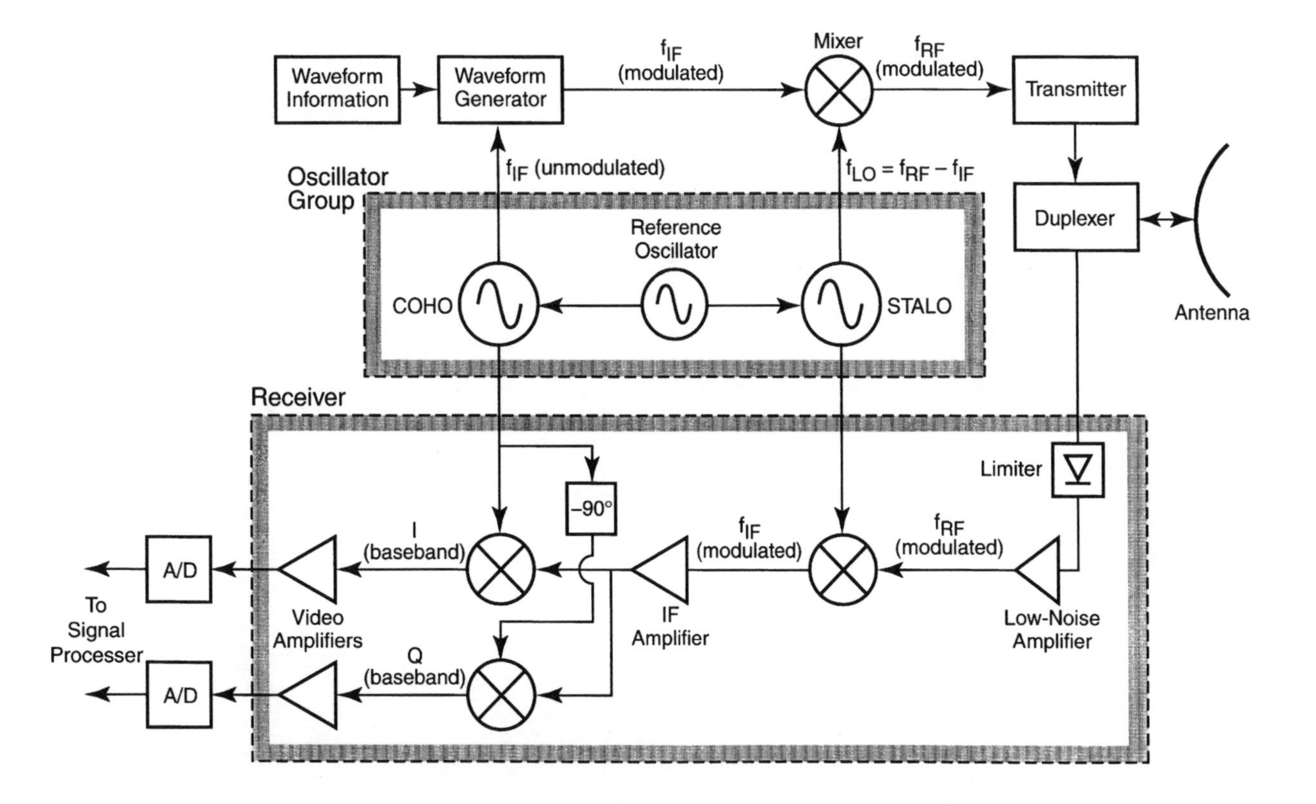

Figure 2.1 Simplified block diagram of typical coherent radar.

details. Frequently the detailed analog waveform information is produced directly from the ouput of a computer via direct digital synthesis (DDS).

The *waveform generator* (or *exciter*) receives the analog waveform information, plus the IF from the COHO, and via a mixer produces a version of the transmitted waveform at IF and low power.

2.2.3 Mixers

A mixer is a device that shifts the carrier frequency of signals, a procedure known as a heterodyne process (from the Greek for *different strength*). As the symbol on Figure 2.1 implies, a mixer is a multiplier. An ideal mixer can be visualized as a device that multiplies two signals, with angular frequencies ω_1 and ω_2. Then, because

$$\cos(\omega_1 t) \cdot \cos(\omega_2 t) = \frac{1}{2}[\cos(\omega_1 + \omega_2)t + \cos(\omega_1 - \omega_2)t] \qquad (2.6)$$

the mixer produces output signals at both the sum and the difference of the two input signals. The output signal is passed through a filter to remove the unwanted signal. Thus, the overall mixer output signal has a frequency that is either the sum or the difference of the frequencies of the two input signals, depending on the application.

2.2.4 Modulation

The IF signal from the COHO (frequency = f_{IF}) is mixed with the analog waveform signal at baseband (average frequency zero), with bandwidth B_{RF}, to produce a signal with average frequency f_{IF} but having been subject to modulation (i.e., multiplication) by the waveform signal. Thus, the upconverted IF signal now occupies the frequency band $f_{IF} \pm B_{RF}/2$. Later in the RF chain are situations in which a modulated carrier experiences demodulation, that is, a mixer is used to discard the carrier frequency, and the modulation signal is downconverted to baseband.

The next mixer combines the waveform generator output with the STALO output to produce a low-power version of the transmitted waveform at the carrier frequency. The process of heterodyning using an IF frequency is called superheterodyning. If no IF frequency is used, the circuit is said to be homodyne (from the Greek for *same strength*).

2.2.5 Transmitter

The low-power (milliwatts) modulated carrier enters the transmitter, which amplifies it to a relatively high power (typically kilowatts). Many types of transmitters are used. The traveling-wave tube (TWT) is widely employed because it is both coherent and relatively wideband, that is, it amplifies the input power with high gain over a relatively wide range of frequencies, typically hundreds of megahertz at X band. Other types of transmitters include the magnetron (noncoherent), klystron, crossed-field amplifier (CFA), and solid-state transmitters [2].

2.2.6 Waveguide

After leaving the transmitter, the high-power microwaves must be "guided" to the antenna with low losses. For that purpose, a *waveguide* usually is used. A waveguide is a hollow metal pipe with a cross-section that usually is rectangular but may be circular or elliptical. The interior may be vacuum, air, or some other dielectric. To find the propagation characteristics inside the waveguide, we solve Maxwell's equations with appropriate boundary conditions. The solution [6, 7] reveals a number of possible modes of increasing frequencies.

In rectangular waveguide, modes are either transverse electric (TE) or transverse magnetic (TM), indicating that the electric (for TE) or the magnetic (for TM) field is entirely transverse to the propagation direction. Figure 2.2 illustrates the cross-section of a rectangular waveguide; the interior dimensions are denoted as a and b, with $a > b$.

For the most common propagation mode in a rectangular waveguide (the TE_{10} mode) , if c_D is the speed of the EM waves in the dielectric ($c_D \leq c$), the minimum frequency that will propagate is the cutoff frequency:

$$f_{\min} = f_c = \frac{c_D}{2a} \tag{2.7}$$

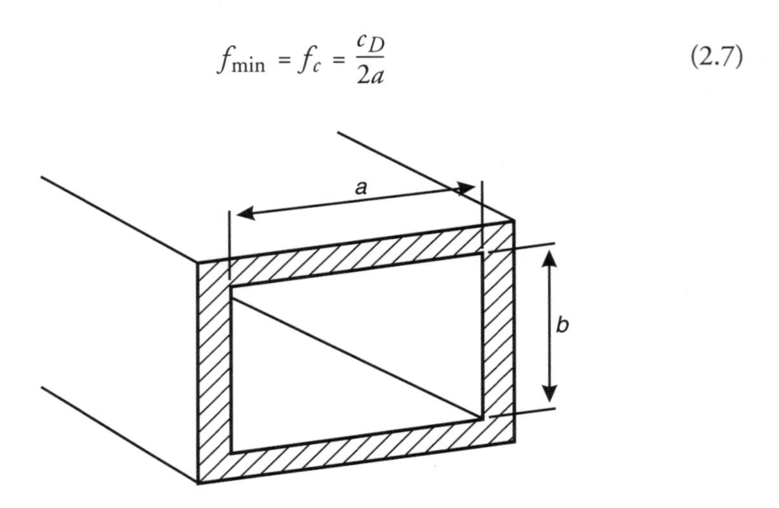

Figure 2.2 Cross-section of rectangular waveguide.

Furthermore, the maximum frequency that will propagate is $f_{max} = c_D/a$. The effective speed of propagation through the guide is the group velocity [6]:

$$c_G = c_D\left[1 - \left(\frac{f_c}{f}\right)^2\right]^{1/2} \tag{2.8}$$

2.2.7 Duplexer

From the transmitter, the high-power RF enters the *duplexer,* which functions as a single-pole, double-throw switch to the antenna. While the pulse is being transmitted, the duplexer connects the antenna to the transmitter. Then, when the returned echo radiation is entering the antenna, the duplexer connects the antenna to the components leading to the receiver. A key function of the duplexer is to keep leakage from the transmitter to the receiver very low. Duplexers usually are either circulators or transmit/receive (T/R) tubes [2].

2.2.8 Antenna

Next, the radiation is emitted from the antenna and scatters from objects in the field of view; some radiation is returned to the antenna. Antennas are discussed in detail in Sections 2.3 and 2.4.

2.2.9 Limiter

The low-power returned radiation first encounters a *limiter,* a nonlinear device that blocks passage of radiation if its power is greater than a level that might damage the delicate receiver components. More generally, in any radar there are a number of points at which strong signals above a certain level, L, are transformed to signals at L, resulting in loss of information about the original signal level. Such signals are said to be saturated. Input signals of too high a power for the overall radar may be saturated at the limiter or possibly at other points downstream in the RF chain.

2.2.10 Low-Noise Amplifier

The signals of most interest frequently are of very low power, and we want to retrieve as much information about them as possible. Therefore, as the returned signal is passed into the receiver, it first encounters a *low-noise amplifier* (LNA).

In the frequency bands above 1 GHz, it generally is possible to obtain receivers for which the sensitivity (minimum detectable signal) in many applications is limited by internal thermal noise. Good radar receiver design aims at ensuring that that sensitivity is minimized (i.e., optimized) and set by the quality of the front-end LNA.

It is convenient to model a real amplifier as an ideal, "noise-free" amplifier with a noise source in parallel with the signal input (Figure 2.3).

The gain of an amplifier is the ratio of the power at the output to the power at the input:

$$G = P_{\text{out}} / P_{\text{in}} \tag{2.9}$$

2.2.11 System Noise

Noise in communications systems, including radar systems, is discussed in [4] and [8–10]. This subsection presents a brief summary. All parameters are assumed to be essentially constant over the applicable range of frequencies (bandwidth).

Consider a radar system that is observing an isotropic scene that, for simplicity, is assumed to be at a constant temperature T_{scene} and constant emissivity ϵ. $T_{\text{scene}}\epsilon$ is known as the antenna temperature, T_{ant}. The antenna and other radar components, up to the input terminals of the LNA, are assumed to be at a constant temperature, T_{radar}, and to produce losses L_{radar} (1 for zero loss, greater than 1 for nonzero loss). T_{rcvr} is defined as the effective input temperature to the LNA if T_{ant} and T_{radar} are zero; T_{rcvr} accounts for noise

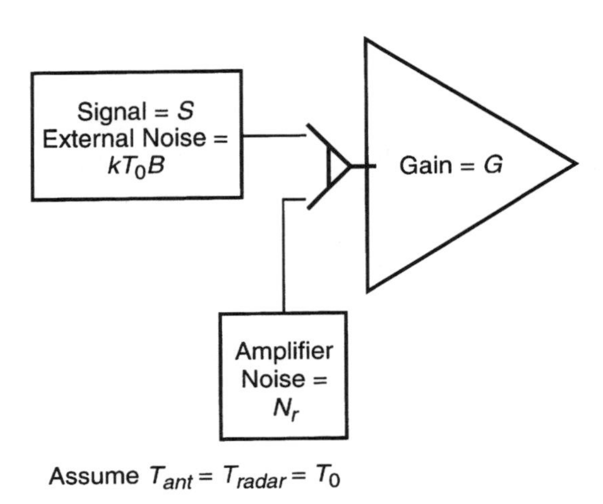

Figure 2.3 Low-noise amplifier.

generated within the LNA itself. Then we define T_{sys}, the system temperature, as the temperature of a matched termination at the input of the LNA that accounts for all the noise at the LNA output. Specifically,

- The noise density (watts per hertz) from the scene is kT_{ant}/L_{radar}, independent of the antenna pattern (see Section 3.3).
- The noise density from the radar (prior to LNA) is $kT_{radar}(1 - 1/L_{radar})$ [10, Eq. 13].
- The noise density from the LNA itself is kT_{rcvr}.
- Total effective noise density at LNA input is

$$kT_{sys} = k(T_{ant}/L_{radar} + T_{radar}(1 - 1/L_{radar}) + T_{rcvr}) \quad (2.10)$$

The noise figure (also called the noise factor), F, is defined in terms of the standard temperature, $T_0 = 290\text{K}$:

$$T_{rcvr} = (F - 1)T_0,\ F > 1 \quad (2.11)$$

It is often assumed that $T_{ant} = T_{radar} = T_0$. If that is the case, the noise density input to the LNA is

$$kT_{sys} = k(T_0/L_{radar} + T_0(1 - 1/L_{radar}) + T_0(F - 1)) = kT_0F \quad (2.12)$$

If, however, those temperatures are different, (2.10) should be used instead (see Problems 2.1 to 2.3). The noise input to the LNA is $kT_{sys}B$, where B is the LNA input bandwidth.

Now consider a network of two amplifiers in cascade (Figure 2.4), and assume that $kT_{sys}B = kT_0FB = kT_0B + N_{r1}$. The second amplifier represents the rest of the receiver, following the initial LNA. The noise at the output of the cascade is

$$\begin{aligned} N_0 &= G_1G_2(kT_0B + N_{r1}) + G_2N_{r2} \\ &= kT_0BG_1G_2 + N_{r1}G_1G_2 + N_{r2}G_2 \end{aligned} \quad (2.13)$$

Dividing by $kT_0BG_1G_2$ (the total power at the output due to source noise) yields the effective system noise figure:

$$F = 1 + \frac{N_{r1}}{kT_0B} + \frac{N_{r2}}{kT_0BG_1} = F_1 + \frac{F_2 - 1}{G_1} \quad (2.14)$$

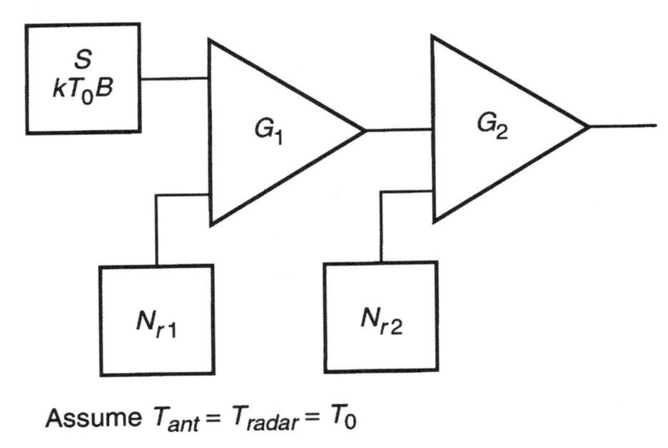

Figure 2.4 Cascaded low-noise amplifiers.

Thus, the system noise figure is dominated by the noise figure of the first LNA. The designer attempts to ensure that internally generated noise of the first stage is, after amplification, much greater than the locally generated noise of subsequent stages, in which case $F \cong F_1$ (see Problem 2.4).

2.2.12 Demodulation

The amplified signal is next mixed with the STALO output to produce a signal with the same modulation but a central frequency of f_{IF}. The latter is typically (but not always) much less than f_{RF}; for example, f_{RF} might be 10 GHz (X band) and f_{IF} might be 500 MHz. The IF generally is chosen to optimize the performance of receiver components. It usually is easier for receivers to operate at megahertz IF frequencies than at either gigahertz carriers or submegahertz frequencies.

The power of the IF signal is further increased via an IF amplifier.

We are interested only in the modulation, not the IF carrier. Therefore, the signal eventually must be demodulated down to baseband. The baseband signal is also known as the video, a term from early radar, referring to the displayed signal.

Assume that the baseband signal is band limited between $\pm f_1$. Then (Section 2.1) sampling can retain all the information in the signal as long as the sampling frequency is at least $2f_1$ (and the sampling accuracy is perfect). The simplest next step after the IF amplifier would be to mix the IF signal to baseband and sample it with an *analog-to-digital (A/D) converter* operating at frequency $2f_1$. Indeed, that is done in some radars. However, $2f_1$ may be several hundred megahertz, and until recently it was difficult to manufacture A/D converters that would operate reliably at such high frequencies.

2.2.13 Quadrature Mixer

It is common for a coherent radar to employ a quadrature mixer, in which the IF signal is mixed with two signals: (1) the IF from the COHO and (2) the IF from the COHO shifted in phase by 90 degrees. The baseband outputs of the two mixers are referred to as the in-phase signal (I) and the quadrature signal (Q), respectively. If the I and Q are each sampled at a rate at least f_1, then again the full signal may be recovered. Thus, by using a quadrature mixer, we can use two A/Ds, each at f_1, instead of one at $2f_1$. Lowering the sampling frequency by a factor of 2 often makes it much easier to find reliable A/Ds.

Consider a quadrature mixer in a homodyne radar; the radar is observing a stationary point target at range R. By the time the transmitted signal has returned from the point target, the phase of the COHO has advanced by an amount $4\pi R/\lambda$ relative to the returned signal. In addition, another arbitrary phase difference, ϕ, typically will be associated with miscellaneous time delays in the circuitry. The I channel of the quadrature mixer then produces the product of the phase-advanced COHO signal (angular frequency ω_0) and the received signal:

$$\begin{aligned} V_I &= \cos\left(\omega_0 t + 4\pi\frac{R}{\lambda} + \phi\right) \cdot \cos(\omega_0 t) \\ &= \frac{1}{2}\cos\left(2\omega_0 t + 4\pi\frac{R}{\lambda} + \phi\right) + \frac{1}{2}\cos\left(4\pi\frac{R}{\lambda} + \phi\right) \end{aligned} \tag{2.15}$$

The Q channel produces the product of phase-advanced COHO signal with a phase shift of –90 degrees, $\sin(\omega_0 t + 4\pi R/\lambda + \phi)$, and the received signal:

$$\begin{aligned} V_Q &= \sin\left(\omega_0 t + 4\pi\frac{R}{\lambda} + \phi\right) \cdot (\cos\omega_0 t) \\ &= \frac{1}{2}\sin\left(2\omega_0 t + 4\pi\frac{R}{\lambda} + \phi\right) + \frac{1}{2}\sin\left(4\pi\frac{R}{\lambda} + \phi\right) \end{aligned} \tag{2.16}$$

For both I and Q channels, the low-pass filter passes only the second term near dc (direct current). Normalizing to unit amplitude,

$$\begin{aligned} V_I &\rightarrow \cos\left(\frac{4\pi R}{\lambda} + \phi\right) \\ V_Q &\rightarrow \sin\left(\frac{4\pi R}{\lambda} + \phi\right) \end{aligned} \tag{2.17}$$

Those I and Q signals then can be considered as the real and imaginary parts of a complex phasor (where $\theta = 4\pi R/\lambda + \phi$):

$$\begin{aligned} P &= \cos\theta + j\sin\theta \\ &= e^{j\theta} = e^{j\frac{4\pi R}{\lambda}} e^{j\phi} \end{aligned} \tag{2.18}$$

Ideally, as R varies over $\lambda/2$, P should trace out an IQ curve that is a circle in the complex plane. Any imbalance between the I and Q channels will result in a distortion in the resulting signals, leading to lack of ideal IQ circularity. Some types of distortion that can occur include the following:

- DC offset: I, Q, or both are nonzero when the corresponding input is zero; the IQ curve is a circle with center displaced from origin.
- Channel imbalance: I and Q amplification factors are not the same, resulting in an IQ curve that is an ellipse with axes parallel to the I and Q axes.
- Nonorthogonality: The phase difference between the I and Q mixers is not exactly 90 degrees; the IQ curve is an ellipse with axes nonparallel to (rotated with respect to) the I and Q axes.
- Nonlinearity over dynamic range of signals: The IQ curve is nonelliptical.

Scheer [3] discusses the consequences of such distortions; a summary is given in Figure 2.5.

2.2.14 A/D Converters

The outputs of the baseband mixer are routed through video amplifiers and then sampled via A/Ds; the resulting digital signals, often called the video phase history (VPH), are used for subsequent signal processing. Frequently, the VPH is recorded on tape or disk. Signal processing may occur immediately or at a later time.

A/Ds are characterized primarily by their number of bits and their maximum sampling frequency. For example, a typical low-cost A/D converter might be 8-bit, 200 MHz. Usually, the greater the number of bits, the longer is the sampling time, and therefore the lower is the sampling frequency. However, demand for more bits and higher sampling frequencies is currently producing a supply of ever more capable A/Ds. Special-purpose A/Ds currently are available with sampling frequency up to 2 GHz and 8-bit resolution.

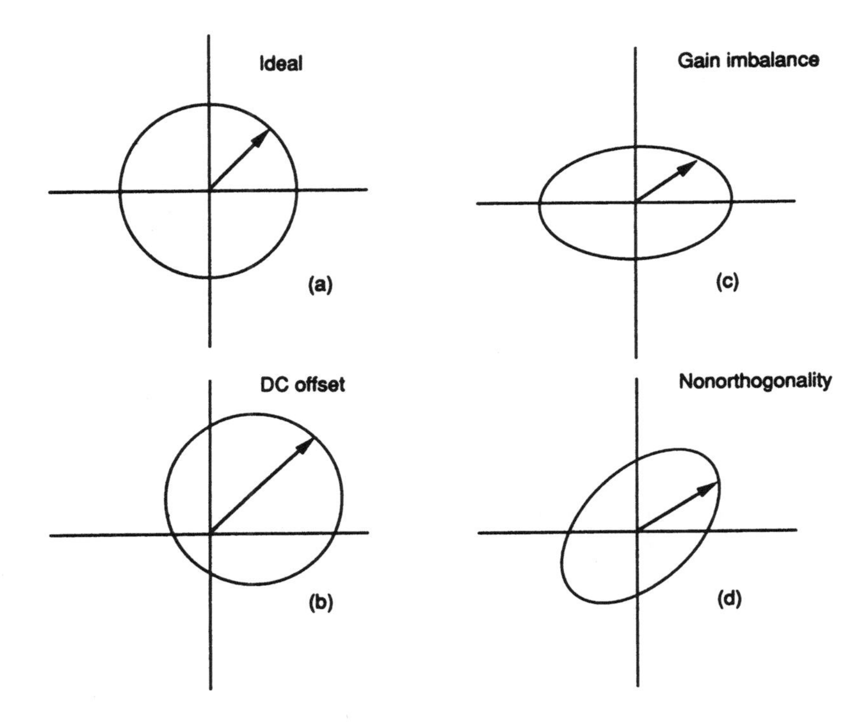

Figure 2.5 IQ distortions. (Reproduced from [3], p. 23, with permission of Artech House.)

The dynamic range of a variable signal is the ratio of its maximum power to its minimum power. Applied to a radar, dynamic range refers to the ratio of maximum power to minimum power that can be detected without being either saturated or lost in the noise. Applied to an A/D, dynamic range is the ratio of the maximum power (P) that can be expressed at full scale to the power represented by the least significant bit. Because A/Ds quantify voltage (V), the maximum voltage value expressible by an A/D converter using n bits to encode amplitude is $2^n - 1$, since one configuration is used for zero. Because $P \sim V^2$, the dynamic range in decibels is

$$\text{dynamic range } (dB) = 10\log[(2^n - 1)]^2 = 20\log(2^n - 1) \cong 20n \log(2) \cong 6n \tag{2.19}$$

Typically, one bit is used for the algebraic sign; thus, an "n-bit" A/D converter uses $n - 1$ bits for signal amplitude, and its dynamic range $\cong 6(n - 1)$.

Because A/D converter resolution is finite, measurements of real voltage values are subject to a round-off error, ϵ, equal to the difference between the exact analog value and its digital representation: $\epsilon = V_{\text{analog}} - V_{\text{digital}}$. The standard deviation (voltage) of the quantization error, in terms of the voltage corresponding to the least significant bit (LSB), is (Problem 2.5)

$$\sigma_{\epsilon} \sim \frac{LSB}{\sqrt{12}} \tag{2.20}$$

The quantization error is thus equivalent to a quantization noise power proportional to

$$P_{nq} \sim \frac{(LSB)^2}{12} \tag{2.21}$$

or about 10.8 dB below the power represented by one LSB.

Quantization error does not establish a fundamental lower limit to the precision of measurements possible with an A/D converter of finite resolution. Provided that quantization error is uncorrelated between samples, it can be reduced (and any bias measured) by coherent averaging of successive samples. To ensure that quantization errors are uncorrelated, the signal applied to the A/D converter should have a random noise component with standard deviation $\cong$ 1 LSB. This is referred to as A/D whitening.

To see why some noise is necessary to measure the bias in an LSB, consider the analogy between the LSB and a coin resting on a table. Suppose we want to use repeated observations to measure the bias (or "trueness") of the coin (probability of obtaining heads by flipping the coin, ideally 1/2). If there is no "noise," that is, if the coin is lying on the table undisturbed, every measurement will yield the same result, and the bias cannot be measured. However, if we add "noise" by banging on the table so that the coin continues to assume new, random states, then with enough measurements we can measure the bias to any desired accuracy.

From the A/D(s), the digitized signal proceeds to the signal processing, which is discussed in subsequent chapters.

2.3 Aperture Antennas

An antenna is a device that serves as the interface between the user's hardware and the "outside world" (the propagation medium). This book refers to the outside world as "free space," with the understanding that the term may include air or occasionally another medium, such as water. The antenna produces an

EM wave in either (1) free space from an EM wave in the hardware (called transmission or transmitting) or (2) the hardware from an EM wave in free space (called reception or receiving).

A number of texts present detailed analyses of the determination of an antenna pattern from Maxwell's equations and other fundamental principles (e.g., [11, 12]). This section is a brief summary.

2.3.1 Introduction to Aperture Antennas

Many antennas can be thought of as flat apertures of a certain shape; common shapes are circular and rectangular. The radar hardware causes time-varying electric (and therefore magnetic) fields to be set up on the aperture. The distribution of the fields over the face of the aperture is the *aperture illumination function*; it may be uniform (constant) or tapered with a lower field strength near the aperture edges. Typically, the electric field (**E**) and the magnetic field (**H**) are parallel to the plane of the aperture (the aperture plane). We denote the pattern of the fields radiated into free space as $f(\theta, \phi)$, where the antenna is considered to be at the origin of a spherical coordinate system. In general, $f(\theta, \phi)$ is complex; the phase represents the relative phase of the radiation as it is emitted in various directions. The pattern typically is normalized, that is, its peak is unity with zero phase, and other values are expressed relative to the peak. That field pattern is sometimes referred to as the antenna voltage pattern to distinguish it from the power pattern $|f(\theta, \phi)|^2$. The term *radiation pattern* can be used to indicate either the voltage pattern or the power pattern; if not otherwise specified, it refers to the power pattern ([13], p. 721).

Throughout this text, it is assumed that the target is in the far field (or Fraunhofer region) of the antenna, that is, the radar-target distance R is great enough that the antenna may be regarded as a point source. In such a case, the wave fronts striking the target are essentially planar. For a target of characteristic dimension L perpendicular to **k**, the Fraunhofer criterion for the far field is that, over the distance L, the wave front deviates from a plane by no more than $\lambda/16$. (Assume that L is greater than the maximum antenna diameter.) That implies that the far field exists for $R > R_{far} = 2L^2/\lambda$ (Problem 2.6.) Regions closer to the radar than R_{far} are known as the near field, which is sometimes divided into the Fresnel region, where the EM fields may be characterized by spherical waves, and the very near field, where the EM fields are quite complex [14].

Long before Maxwell developed his seminal equations, Christiaan Huygens (1629–1695), in his *Treatise on Light* [15], reported his discovery that the future behavior of a wave front can be predicted by considering each instantaneous wave front as a set of sources for new waves. Huygens assumed the existence of "ether" particles everywhere and stated:

> There is the further consideration in the emanation of these waves, that each particle of matter in which a wave spreads, ought not to communicate its motion only to the next particle which is in the straight line drawn from the luminous point, but that it also imparts some of it necessarily to all the others which touch it and which oppose themselves to its movement. So it arises that around each particle there is made a wave of which that particle is the centre . . . And all this ought not to seem fraught with too much minuteness of subtlety, since we shall see in the sequel that all the properties of Light, and everything pertaining to its reflection and its refraction, can be explained in principle by this means. [15]

Indeed, Huygens went on to derive the principles of reflection, refraction (Problem 2.7), and the "strange refraction of Icelandic Crystal" (calcite, $CaCO_3$).

It is now known that a mathematical generalization of Huygens' principle can be derived from Maxwell's equations and is of fundamental importance in EM theory [16]. As shown in Figure 2.6, Huygens' principle can account

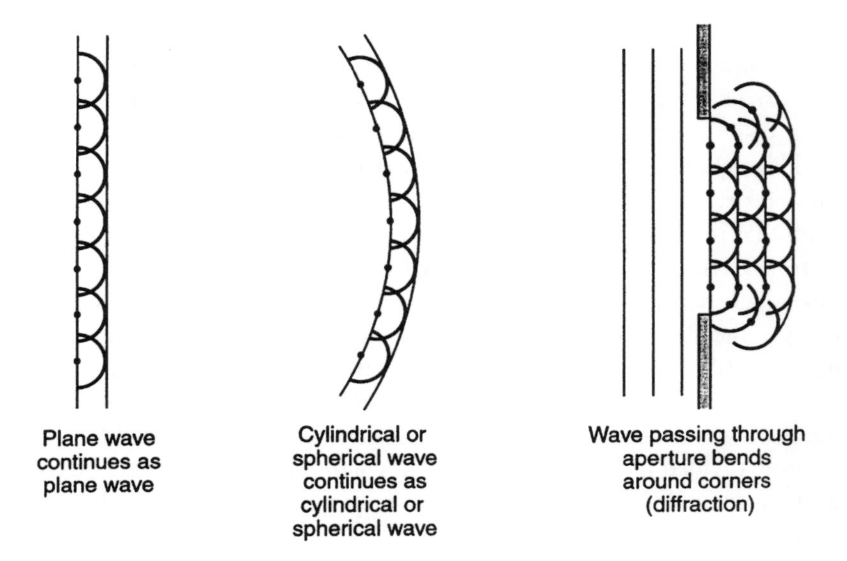

Figure 2.6 Huygens' principle: (a) plane wave continues as plane wave; (b) cylindrical or spherical wave continues as cylindrical or spherical wave; (c) wave passing through aperture bends around corners (diffraction).

for the facts that (1) a plane wave will continue as a plane wave, (2) a cylindrical wave or spherical wave will continue as a cylindrical wave or a spherical wave, respectively, and (3) when a plane wave encounters and passes through an aperture, to some extent it spreads out beyond the aperture, appearing to bend around corners, a phenomenon known as *diffraction* [17].

2.3.2 Radiation Patterns

In discussing antenna radiation patterns, we have a choice of at least two different spherical coordinate systems, as summarized in Figure 2.7. In both cases, the antenna is located at the origin, and we also refer to a Cartesian (x-y-z) system for comparison. Either spherical system is analogous to the latitude-longitude system used to describe the surface of the Earth.

Experimentalists, especially those using antennas in remote locations, generally prefer the azimuth-elevation (AZ-EL) system. The polar axis (z) of the coordinate system is parallel to the aperture plane (in fact, in the aperture plane), and the direction $AZ = EL = 0$ is taken as the boresight direction (perpendicular to the aperture). Assume that the aperture is perpendicular to the Earth's surface; then the equatorial (x-y) plane of the coordinates is horizontal. Azimuth (analogous to longitude) is usually measured from –90 degrees to +90 degrees in the forward hemisphere. If the pattern behind the aperture is of interest (back lobes), azimuth would extend from −180 degrees to +180 degrees. Elevation (analogous to latitude) extends from –90 degrees (nadir) to +90 degrees (zenith).

Theorists generally prefer the theta-phi coordinate system to express antenna patterns, because of its symmetry. In the theta-phi coordinate system, the polar axis (z) is perpendicular to the aperture. In a direction (θ, ϕ), a unit vector is expressed as $\mathbf{r} = u\mathbf{x} + v\mathbf{y} + w\mathbf{z}$, where $u = \sin\theta\cos\phi$, $v = \sin\theta\sin\phi$, and $w = \cos\theta$. It is straightforward to transform from one spherical system to the other via orthogonal rotation matrices [18] (Problem 2.8).

The mathematical formulation of Huygens' principle may be used to derive the voltage radiation pattern from the aperture illumination function, using the theta-phi system. Thus, we make the *physical optics approximation,* which essentially involves assuming the wavelength to be much less than any aperture dimension.

The electric field in the aperture S_a is $\mathbf{E}(x, y, 0)e^{j\omega t} = \mathbf{E_a}(\mathbf{r}')e^{j\omega t}$ where $\mathbf{r}' = x\mathbf{x} + y\mathbf{y}$. As shown in [11, p. 382], we integrate $\mathbf{E_a}$ over S_a and incorporate a specific direction $\hat{\mathbf{r}}(\theta, \phi)$:

$$\mathbf{P}(\hat{\mathbf{r}}) = \mathbf{P}(\theta, \phi) \equiv \int_{S_a} dS'\mathbf{E_a}(\mathbf{r}')e^{jk\hat{\mathbf{r}}\cdot\mathbf{r}'} \tag{2.22}$$

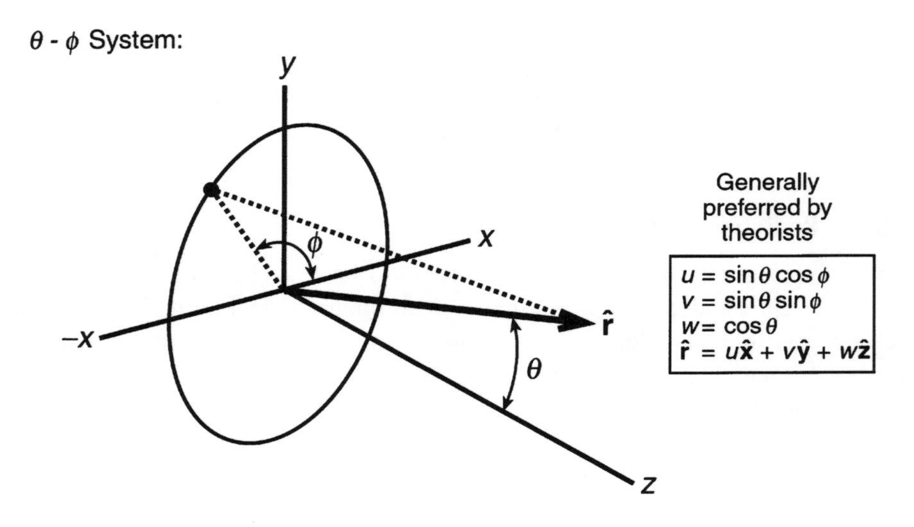

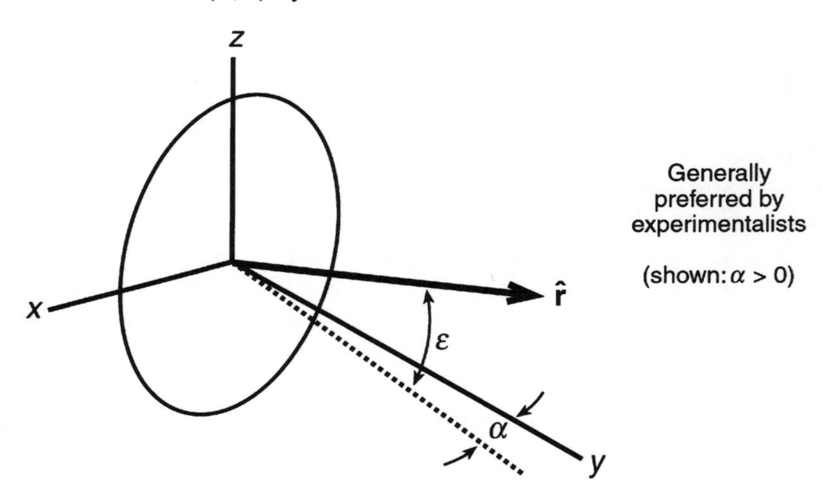

Figure 2.7 Spherical coordinate systems: (a) θ-ϕ system, generally preferred by theorists; (b) azimuth-elevation (α, ϵ) system, generally preferred by experimentalists.

$\mathbf{P}(\hat{\mathbf{r}})$ is an FT of $\mathbf{E_a}(\mathbf{r}')$ with components $P_x(\hat{\mathbf{r}})$ and $P_y(\hat{\mathbf{r}})$. $P_z(\hat{\mathbf{r}}) = 0$ because $E_z = 0$. More specifically,

$$P_{\substack{x\\y}}(\theta, \phi) = \iint_{S_a} dx'dy' E_{\substack{ax\\y}}(x', y') e^{jk(x'\sin\theta\cos\phi + y'\sin\theta\sin\phi)} \qquad (2.23)$$

The radiated electric field pattern is then

$$E_\theta(r, \theta, \phi) = jk\frac{e^{-jkr}}{2\pi r}[P_x(\theta, \phi)\cos\phi + P_y(\theta, \phi)\sin\phi] \qquad (2.24)$$

$$E_\phi(r, \theta, \phi) = jk\frac{e^{-jkr}}{2\pi r}\cos\theta[P_y(\theta, \phi)\cos\phi - P_x(\theta, \phi)\sin\phi] \qquad (2.25)$$

Equations (2.23)–(2.25) can be used to calculate the transmitted electric field from an aperture of any shape and illumination pattern, under the assumptions of $r >> D$ and $\lambda << D$, where D is the characteristic aperture dimension. The equations also describe the gain pattern describing the received electric field.

Using these equations, specific radiation patterns can be derived. In each case, illumination is uniform, and the **E** vector is in the aperture plane. For an infinite slit of width L parallel to the x axis, with $\mathbf{E_{incident}}$ perpendicular to the slit edges, that is, $\mathbf{E_{incident}} = (0, E_y, 0)$, in the yz plane, $E_x = 0$, and (Problem 2.9)

$$E_\theta(r, \theta) = jk\frac{e^{-jkr}}{2\pi r}E_y L \operatorname{sinc}\left(\frac{kL\sin\theta}{2}\right) \qquad (2.26)$$

where $\operatorname{sinc}(x) \equiv \sin(x)/x$. The angular interval from peak to first null is $\theta_{pn} = \lambda/L$.

For a finite rectangular aperture, L_x by L_y, with $\mathbf{E_{incident}} = \mathbf{E_0}$ parallel to **y** (Problem 2.10):

$$E_\theta(r, \theta, \phi) = jk\frac{e^{-jkr}}{2\pi r}E_0 L_x L_y \underline{\sin\phi}\operatorname{sinc}\left(\frac{kL_x}{2}u\right)\operatorname{sinc}\left(\frac{kL_y}{2}v\right) \qquad (2.27)$$

$$E_\phi(r, \theta, \phi) = jk\frac{e^{-jkr}}{2\pi r}E_0 L_x L_y \underline{\cos\theta\cos\phi}\operatorname{sinc}\left(\frac{kL_x}{2}u\right)\operatorname{sinc}\left(\frac{kL_y}{2}v\right)$$

where $u = \sin\theta\cos\phi$, $v = \sin\theta\sin\phi$. The underlined factors are called obliquity factors. In the E plane (yz plane), ϕ = 90 degrees, and

$$E_\theta(r, \theta) = jk\frac{e^{-jkr}}{2\pi r}E_0 L_x L_y \operatorname{sinc}\left(\frac{kL_y}{2}\sin\theta\right),\ E_\phi = 0 \tag{2.28}$$

In the H plane (xz plane), $\phi = 0$ degrees, and

$$E_\theta(r, \theta) = jk\frac{e^{-jkr}}{2\pi r}E_0 L_x L_y \underline{\cos\theta} \operatorname{sinc}\left(\frac{kL_x}{2}\sin\theta\right),\ E_\phi = 0 \tag{2.29}$$

For a circular aperture of radius a, with $\mathbf{E_{incident}} = \mathbf{E_0}$ parallel to the aperture plane (Problem 2.11):

$$\mathbf{E} = (\hat{\boldsymbol{\theta}}\cos\phi - \hat{\boldsymbol{\phi}}\sin\phi\cos\theta)jk\frac{e^{-jkr}}{2\pi r} \cdot E_0 2\pi a^2 \frac{J_1(ka\sin\theta)}{ka\sin\theta}$$
$$f(\theta, \phi) = f_1(\theta) = \frac{2J_1(ka\sin\theta)}{ka\sin\theta} \tag{2.30}$$

where $J_n(x)$ is the nth-order ordinary Bessel function. This pattern has its peak at $\phi = 0$; the angular distance from peak to first null is $\phi_{pn} = (1.22)(\lambda/2a)$ for $\lambda << a$ [11, 17].

For the infinite slit and its characteristic sinc(x) pattern, the first sidelobe is −13.3 dB below the peak; for the circular aperture, the first sidelobe is −17.6 dB below the peak (Problem 2.12).

The effective width of a power-pattern mainlobe is usually defined as the beam width at the half-power points, or 3-dB width θ_{3dB}, of the pattern. Consider the infinite slit with θ small. Then $|f(\theta)|^2 \sim |E_\theta|^2 \sim \operatorname{sinc}^2(kL\theta/2)$. Setting $x = (kL\theta/2)$, we want to find $x = x_{1/2}$ such that

$$\frac{\sin^2 x_{1/2}}{x_{1/2}^2} = \frac{1}{2},\ \text{or}\ \frac{\sin x_{1/2}}{x_{1/2}} = \frac{1}{\sqrt{2}} \tag{2.31}$$

Equation (2.31), although transcendental, can be easily solved iteratively with a hand calculator, yielding

$$x_{1/2} = 1.391557 = \frac{\pi L}{\lambda}\theta_{1/2}$$
$$\theta_{1/2} = \frac{1.391557}{\pi} \cdot \frac{\lambda}{L} \tag{2.32}$$
$$\theta_{3dB} = 2\theta_{1/2} = (0.885893)\frac{\lambda}{L} \cong (0.886)\frac{\lambda}{L}$$

In general, for a sinc function, $\theta_{3dB} = (0.886)\theta_{pn}$, and the first sidelobe peak is −13.3 dB below the mainlobe peak.

The expressions just derived correspond to a one-way power pattern. We are also frequently interested in the two-way pattern. For that case,

$$\frac{\sin^4 x_{1/2}}{x_{1/2}^4} = \frac{1}{2}, \text{ or } \frac{\sin x_{1/2}}{x_{1/2}} = \frac{1}{2^{1/4}}$$

$$x_{1/2} = 1.001906 = \frac{\pi L}{\lambda}\theta_{1/2}$$

$$\theta_{1/2} = \frac{1.001906}{\pi} \cdot \frac{\lambda}{L} \tag{2.33}$$

$$\theta_{3dB,\,2\text{-way}} = 2\theta_{1/2} = (0.637833)\frac{\lambda}{L} \cong (0.634)\frac{\lambda}{L}$$

$$\frac{\theta_{3dB,\,1\text{-way}}}{\theta_{3dB,\,2\text{-way}}} = 1.388911$$

For the circular aperture, if **E** is constant over the aperture, the polarization at any (θ, ϕ) is the projection of **E** onto the sphere at (θ, ϕ).

A common type of circular aperture is the parabolic reflector. The source of microwave radiation, called the feed, is positioned at the focal point of a paraboloid of revolution. The paraboloid is shaped like a dish; the planar surface (usually perpendicular to the axis of revolution) located within the rim of the dish is regarded as the aperture. If we approximate the feed as a point, then radiation emitted from it will reflect from the parabolic dish in such a way as to form a plane wave of constant phase as it leaves the aperture (Problems 2.13 and 2.14).

2.3.3 Gain

From Section 1.6, the gain of an antenna is the ratio of the average power per steradian at the peak of the pattern to the average power per steradian of an isotropic antenna that emits the same overall power. We want to derive the gain, G, of a uniformly illuminated lossless aperture antenna of arbitrary shape, and we assume the pattern peak is in the boresight direction. In the (θ, ϕ) coordinate system, the boresight direction is (0, 0). In that direction, $\hat{\boldsymbol{\theta}}$ and $\hat{\boldsymbol{\phi}}$ are not defined. To alleviate that singularity, we consider a direction slightly off boresight: $\theta = \epsilon$, $\phi = 0$. Then $\hat{\boldsymbol{\theta}} = \hat{\mathbf{x}}$, $\hat{\boldsymbol{\phi}} = \hat{\mathbf{y}}$, $\sin\theta \cong 0$, and $\sin\phi = 0$. With no loss of generality, we take $E_{\text{aperture}} = \hat{\mathbf{x}}E_0$. Then, from (2.23) to (2.25),

$$P_x(\epsilon, 0) = \iint_{S_a} dx'dy' \, E_0 = E_0 A$$

$$P_y(\epsilon, 0) = 0$$

$$E_\theta(r, \epsilon, 0) = E_x(r, \epsilon, 0) = jk\frac{e^{-jkr}}{2\pi r}E_0 A = j\frac{e^{-jkr}}{\lambda r}E_0 A \quad (2.34)$$

$$E_\phi(r, \epsilon, 0) = E_y(r, \epsilon, 0) = 0$$

$$\mathbf{E}(r, \epsilon, 0) = \hat{\mathbf{x}}j\frac{e^{-jkr}}{\lambda r}E_0 A$$

The flux density (W/m^2) $F(r, \epsilon, 0)$ is

$$F(r, \epsilon, 0) = K|E|^2 = \frac{KE_0^2A^2}{\lambda^2 r^2} \quad (2.35)$$

where K is a constant.

Consider a small solid angle $\Delta\Omega$ surrounding the direction $(\epsilon, 0)$ intersecting a sphere of radius r, the center of which is at the origin. The area of the surface of the sphere that is bounded by $\Delta\Omega$ is $r^2\Delta\Omega$. The power incident on this region is $Fr^2\Delta\Omega$, and the watts per steradian is

$$\Phi(r, \epsilon, 0) = \frac{Fr^2\Delta\Omega}{\Delta\Omega} = Fr^2 = \frac{KE_0^2A^2}{\lambda^2} \quad (2.36)$$

Then, from (1.9),

$$G = \frac{\Phi(\text{W/sr})}{P_{\text{trans}}/4\pi} = \frac{KE_0^2A^2/\lambda^2}{KE_0^2A/4\pi} = \frac{4\pi A}{\lambda^2} \quad (2.37)$$

Thus, the gain is independent of the aperture shape.

2.3.4 Tapered Apertures

If the aperture illumination function is uniform, the resulting radiation pattern typically will have relatively high sidelobes compared to patterns formed using tapered illumination functions. For that reason, the illumination function usually is chosen to be tapered. With tapering, we pay the price of a broader mainlobe and lower gain to receive the benefit of lower sidelobes. The radiation pattern for the tapered illumination function usually results in a broader mainlobe and lower sidelobes than the radiation pattern for the uniform illumination function. Typical tapering, or weighting, functions involve cosine functions raised to the nth power, sometimes including a "pedestal" such that they go to a constant at the aperture edge rather than zero. Common tapering functions include Hamming, Hann ("Hanning"), and Taylor. Harris [19] presents a useful summary of tapering functions. The Chebyshev function is ideal in that all its sidelobes are of uniform height [20, p. 4–9].

For a tapered aperture function, θ_{3dB} is broader than for a nontapered aperture; a typical value is $\theta_{3dB} = \beta\lambda/L$, where $\beta \sim 1.2$. Figure 2.8 summarizes the tradeoff between coherent gain (1.0 for a uniform aperture function) and β for various tapering functions (applicable to a one-dimensional aperture, one-way pattern). For uniform tapering, Taylor tapering (–35 dB,

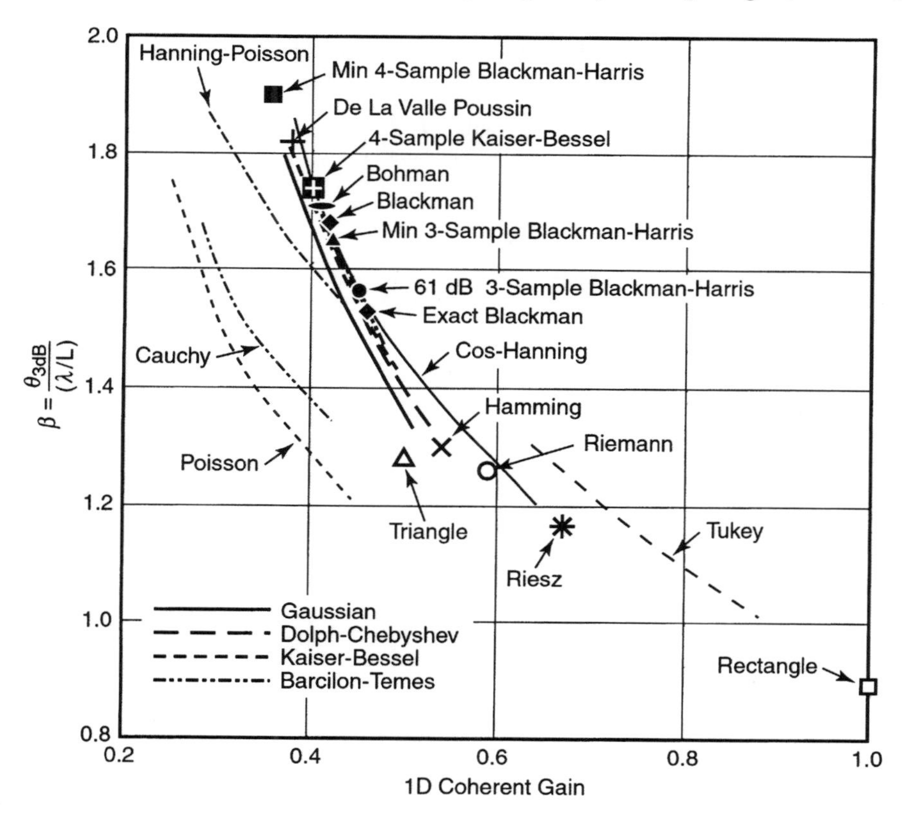

Figure 2.8 Parameters of tapering functions. (Graph prepared by Dr. Kent Haspert of the Institute for Defense Analyses, based on [19].)

nbar = 5), and Hann tapering, θ_{3dB} is 0.886, 1.19, and 1.43 times λ/L, respectively [21 (Also see Section 7.3.1)].

By judicious use of tapering, antenna sidelobes can be made quite low. Table 2.1 summarizes the definitions of low, very low, and ultralow antenna sidelobes.

2.3.5 Cosecant-Squared Antenna Pattern

Referring to Figure 2.9, consider an airborne radar at point A, at altitude h, observing the point P on the surface of the earth (here we make the "flat-earth" assumption). The subradar point is O, $\overline{AP}$ = slant range = R_s, $\overline{OP}$ = ground range = R_g. The angle between $\overline{AP}$ and the ground is the grazing angle ψ for flat earth:

$$\psi(\text{flat-earth}) = \tan^{-1}(h/R_g) = \sin^{-1}(h/R_s) \qquad (2.38)$$

Table 2.1
Definitions of Sidelobe Levels

Antenna Descriptions	Sidelobe Levels dB Below G_m Peak	Sidelobe Levels dB Below G_m Average	dB Below Isotropic Average*
Normal	>–25	>–30	>–3
Low-sidelobe	–25 to –35	–35 to –45	–3 to –10
Very-low-sidelobe	–35 to –45	–45 to –55	–10 to –20
Ultralow-sidelobe (ULSA)	<–45	<–55	<–20

*Averages apply to the worst 30-degree sector starting 10 degrees from the mainlobe.
Source: Reproduced from [9], p. 188, with permission of Artech House.

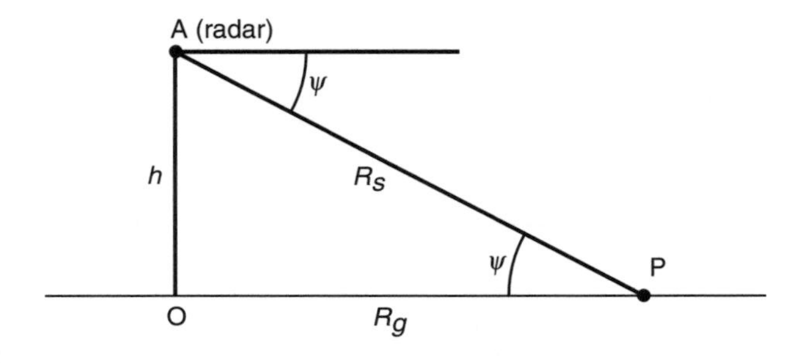

Figure 2.9 Geometry for cosecant-squared antenna pattern.

and the depression angle (angle between AP and the horizontal at A) is equal to ψ. The received power from P as a function of ψ is

$$P_{\text{recd}}(\psi) = \frac{P_{\text{trans}} G^2 \lambda^2 \sigma}{(4\pi)^3 R^4 L} \tag{2.39}$$

If the radar is observing a target with RCS independent of ψ, then

$$P_{\text{recd}}(\psi) = K \cdot \frac{G^2(\psi)}{R^4(\psi)} \tag{2.40}$$

where K is a constant.

In many applications, we want an antenna pattern such that $P_{\text{recd}}(\psi)$ is constant over a wide range of values of ψ (though clearly not extending to $\psi = 0$). For such a pattern,

$$G(\psi) \sim R^2(\psi) = \frac{h^2}{\sin^2 \psi} = h^2 \csc^2 \psi \tag{2.41}$$

That is referred to as a cosecant-squared ($\csc^2$) antenna pattern. Ground-based radars designed to search the sky for aircraft flying at some maximum altitude also often use antennas with patterns approximating the $\csc^2$ pattern.

Many real antennas are designed so that their patterns will approximate the $\csc^2$ pattern. L. J. Chu designed a specially shaped reflector to achieve (approximately) such a $\csc^2$ pattern [22, pp. 494–509].

2.4 Phased-Array Antennas

"Old radar types never die; they just phase array."

Anonymous, quoted by Fowler [23]

So far, we have considered aperture antennas such that, at a particular time, the entire aperture is radiating with a uniform phase. It is frequently advantageous to use an aperture such that different parts are radiating with different variable phases; in that way, a beam can be steered, over times on the order of milliseconds, without any motion of the physical antenna. The most effective method for producing such an antenna is to use a relatively large number of small radiating elements. The phase of each is separately controlled to achieve the desired beam direction. A phased array antenna, often called an electronically scanned array (ESA), is a configuration of coordinated radiating elements designed to achieve that effect [24].

In a *passive array,* the microwave radiation is produced by a central transmitter and delivered via waveguide or through space (a space-fed array) to each element, which contains a phase shifter [25]. In an *active array,* each element contains its own transmit-receive (T/R) module [26], which produces the microwave radiation locally within the element (this is a non-MOPA radar). In either case, given the desired (θ, ϕ) versus time of the beam, a beam-steering computer (BSC) computes the phase versus time of each element.

Again following [11], we denote the voltage radiation pattern from a single element as the element factor, which results from the illumination function of the individual element (element function). We also consider the voltage radiation pattern that would result from a set of isotropic point emitters located at the configuration of element positions; this is the array factor. We can calculate the array factor from the configuration of element locations and their amplitudes and phases.

As with aperture antennas, the voltage radiation pattern (in "beam space") is the FT of the aperture illumination function. From FT theory, a convolution in one domain transforms to a product in the other domain [1]. Because the aperture function can be considered the convolution of the element function and the array function, the voltage radiation pattern is the product of the element factor and the array factor. We assume that the element function is real and that the element factor approximates an isotropic antenna, a very simple pattern. For that reason, the remainder of this section is concerned only with the array factor.

2.4.1 One-Dimensional Line Array

We first consider a line array (a line of point isotropic sources in space) containing N elements, numbered with $0, 1, \ldots, N-1$, with spacing d between them. The array factor $F(\theta)$ is determined by computing the relative strength of the radiated field at angle θ from the array, by adding the contributions of all elements, each with the appropriate phase (Figure 2.10):

$$F_A(\theta) = \sum_{n=0}^{N-1} I_n e^{jknd\cos\theta} \tag{2.42}$$

We now make two assumptions:

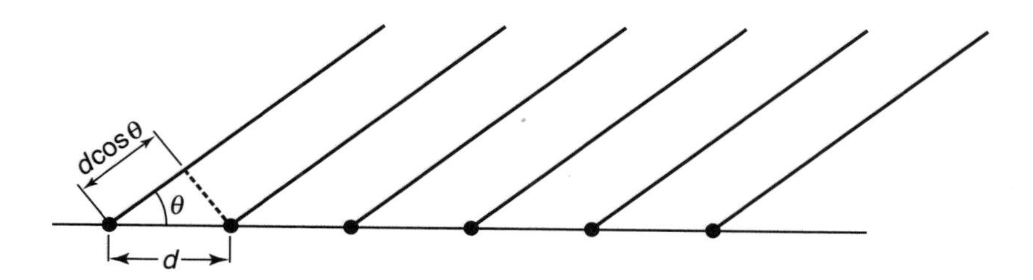

Figure 2.10 The array principle.

1. Phase progression is linear: $I_n = C_n e^{jn\alpha}$, C_n is real,
2. The array is uniformly excited: $C_n = C$, all n.

Then

$$F_A = C \sum_{n=0}^{N-1} e^{jn(kd\cos\theta + \alpha)} \equiv C \sum_{n=0}^{N-1} e^{jn\psi(\theta)} \tag{2.43}$$

where

$$\psi(\theta) \equiv kd\cos\theta + \alpha \tag{2.44}$$

Because the sum of a geometric series is

$$\sum_{n=0}^{N-1} c^n = \frac{c^N - 1}{c - 1} \tag{2.45}$$

then

$$\begin{aligned}\sum_{n=0}^{N-1} e^{jn\psi} &= \frac{e^{jN\psi} - 1}{e^{j\psi} - 1} = \frac{e^{jN\psi/2}}{e^{j\psi/2}} \cdot \left[\frac{e^{jN\psi/2} - e^{-jN\psi/2}}{e^{j\psi/2} - e^{-j\psi/2}}\right] \\ &= e^{j(N-1)\psi/2} \cdot \frac{\sin(N\psi/2)}{\sin(\psi/2)}\end{aligned} \tag{2.46}$$

and

$$|F_A| = C\frac{\sin(N\psi/2)}{\sin(\psi/2)} \tag{2.47}$$

We normalize the array factor so that its peak is 1:

$$f_A(\psi) = \frac{\sin(N\psi/2)}{N\sin(\psi/2)} \tag{2.48}$$

As with an aperture antenna, the voltage radiation pattern of an array antenna $f_A(\psi)$ is characterized by mainlobes and sidelobes or, more generally, by major and minor lobes.

- As N increases, the mainlobe narrows.
- As N increases, the number of sidelobes (minor lobes) between major lobes increases $(N - 2)$.
- The width of a minor lobe is $2\pi/N$.
- The width of a major lobe is $4\pi/N$.
- The height of minor lobes decreases with increasing N.

If we set $\alpha = 0$ so that all elements have the same phase, the array factor describes the pattern generated by a diffraction grating [17]. With an array antenna, we can control α and thus steer the peak of the beam to a desired θ.

The beam direction θ varies from 0 to π. Therefore,

- $\cos\theta$ varies from -1 to $+1$.
- $kd\cos\theta$ varies from $-kd$ to $+kd$.
- ψ varies from $(-kd + \alpha)$ to $(kd + \alpha)$.

As θ varies from 0 to π, the limits between which ψ varies, known as the visible region, will depend on k, d, and α. When $\psi = \pm 2\pi m$ (m is a positive integer), a major lobe occurs in $f(\psi)$. If more than one such lobe occurs in the visible region, we refer to the additional major lobes as grating lobes, since an analogous phenomenon occurs in the diffraction grating. It is left as a problem (Problem 2.17) to show that

- $kd < \pi \Rightarrow d < \lambda/2$ (no grating lobe occurs);
- $\pi < kd < 2\pi \Rightarrow \lambda/2 < d < \lambda$ (a grating lobe may occur);
- $kd > 2\pi \Rightarrow d > \lambda$ (one or more grating lobes occur).

Furthermore (Problem 2.18), for a particular chosen beam direction θ, the criterion for no grating lobe peaks in the visible region is

$$\frac{d}{\lambda} < \frac{1}{1 + |\cos\theta|} \tag{2.49}$$

If a broadside beam (a beam pointing perpendicular to the array) is desired, we choose $\alpha = 0$; for an element spacing less than the wavelength ($d < \lambda$), no grating lobe peaks appear. If endfire operation (beam pointing parallel to array, emitted from each end) is desired, we choose $\alpha = \pm kd$ and no grating lobe occurs as long as $d < \lambda/2$.

2.4.2 Two-Dimensional Phased Array

Most ESAs have a two-dimensional array of elements. The beam equations are straightforward generalizations of those just discussed [11]. For an array of N by M elements, where $N >> 1$, $M >> 1$, and $d < \lambda$ as with the aperture antenna, the boresight gain is

$$G = \frac{4\pi A}{\lambda^2} \tag{2.50}$$

where A = area = $(M - 1)(N - 1)d^2$.

Suppose a particular array function (relative phases of the elements) produces a particular radiation pattern with its peak in a particular direction, and a second array function produces a different radiation pattern with its peak in a different direction. Because of the linearity of the equations, if the total array function is selected to be the sum of the two array functions, the total radiation pattern also will be the sum of the two radiation patterns; that is, the antenna would produce two beams. More generally, an array may produce multiple beams. Of course, the transmitted power is divided among the beams. It often is advantageous to transmit on one broad beam and receive simultaneously on several narrower beams. Furthermore, the beam direction may be switched after a dwell time of only a few milliseconds (the agile-beam technique; see Section 7.5.7).

Consider an ESA operating at a particular frequency, with the phase of each element controlled by a phase shifter. If the frequency is changed, for example, swept over a band, and the phase shifts are not changed, the mainlobe direction will change, which is generally an undesirable effect. For that reason, engineers are developing time-delay shifters to replace phase shifters in such antennas. Time-delay shifters can result in a mainlobe direction that does not vary as the frequency is swept over the band.

References

[1] Brigham, E. O., *The Fast Fourier Transform and Its Applications*, Englewood Cliffs, NJ: Prentice Hall, 1988.

[2] Edde, B., *Radar: Principles, Technology, Applications*, Upper Saddle River, NJ: Prentice Hall, 1993.

[3] Scheer, J. A., and J. L. Kurtz, *Coherent Radar Performance Estimation*, Norwood, MA: Artech House, 1993.

[4] Skolnik, M. I., *Introduction to Radar Systems*, New York: McGraw-Hill, 1980.

[5] Robins, W. P., *Phase Noise in Signal Sources*, London: Peregrinus, 1984.

[6] Jackson, J. D., *Classical Electrodynamics*, 2nd ed., New York: Wiley, 1975.

[7] Henney, K., *Radio Engineering Handbook*, 5th ed., New York: McGraw-Hill, 1959.

[8] Blake, L. V., *Radar Range Performance Analysis*, Norwood, MA: Artech House, 1986.

[9] Barton, D. K., *Modern Radar System Analysis*, Norwood, MA: Artech House, 1988.

[10] Mumford, W. W., and E. H. Scheibe, *Noise Performance Factors in Communication Systems*, Dedham, MA: Horizon House, 1968.

[11] Stutzman, W. L., and G. A. Thiele, *Antenna Theory and Design*, New York: Wiley, 1981.

[12] Collin, R. E., *Antennas and Radiowave Propagation*, New York: McGraw-Hill, 1985.

[13] Kurpis, G. P., and C. J. Booth (ed.), *The New IEEE Standard Dictionary of Electrical and Electronics Terms*, 5th ed., New York: IEEE, 1993.

[14] Sherman, J. W., "Aperture-Antenna Analysis," Chap. 9 in M. Skolnik, ed., *Radar Handbook*, 1st ed., New York: McGraw-Hill, 1970. (Note: This article is not in the 2nd edition.)

[15] Huygens, C., *Treatise on Light*, New York: Dover, 1962.

[16] Baker, B. B., and E. T. Copson, *The Mathematical Theory of Huygens' Principle*, 2nd ed., Oxford: Clarendon, 1949.

[17] Born, M., and E. Wolf, *Principles of Optics*, 4th ed., New York: Pergamon, 1969.

[18] Goldstein, H., *Classical Mechanics*, Reading, MA: Addison-Wesley, 1959.

[19] Harris, F. H., "On the Use of Windows for Harmonic Analysis With the Discrete Fourier Transform," *Proc. IEEE*, Vol. 66, No. 1, Jan. 1978, pp. 51–83.

[20] *Signal-Processing Toolbox for Use With MATLAB*, Natick, MA: The Math Works, 1998.

[21] Carrara, W. G., R. S. Goodman, and R. M. Majewski, *Spotlight Synthetic Aperture Radar*, Norwood, MA: Artech House, 1995.

[22] Silver, S., *Microwave Antenna Theory and Design*, New York: McGraw-Hill, 1949.

[23] Fowler, C. A., "Old Radar Types Never Die; They Just Phased Array," *IEEE-AES Systems Magazine*, Sept. 1998, pp. 24A–24L.

[24] Brookner, E. (ed.), *Practical Phased-Array Antenna Systems*, Norwood, MA: Artech House, 1991.

[25] Koul, S. K., and B. Bhat, *Microwave and Millimeter-Wave Phase Shifters*, Norwood, MA: Artech House, 1991.

[26] McQuiddy, D. N., et al., "Transmit/Receive Module Technology for X-Band Active Array Radar," *Proc. IEEE*, Vol. 79, No. 3, March 1991, pp. 308–341.

[27] Cook, J. H., "Earth Station Antennas," Chap. 36 in R. C. Johnson and H. Jasik, *Antenna Engineering Handbook*, 3rd ed., New York: McGraw-Hill, 1993.

Problems

Problem 2.1

Show that $F = (S/N)_{in}/(S/N)_{out} = (N_{out}/N_{in})(S_{in}/S_{out}) = (N_{out}/N_{in})(1/G_{LNA})$ if and only if $T_{ant} = T_{radar} = T_0$.

Problem 2.2

Show that, if we define $F2 = (S/N)_{in}/(S/N)_{out}$ and if $T_{ant} = T_{radar}$, then $T_{rcvr} = (F2 - 1)T_{ant}$.

Problem 2.3

Consider a sensitive radar observing targets against deep space (T_{ant} = 3K), with an LNA cooled with liquid helium to T_{rcvr} = 4.2K. If $L_{radar} = 1$, what is the noise figure in decibels? What is T_{sys}?

Problem 2.4

In the front-end circuit shown in Figure 2.11, the designer has put a system gain control, in the form of a variable attenuator, after the first LNA. Sketch the variation of overall system gain and system noise figure as this attenuator is varied over its full range. (Assume that $T_{ant} = T_{radar} = T_0$.)

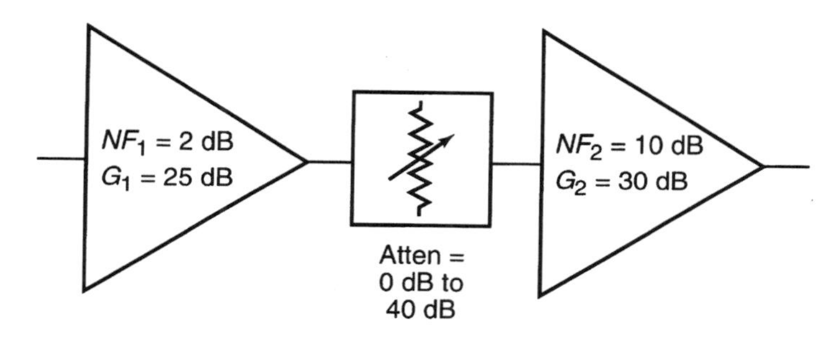

Figure 2.11 LNA circuit.

Problem 2.5

Show that the standard deviation of the quantization error of an A/D converter is

$$\sigma_\epsilon = \frac{\textit{least significant bit}}{\sqrt{12}}$$

Problem 2.6

Consider a spherical wave emitted from a point antenna. For the far field, we want the wave to be essentially planar over an aperture of width D at range R. If the far-field criterion is that the wavefront deviate no more than $\lambda/16$ from a plane (the Fraunhofer criterion), show that R (far field) $= 2D^2/\lambda$. Calculate the near- and far-field "boundaries" ($R = 2D^2/\lambda$), for $D = 1\text{m}$, at L, S, X, Ku, and Ka bands.

Problem 2.7

Using Huygens' principle, derive Snell's law of refraction between two media: $n_1 \sin\theta_1 = n_2 \sin\theta_2$, where n is the index of refraction and θ is the incidence angle (zero for normal incidence).

Problem 2.8

Calculate the coordinate-transformation matrices between the (θ, ϕ) coordinates and the (α, ϵ) coordinates. Suppose that $\alpha = -30$ degrees and $\epsilon = 45$ degrees; find θ, ϕ. Suppose that $\theta = 45$ degrees and $\phi = 30$ degrees; find α, ϵ.

Problem 2.9

For an infinite slit aperture of width L parallel to the x axis, with $\mathbf{E}$ perpendicular to the slit edges—$\mathbf{E} = (0, E_y, 0)$—in the yz plane, show that $E_x = 0$ and in the yz plane, in polar coordinates,

$$E_\theta(r, \theta) = jke\frac{e^{-jkr}}{2\pi r}E_y L \,\text{sinc}\left(\frac{kL\sin\theta}{2}\right)$$

Problem 2.10

For a finite rectangular aperture, L_x by L_y, with $\mathbf{E} = \mathbf{E_0}$ parallel to $\mathbf{y}$, show that

$$E_\theta(r,\ \theta,\ \phi) = jk\frac{e^{-jkr}}{2\pi r}E_0 L_x L_y\ \underline{\sin\phi}\ \text{sinc}\left(\frac{kL_x}{2}u\right)\text{sinc}\left(\frac{kL_y}{2}v\right)$$

$$E_\phi(r,\ \theta,\ \phi) = jk\frac{e^{-jkr}}{2\pi r}E_0 L_x L_y\ \underline{\cos\theta\cos\phi}\ \text{sinc}\left(\frac{kL_x}{2}u\right)\text{sinc}\left(\frac{kL_y}{2}v\right)$$

where $u = \sin\theta\cos\phi$, $v = \sin\theta\sin\phi$. (The underlined factors are the obliquity factors.) In the E plane (*yz* plane), $\phi = 90$ degrees; show that

$$E_\theta(r,\ \theta) = jk\frac{e^{-jkr}}{2\pi r}E_0 L_x L_y\ \text{sinc}\left(\frac{kL_y}{2}\sin\theta\right),\ E_\phi = 0$$

In the H plane (*xz* plane), $\phi = 0$ degrees; show that

$$E_\theta(r,\ \theta) = jk\frac{e^{-jkr}}{2\pi r}E_0 L_x L_y\ \underline{\cos\theta}\ \text{sinc}\left(\frac{kL_x}{2}\sin\theta\right),\ E_\phi = 0$$

Problem 2.11

For a circular aperture of radius a, with $\mathbf{E} = \mathbf{E_0}$ parallel to the aperture plane, show that

$$\mathbf{E} = (\hat{\boldsymbol{\theta}}\cos\phi - \hat{\boldsymbol{\phi}}\sin\phi\cos\theta)jk\frac{e^{-jkr}}{2\pi r}\cdot \mathbf{E}_0 2\pi a^2\frac{2J_1(ka\sin\theta)}{ka\sin\theta}$$

$$f(\theta,\ \phi) = f_1(\theta) = \frac{2J_1(ka\sin\theta)}{ka\sin\theta}$$

where $J_n(x)$ is the nth-order ordinary Bessel function. Show that the angle from peak to first null is $(1.22)(\lambda/2a)$.

Problem 2.12

Show that for the infinite slit and its characteristic $\sin(x)/x \equiv \text{sinc}(x)$ pattern, the first sidelobe peak is -13.3 dB below the mainlobe peak; show that for the circular aperture, the first sidelobe peak is -17.6 dB below the mainlobe peak.

Problem 2.13

From the fundamental properties of a parabola, show that for a parabolic dish antenna with a point-source feed at the focal point and the aperture (planar surface within rim of dish) perpendicular to the axis of revolution, all radiation leaving the aperture has the same phase.

Problem 2.14

For an axially symmetric parabolic reflector with a feed at the focus radiating isotropically with power P a distance d from the closest point on the reflector surface, show that the aperture illumination function $F(r)$ (W/m^2) is

$$F(r) = \frac{P}{4\pi d^2} \cdot \frac{1}{\left(1 + \frac{r^2}{4d^2}\right)^2}$$

Problem 2.15

Show that, for an ideal rectangular or circular aperture where $\lambda >>$ aperture dimension, the sidelobe envelope (smooth line connecting the sidelobe peaks), measured in dBi, is independent of λ or the aperture dimension. (The International Radio Consultative Committee [CCIR] recommends that sidelobes for large circular apertures [D > 100λ] be below $G = 29 - 25\log_{10}(\theta_{\text{degrees}})$ dBi [27].)

Problem 2.16

Consider a long ($N >> 1$) linear array of length L. Suppose that the center $N/4$ elements burn out. Develop a revised expression for the radiation pattern, for the mainlobe at 90 degrees (broadside), 45 degrees, and 0 degrees (endfire).

Problem 2.17

Show that

- $kd < \pi \Rightarrow d < \lambda/2$ (no grating lobe occurs);
- $\pi < kd < 2\pi \Rightarrow \lambda/2 < d < \lambda$ (a grating lobe may occur);
- $kd > 2\pi \Rightarrow d > \lambda$ (one or more grating lobes occur).

Problem 2.18

Show that the criterion for no grating lobes (peak) is

$$\frac{d}{\lambda} < \frac{1}{1 + |\cos\theta|}$$

Problem 2.19

If θ (mainlobe) = 30 degrees, at what element spacing (d/λ) do the peaks of the first, second, and third grating lobes appear?

3

Interaction of Radar Systems with the External Environment

After the EM radiation leaves the radar antenna, it propagates into the external environment. Some of the radiation scatters from one or more objects in the environment, returns to the antenna, and enters the radar receiver. The radar may also receive some external noise simultaneously. This chapter discusses those effects.

3.1 Radar Cross-Section

The RCS of an object, denoted as σ, is given by [1]

$$\sigma \equiv \lim_{R \to \infty} 4\pi R^2 \frac{|\mathbf{E}_s|^2}{|\mathbf{E}_i|^2} \tag{3.1}$$

where $\mathbf{E_i}$ is the incident electric field (a plane wave) and $\mathbf{E_s}$ is the scattered electric field (a spherical wave as $R \to \infty$). From Section 1.3, the average flux density S_{inc}(W/m^2) incident on the target is

$$S_{inc} = \frac{c\epsilon_0}{2}|\mathbf{E}_i|^2 \tag{3.2}$$

and the scattered flux density is

$$S_{\text{scatt}} = \frac{c\epsilon_0}{2}|\mathbf{E}_s|^2 \tag{3.3}$$

From Section 1.7, Φ represents the scattered intensity in watts per steradian (sr). Thus,

$$\Phi_{\text{scatt}} = S_{\text{scatt}} \cdot R^2 = \frac{c\epsilon_0}{2}|\mathbf{E}_s|^2 R^2 \tag{3.4}$$

Then

$$\frac{\Phi_{\text{scatt}}}{S_{\text{inc}}} = \frac{|\mathbf{E}_s|^2 R^2}{|\mathbf{E}_i|^2} \tag{3.5}$$

and, for $R \to \infty$,

$$\sigma = 4\pi \frac{\Phi_{\text{scatt}}}{S_{\text{inc}}} \tag{3.6}$$

as discussed in Section 1.7.

The RCS has the dimensions of area and can be high or low with respect to the projected area of the target normal to the line of sight. The concept applies only if the radar and the target are in the far field with respect to each other.

It is useful to consider a spherical coordinate system fixed in the target (assumed to be a rigid body). The RCS is, in general, a strong function of azimuth and elevation in this coordinate system and a strong function of frequency and polarization as well.

3.1.1 RCS of Simple Targets

The RCS of simple targets can be calculated from EM theory. Because the calculations have been extensively documented, for example, by Knott, Shaeffer, and Tuley [1], Ruck et al. [2], and Crispin and Siegel [3], only a few key results are summarized here. Table 3.1 summarizes the RCS for some simple shapes. Most are discussed in more detail in the following sections.

3.1.1.1 Perfectly Conducting Sphere

For a sphere of radius a, it can easily be shown (Problem 3.1) that, assuming geometrical optics ($\lambda << a$) and a perfectly reflecting sphere, the incident

Table 3.1
RCS of Perfectly Conducting Simple Targets

Target	RCS	Definition of Symbols
Sphere	$\sigma = \pi a^2$	a = radius ($a >> \lambda$)
Cylinder	$\sigma = \frac{a\lambda}{2\pi} \cdot \frac{\cos\theta \sin^2(kL\sin\theta)}{\sin^2\theta}$ $= \frac{2\pi a L^2}{\lambda}, \theta = 0$	a = radius; L = length; θ = angle off broadside
Flat plate, any shape (normal to plate)	$\sigma = \frac{4\pi A^2}{\lambda^2}$	A = area ($A >> \lambda^2$)
Triangular trihedral (boresight direction)	$\sigma = \frac{4\pi a^4}{3\lambda^2}$	a = distance from trihedral vertex to dihedral vertex ("short side")
Square trihedral (boresight direction)	$\sigma = \frac{12\pi a^4}{\lambda^2}$	a = distance from trihedral vertex to dihedral vertex ("short side")
Luneburg lens	$\sigma = \frac{4\pi^3 a^4}{\lambda^2}$	a = radius

radiation is scattered uniformly into all 4π steradians (isotropic scattering), and the RCS is

$$\sigma = \pi a^2 \tag{3.7}$$

When the wavelength is not small compared with the sphere radius, the calculation is much more complicated. It was first performed by Mie [4] and later developed by Stratton [5] and Kerr [6]. According to Blake [7], the solution is

$$\frac{\sigma}{\pi a^2} = \frac{1}{\rho^2}\left|\sum_{n=1}^{\infty}(-1)^n(2n+1)(a_n+b_n)\right|^2 \tag{3.8}$$

where $\rho = 2\pi a/\lambda$; a_n and b_n (for a perfectly conducting sphere) are given by

$$a_n = \frac{j_n(\rho)}{h_n^{(2)}(\rho)} \tag{3.9}$$

$$b_n = \frac{-[\rho j_n(\rho)]'}{[\rho h_n^{(2)}(\rho)]'} \tag{3.10}$$

where j_n and $h_n^{(2)}$ are the spherical Bessel function of the first kind and the spherical Hankel function of the second kind, respectively. The primes denote differentiation with respect to ρ. A graph of (3.8) is given in Figure 3.1.

The radius-to-wavelength ratio is divided into three regions:

- Rayleigh scattering, $a << \lambda$: The RCS is much less than πa^2 and becomes progressively smaller for longer wavelengths (lower frequencies); that explains why lower frequencies penetrate rain better than higher ones (see Sections 3.2.4, 14.1). The RCS is equal to $(144)\pi^5 a^6/\lambda^4$ [2, p.150].

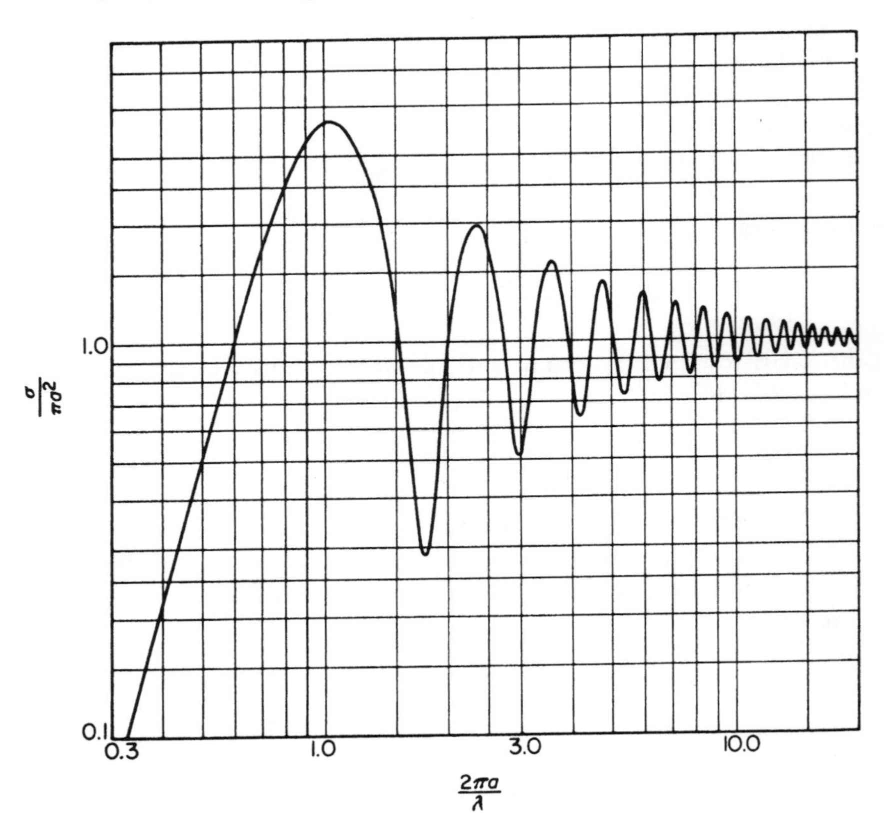

Figure 3.1 RCS of perfectly conducting sphere. (Reproduced from [13], Figure 3.2.3, with permission of Artech House.)

- Mie scattering, $a \sim \lambda$: The RCS fluctuates above and below πa^2, depending on the exact wavelength.
- Optical scattering, $a >> \lambda$: The RCS is equal to πa^2.

3.1.1.2 Flat Plate

When radar energy strikes a target, the target reradiates energy. Thus, the target behaves as an antenna. As a result, a strong analogy exists between the gain of a flat-plate antenna and the RCS of a flat-plate target. The application of Huygens' principle (Section 2.2) is summarized in Figure 3.2, and the results are summarized in Table 3.2, where the information on the left side is from an antenna text [8] and that on the right side is from an RCS text [2]; *the sinc function* is defined in Section 2.3.2. We observe the following observations (λ << plate dimensions):

- For an arbitrary shape, the broadside (i.e., normal incidence) RCS equals the boresight gain multiplied by the area.
- For both rectangular and circular shapes, the functional form of the lobing pattern is the same for the antenna gain and the RCS, except that the lobes for the RCS are half the angular width of the lobes for the antenna. The difference occurs because RCS represents a two-way phenomenon, whereas antenna gain is one-way.

3.1.1.3 Corner Reflector

As a standard target of known RCS, the trihedral corner reflector (or, simply, trihedral) is widely used to calibrate radars (see Section 1.7). Figure 3.3 illustrates

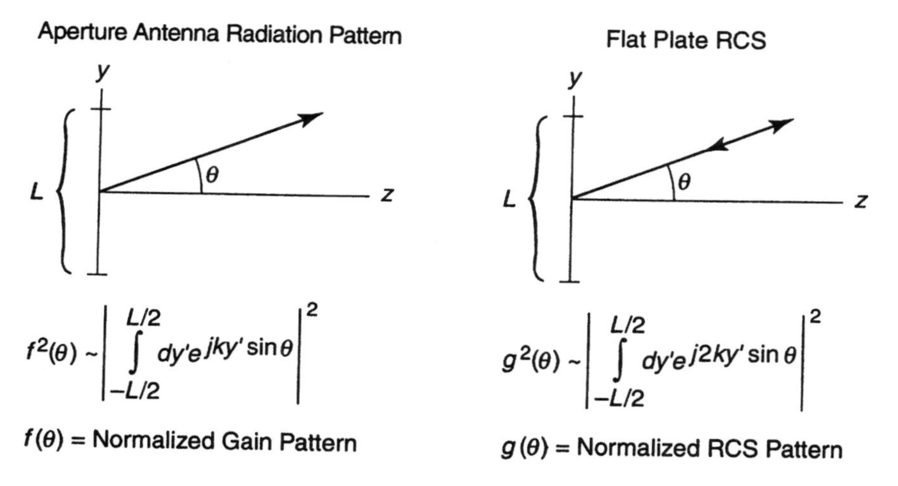

Figure 3.2 Comparison of (a) aperture antenna radiation pattern and (b) flat-plate RCS.

Table 3.2
Comparison of Aperture Antenna Peak Gain and Flat-Plate RCS Near Broadside (Physical Optics Approximation; Neglect Obliquity factors; $u = \sin\theta\cos\phi$, $v = \sin\theta\sin\phi$, z Axis Perpendicular to Aperture)

Aperture Antenna[a] **Directivity = $4\pi A/\lambda^2$**	**Flat-Plate RCS**[b] **Broadside RCS = $4\pi A^2/\lambda^2$**
Rectangular aperture (L_x by L_y)	Rectangular plate (L_x by L_y)
$G(u, v) = \frac{4\pi L_x L_y}{\lambda^2}\text{sinc}^2\left(\frac{kL_x u}{2}\right)\text{sinc}^2\left(\frac{kL_y v}{2}\right)$	$\sigma(u, v) = \frac{4\pi L_x^2 L_y^2}{\lambda^2}\text{sinc}^2(kL_x u)\,\text{sinc}^2(kL_y v)$
Circular aperture (radius = a)	Circular plate (radius = a)
$G(\theta) = \frac{4\pi(\pi a^2)}{\lambda^2}\left(\frac{2J_1(ka\sin\theta)}{ka\sin\theta}\right)^2$	$\sigma(\theta) = \frac{4\pi(\pi a^2)^2}{\lambda^2}\left(\frac{2J_1(2ka\sin\theta)}{2ka\sin\theta}\right)^2$

[a]Source: Stutzman and Thiele [8].
[b]Source: Ruck et al. [2].

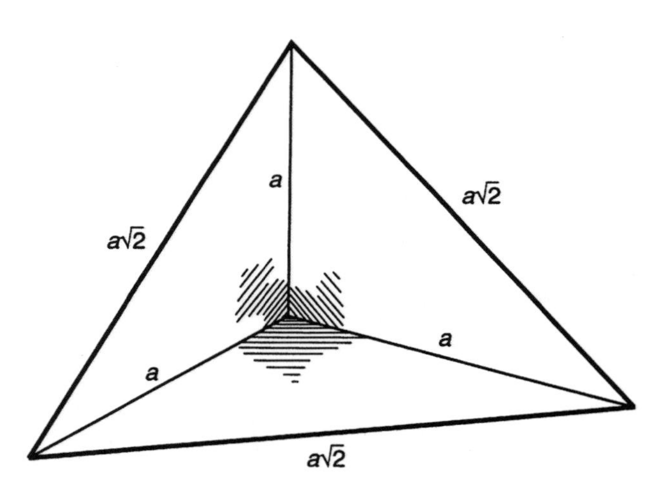

Figure 3.3 Triangular trihedral corner reflector.

the most common type of corner reflector, the triangular trihedral, which is often used for calibration in field tests of radars because of its simplicity, low cost, ruggedness, and relative insensitivity to alignment errors.

An approximate (ray-based) theory of corner reflector RCS was worked out by Spencer [9]. The following is a brief summary. A corner reflector is a retroreflector, that is, a ray of light (or other EM radiation) entering the corner reflector is reflected back in the direction from which it came. Furthermore, the reflected ray, as seen from the direction of the incident ray, appears to be emitted from a point reflected across the vertex (Figure 3.4). Consider a corner reflector with its three planar sides in the xy, yz, and xz planes and consider a source point (in the far field) x, y, z. By the method of images, after one reflection the apparent source point is at $-x$, y, z; after two reflections, it is at $-x$, $-y$, z; and after three reflections, it is at $-x$, $-y$, $-z$.

As Figure 3.5 shows, because of that effect, when radiation enters a triangular corner reflector exactly along its boresight, making equal angles with each of the three axes, reflected energy is received, not from the total projected area of the corner reflector, but rather from a subset of that area, which has

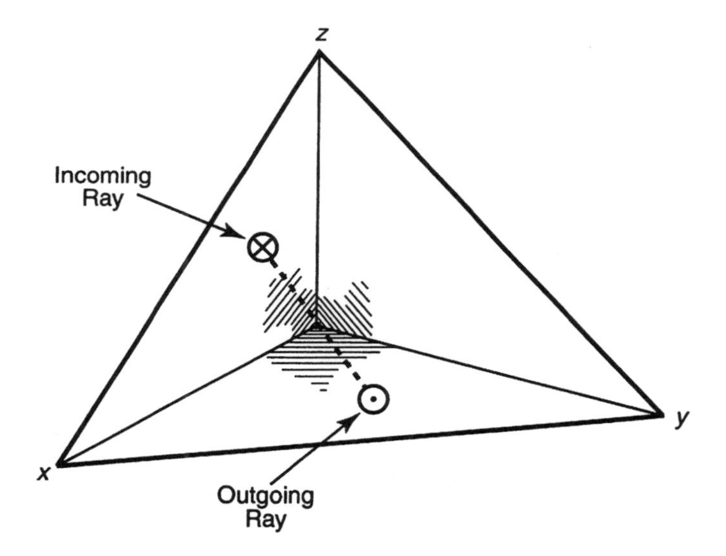

Figure 3.4 Reflection from trihedral corner reflector.

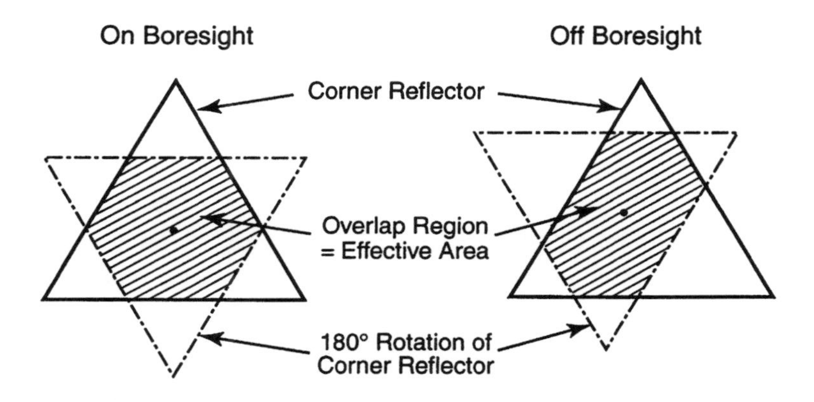

Figure 3.5 Effective area of triangular trihedral.

the shape of a regular hexagon. Furthermore, if the ray enters off-boresight, the subarea contributing to the retroreflection is an irregular hexagon formed by the intersection of the projected area of the corner reflector and its rotation 180 degrees about the incident direction. Portions of the projected area not in the hexagon do not contribute because for at least one of the three reflections, the physical corner reflector does not extend far enough along the appropriate plane (e.g., the *xy* plane).

Finally, Spencer showed that, in complete analogy to the broadside flat plate, the RCS is $4\pi A^2/\lambda^2$, where A is the area contributing to the retroreflection. For a triangular corner, that value is (Problem 3.5)

$$\sigma \text{ (triangular } CR) = \frac{4\pi a^4}{3\lambda^2} \tag{3.11}$$

where, as shown in Figure 3.3, a is the distance from the trihedral vertex (i.e., where all three planes come together) to one of the dihedral vertices. a is sometimes referred to as the "short side," $a\sqrt{2}$ as the "long side."[1]

For a triangular corner reflector, Spencer also showed (summarized in [3]) that if l, m, and n are the direction cosines with respect to the x-, y-, and z-axes (defined such that $l \le m \le n$), then the effective projected area is

$$\begin{aligned} A &= 4\frac{lm}{l+m+n}a^2 \quad (l+m \le n) \\ A &= \left(l+m+n-\frac{2}{l+m+n}\right)a^2 \quad (l+m \ge n) \end{aligned} \tag{3.12}$$

3.1.1.4 Luneburg Lens

A Luneburg lens (LL) [3] consists of a sphere having an index of refraction, n, that is a function of its radius, a:

$$n(r) = \left[2 - \left(\frac{r}{a}\right)^2\right]^{1/2} \tag{3.13}$$

n is a maximum ($\sqrt{2}$) at the center and decreases monotonically to a minimum (unity) at the surface. Commercial LLs typically are built of concen-

1. The formula on page 591 of Ruck [2] for the RCS of a triangular corner reflector contains a typographical error: The "3" in the denominator was omitted. The same error occurs in Mott [10], equations (6.149) and (6.151).

tric spherical shells, which provide a stepped approximation to (3.13). The effect is that a plane EM wave incident on the sphere will be focused to a point on the far side. If that portion of the sphere is coated with a reflecting material, the EM radiation is again reflected, so as to emerge as a plane wave traveling in the direction opposite to that of the incident wave. Thus, the lens acts as a retroreflector. The RCS of such an LL is

$$\sigma = \frac{4\pi A^2}{\lambda^2} = \frac{4\pi}{\lambda^2}(\pi a^2)^2 = \frac{4\pi^3 a^4}{\lambda^2} \tag{3.14}$$

LLs can be used as retroreflectors for radar calibration. LLs possess an RCS that may be relatively constant over a fairly wide range of incident aspect and elevation angles. On the other hand, the corner reflector has the advantages of simplicity. The choice of retroreflector depends on the specific application.

3.1.2 RCS of Complex Targets

In general, the measured RCS of a complex target, such as an aircraft or a ship, is highly variable with respect to azimuth, elevation, frequency, and polarization. It is convenient to consider a spherical coordinate system fixed within the target and to discuss RCS relative to azimuth (0 to 360 degrees) and elevation (+90 to −90 degrees) defined by that coordinate system. If we consider the target as an antenna reradiating energy back to the radar, then, from Section 3.1.1.2, we see that lobes will occur with 3-dB angular widths of approximately $\lambda/2L$, where L is the cross-range dimension in the plane perpendicular to the axis of rotation. Because $L >> \lambda$, a great many narrow lobes occur; an example is given in Figure 3.6. Table 3.3 summarizes the approximate RCS of some complex targets.

3.1.3 Methods of Calculating RCS

The RCS of relatively complex targets can be calculated according to several different approximations. Elaborate computer codes have been developed. A brief summary of some approximations follows [1].

- *Geometrical optics* (GO) assumes that the radiation travels in straight lines (assuming homogeneous media); use classical ray-tracing.
- *Physical optics* (PO) invokes approximation of tangent plane [1] and calculates RCS via Huygens' principle.
- *Geometric theory of diffraction* (GTD) is a hybrid system based on a combination of GO and the concept of "diffracted rays."

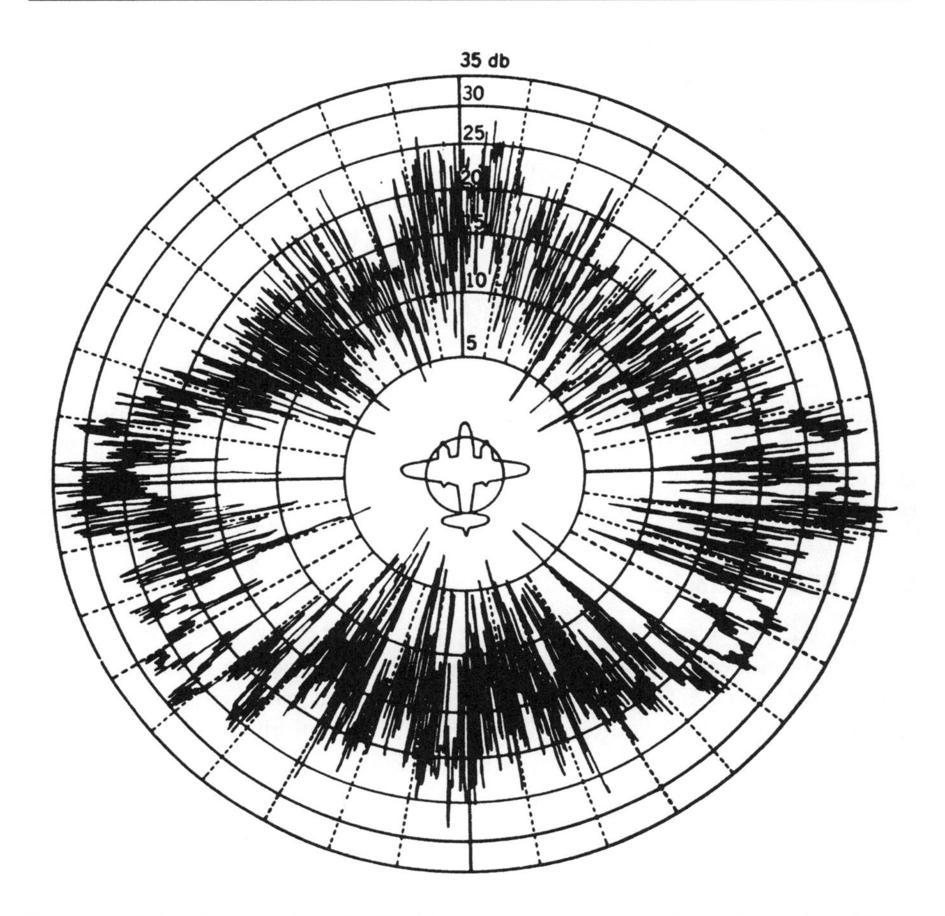

Figure 3.6 Azimuthal variation of RCS of B-25 aircraft at S band. (Reproduced from [7] with permission of Artech House and McGraw-Hill.)

3.1.4 Polarization Considerations in RCS [10]

So far, the discussion of RCS has considered all the reflected energy, regardless of polarization. Most radars, however, observe only one component of polarization at a time, and given the incident polarization, the reflected radiation is a function of polarization.

3.1.4.1 Polarization Scattering Matrix

The relationship between incident polarization and scattered polarization is given by the normalized polarization scattering matrix (PSM), where $\sqrt{\sigma_{HH}}$ is the positive square root of the HH RCS, per (3.1):

Table 3.3
Approximate RCS of Some Complex Targets (Reproduced from [7], Table 3-1, with permission of Artech House.)

Type of Target	Aspect	Cross-Section, m^2
Small jet fighter aircraft or small commercial jet	Nose, tail	0.2–10
	Broadside	5–300
Medium bomber or midsize airline jet such as 727, DC-9	Nose, tail	4–100
	Broadside	200–800
Large bomber or large airline jet such as 707, DC-8	Nose, tail	10–500
	Broadside	300–550
Wooden minesweeper, 144 long, from airborne radar (5 to 10 GHz)	Broadside	10–300
	25° from bow, stern	0.1–300
Small bird, 450 MHz	Average overall	$10^{-5.6}$
Large bird, 9 GHz	Broadside	10^{-2}
Insect (bee), 9 GHz	Average	$10^{-2.8}$
Large insect (5 cm), 9 GHz	Average	$10^{-1.8}$

$$\begin{bmatrix} E^s_H \\ E^s_V \end{bmatrix} = \frac{\sqrt{\sigma_{HH}}}{R\sqrt{4\pi}} \begin{bmatrix} A_{HH} & A_{VH} \\ A_{HV} & A_{VV} \end{bmatrix} \begin{bmatrix} E^i_H \\ E^i_V \end{bmatrix} = \mathbf{A_{HV}E^i} \tag{3.15}$$

The incident wave and the scattered wave are represented by two-component spatial vectors. Each component is complex, indicating the phase of the wave. The two components are expressed here as horizontal (H) and vertical (V), although any other convenient basis can be used. The equation relates the two-component scattered **E** vector to the two-component incident **E**. The 2-by-2 **A** matrix is the most general linear relationship between them (linearity is assumed—an excellent approximation). The factor of $1/R$ indicates the spherical nature of the scattered wave. In the formalism of (3.15), the normalized PSM is unitary, that is, the absolute value of its determinant is unity, and also the magnitude of each of its rows and columns is unity (see Table 8.1).

It is sometimes more convenient to represent the waves with a basis in which the components are circularly polarized (R = right-circular; L = left-circular):

$$\begin{bmatrix} E^s_R \\ E^s_L \end{bmatrix} = \frac{\sqrt{\sigma_{RR}}}{R\sqrt{4\pi}} \begin{bmatrix} A_{RR} & A_{LR} \\ A_{RL} & A_{LL} \end{bmatrix} \begin{bmatrix} E^i_R \\ E^i_L \end{bmatrix} = \mathbf{A_{RL}E_i} \tag{3.16}$$

Given the PSM, the backscatter from any incident polarization can be calculated, via appropriate coordinate transformations. Furthermore, for a monostatic radar operating in free space, the PSM can be shown to be symmetric in any basis.

The PSM components are complex because there generally is a phase difference between incident and scattered components. For monostatic operation, because the PSM is symmetric, it contains three independent complex numbers.

3.1.4.2 PSM for Simple Targets

In the HV basis, the following normalized PSMs for some simple perfectly conducting targets are independent of target orientation.

$$\text{Flat plate (broadside) or trihedral corner (boresight):} \quad \mathbf{A} = \begin{bmatrix} -1 & 0 \\ 0 & -1 \end{bmatrix} \tag{3.17}$$

If H or V polarized radiation strikes either of those targets, the scattered radiation has the same polarization as the incident radiation, and there is a phase change of π(180 degrees).

The dihedral corner reflector (Figure 3.7) is a bit more complicated; its normalized PSM depends on its orientation. When viewed along the dihedral angle bisector, if the fold line is normal to the (horizontal) line of sight (LOS), and if the fold line makes an angle ψ with the vertical, then the normalized PSM in the HV basis is

$$\mathbf{A} = \begin{bmatrix} -\cos 2\psi & \sin 2\psi \\ \sin 2\psi & \cos 2\psi \end{bmatrix} \tag{3.18}$$

The value of $\mathbf{A}$ for several values of ψ is

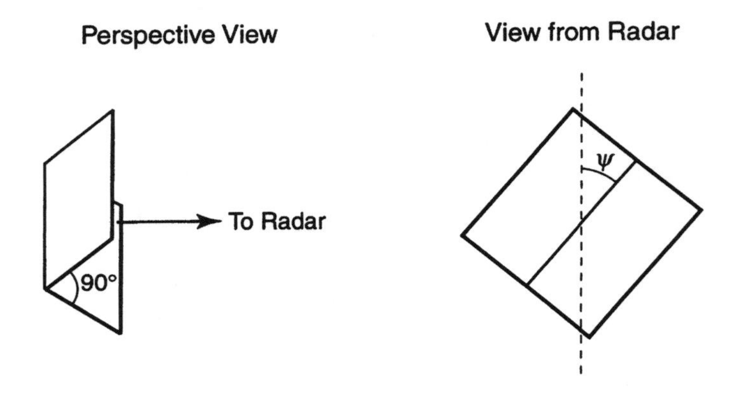

Figure 3.7 Dihedral corner reflector: (a) perspective view; (b) view from radar.

$$
\begin{aligned}
\mathbf{A} &= \begin{bmatrix} -1 & 0 \\ 0 & 1 \end{bmatrix} \quad \psi = 0 \text{ degrees} \\
&= \frac{1}{\sqrt{2}}\begin{bmatrix} -1 & 1 \\ 1 & 1 \end{bmatrix} \quad \psi = 22.5 \text{ degrees} \\
&= \begin{bmatrix} 0 & 1 \\ 1 & 0 \end{bmatrix} \quad \psi = 45 \text{ degrees} \\
&= \frac{1}{\sqrt{2}}\begin{bmatrix} 1 & 1 \\ 1 & -1 \end{bmatrix} \quad \psi = 67.5 \text{ degrees} \\
&= \begin{bmatrix} 1 & 0 \\ 0 & -1 \end{bmatrix} \quad \psi = 90 \text{ degrees}
\end{aligned} \tag{3.19}
$$

Thus, when ψ is 0 degrees or 90 degrees, the scattered polarization is the same as the incident polarization. When ψ is 45 degrees, it is the opposite. When ψ is 22.5 or 67.5 degrees, if the incident radiation is either H or V, the scattered polarization is distributed evenly between H and V.

3.1.4.3 More General Polarization Basis

More generally, an EM wave traveling in the z-direction can be described as follows:

$$
\begin{aligned}
\mathbf{E}(z, t) &= Re[e^{j(\omega t - kz)}(E_{0x}e^{j\theta_x}\hat{\mathbf{x}} + E_{0y}e^{j\theta_y}\hat{\mathbf{y}})] \\
E_x &= E_{0x}e^{j\theta_x} \\
E_y &= E_{0y}e^{j\theta_y}
\end{aligned} \tag{3.20}
$$

Here, θ_x and θ_y determine the relative phase of E_x and E_y and hence the type of polarization.

For the linearly polarized case, the x and y components of $\mathbf{E}$ are in phase; that is, $\theta_x = \theta_y = \theta$.

$$
\mathbf{E}(z, t) = [E_{0x}\hat{\mathbf{x}} + E_{0y}\hat{\mathbf{y}}]e^{(j\omega t - kz + \theta)} \tag{3.21}
$$

For the right-circularly polarized case, the y component lags the x component by 90 degrees, or $\lambda/4$: $|E_x| = |E_y|$,

$$
\theta_x = 0 \quad \theta_y = -90° \tag{3.22}
$$

$$E_x = E_{0x} \quad E_y = -jE_{0y} \tag{3.23}$$

$$\mathbf{E}(z, t) = [E_{0x}\hat{\mathbf{x}} - jE_{0y}\hat{\mathbf{y}}]e^{j(\omega t - kz)} \tag{3.24}$$

For the left-circularly polarized case, $\theta_x = 0$, $\theta_y = +90$ degrees. This formalism can be used to analyze waves with arbitrary elliptical polarization.[2]

To convert from HV to RL, the transformation matrix is [10]

$$\mathbf{C} = \frac{1}{\sqrt{2}}\begin{bmatrix} 1 & 1 \\ -j & j \end{bmatrix}, \; [\;]_{HV} = \mathbf{C}[\;]_{RL} \tag{3.25}$$

$$\mathbf{C}^{-1} = \frac{1}{\sqrt{2}}\begin{bmatrix} 1 & j \\ 1 & -j \end{bmatrix}, \; [\;]_{RL} = \mathbf{C}^{-1}[\;]_{HV} \tag{3.26}$$

We define

$$\hat{\mathbf{H}} = \begin{bmatrix} 1 \\ 0 \end{bmatrix}_{HV} \quad \hat{\mathbf{V}} = \begin{bmatrix} 0 \\ 1 \end{bmatrix}_{HV} \tag{3.27}$$

and then

$$\hat{\mathbf{R}} = \frac{1}{\sqrt{2}}\begin{bmatrix} 1 \\ -j \end{bmatrix}_{HV} = \begin{bmatrix} 1 \\ 0 \end{bmatrix}_{RL} \quad \hat{\mathbf{L}} = \frac{1}{\sqrt{2}}\begin{bmatrix} 1 \\ j \end{bmatrix}_{HV} = \begin{bmatrix} 0 \\ 1 \end{bmatrix}_{RL} \tag{3.28}$$

It is of particular interest that for circularly polarized waves, the PSM of a dihedral corner reflector is (except for phase) independent of the angle ψ, as is evident from symmetry (Problem 3.7).

3.2 Propagation and Clutter

For a radar on or near the surface of the Earth, the EM waves, after leaving the antenna, typically interact with the atmosphere and the Earth's surface before returning to the radar. This section discusses those important interactions.

2. These equations follow the electrical engineering convention defined by the Institute of Electrical and Electronics Engineers (IEEE). The physics community uses a different convention to define right- and left-circular polarization.

3.2.1 Refraction of Radar Waves by Air

The index of refraction (n) of a medium is defined as the ratio of the speed of EM waves in vacuum to their speed in the medium. Because $c_{air} = c/n_{air}$, with n_{air} very close to unity, we define $n_{air} - 1 \equiv \epsilon << 1$. A formula for ϵ is [11, p. 448]

$$\epsilon \cdot 10^6 \cong \frac{77.6p}{T} + \frac{(3.73 \cdot 10^5)e}{T^2} \tag{3.29}$$

where p = pressure (mbar) (1 bar = 10^5 newtons/m^2; 1 atm = 1,013 mbar), T = temperature (Kelvin), and e = partial pressure of water vapor (mbar).

For example, for p = 1 atm, T = 290K (17 C), and e = 0, we find that n = 1.00027. Usually the difference between c_{air} and c can be neglected, but for very precise work it cannot.

3.2.2 Effect of Earth Curvature

For a spherical Earth of radius R_E with no atmosphere, the distance to the horizon d from a vantage point at altitude h is shown in Figure 3.8(a) and can be found as follows:

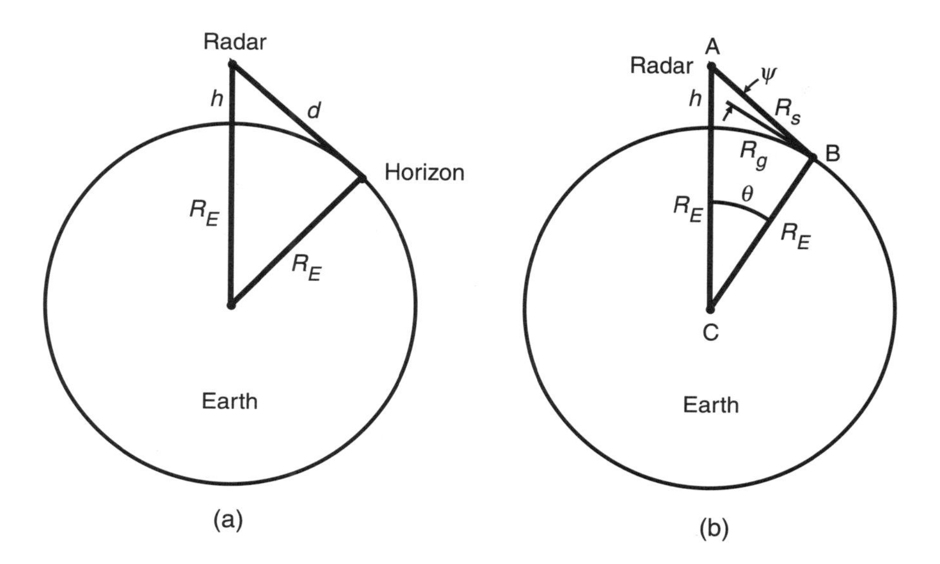

Figure 3.8 Effects of Earth's curvature: (a) distance to horizon, (b) grazing angle.

$$d^2 + R_E^2 = (R_E + h)^2 = R_E^2 + 2R_E h + h^2 \quad (3.30)$$
$$d = [2R_E h + h^2]^{1/2}$$

For $h << R_E$, $d \cong (2R_E h)^{1/2}$

Because n generally decreases with altitude, a radar ray with an upward direction $\hat{\mathbf{k}}$ usually curves down as it passes through the atmosphere. A ray emitted from a radar at altitude h and aimed downward will strike the ground at a ground range less than the corresponding ground range for no atmosphere. Similarly, some rays directed slightly upward will bend downward and strike the Earth at points past the conventional horizon. Thus, the distance to the horizon is greater than for no atmosphere. Although the details are complicated, we often use a simple approximation to calculate the distance to the horizon. We replace the actual Earth radius R_E = (6,371 km) by a greater radius. A commonly used value is $(4/3)R_E$ = 8,495 km. The greater the radar altitude, the less effect the atmosphere has; values of αR_E with $1 < \alpha < 4/3$ are sometimes used.

In computing the grazing angle of the beam from an airborne radar, it is frequently important that spherical-Earth geometry, rather than flat-Earth geometry, be used. Consider a radar at altitude h observing a point P at ground range (length of arc) R_g, as shown in Figure 3.8(b). For flat Earth, ψ (flat Earth) = $\tan^{-1}(h/R_g)$ (see Problem 3.8). For spherical Earth, we first compute slant range = R_s using the law of cosines, with $\theta = R_g/R_E$:

$$R_s^2 = (R_E + h)^2 + R_E^2 - 2(R_E + h)R_E \cos\theta \quad (3.31)$$

Then angle ABC is found from the law of sines:

$$\sin(ABC) = \frac{\sin\theta}{R_s} \cdot (R_E + h),\ ABC > \frac{\pi}{2} \quad (3.32)$$

and

$$\psi \text{ (spherical Earth)} = ABC - \frac{\pi}{2} \quad (3.33)$$

For example, for h = 10 km and R_g = 200 km, using $R_E = R_{4/3}$ = 8,495 km, ψ (spherical Earth) = 0.93 degrees, whereas ψ (flat Earth) = 1.98 degrees; the Earth's curvature cannot be neglected.

3.2.3 Attenuation of Radar Waves by Air

To some extent, microwaves are absorbed by air. Figure 3.9 summarizes the two-way attenuation (decibels per kilometer) at sea level as a function of frequency. Figure 3.10 summarizes the two-way attenuation, as a function of elevation (grazing) angle, of radar waves by the entire troposphere, which extends to an altitude of about 10 km. The latter figure shows that microwaves are strongly absorbed by O_2 resonances at ~60 GHz and by an H_2O line at 22.2 GHz. That produces "windows" (regions of relatively low absorption) around 35 GHz (Ka band) and 95 GHz (W band), both of which are frequently used millimeter-wave bands. Townes and Schawlow [12] provide a detailed explanation of the origin of those absorption bands.

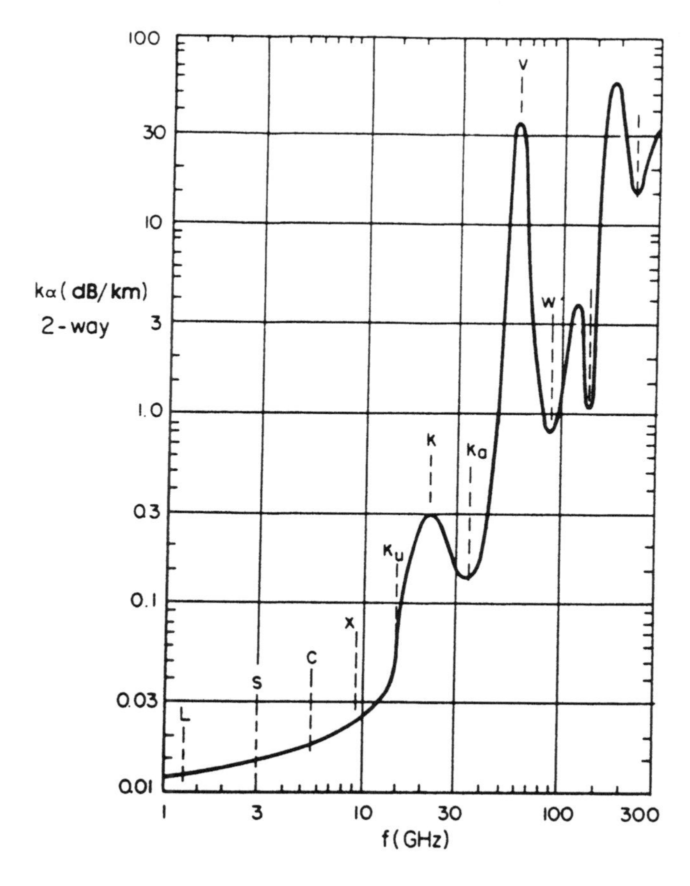

Figure 3.9 Attenuation by clear air at sea level. (Reproduced from [13], Figure 6.1.1, with permission of Artech House.)

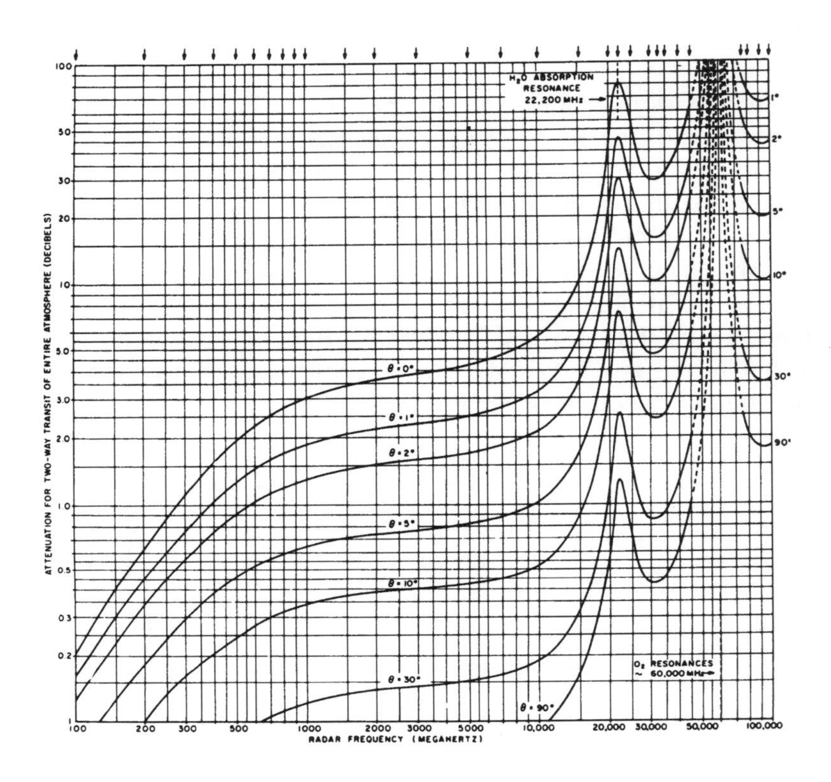

Figure 3.10 Attenuation by entire troposphere. (Reproduced from [13], Figure 6.1.4, with permission of Artech House.)

3.2.4 Attenuation of Radar Waves by Rain [13, 14]

The rainfall rate in $mm^3/(mm^2\text{-hr})$ equates to the accumulation rate in millimeters per hour. It is convenient to characterize various rainfall rates by the typical status of windshield wipers on vehicles moving through the rain, as shown in Table 3.4.

Figure 3.11 summarizes microwave attenuation as a function of frequency band and rainfall rate (and also of attenuation by clouds and fog). For example, if the rainfall rate is 4 mm/hr (a standard assumption used in radar design) and if the one-way path length through the rain is 5 km, then the two-way attenuation a is as given in Table 3.5, based on [7, 13]. The attenuation is negligible for S and C bands, small for X and Ku, considerable for Ka, and overwhelming for W. (For long-range radars, rain rarely is present over the entire path, particularly at the higher rates. Thus, in modeling radar performance, reasonable rain path lengths should be assumed.)

Table 3.4
Rainfall Rate

Rainfall Rate (mm/hr)	Windshield Wiper Status
1	Intermittent
4	Regular
15	High
60	Useless!

Note: 1 inch/hr = 25.4 mm/hr.

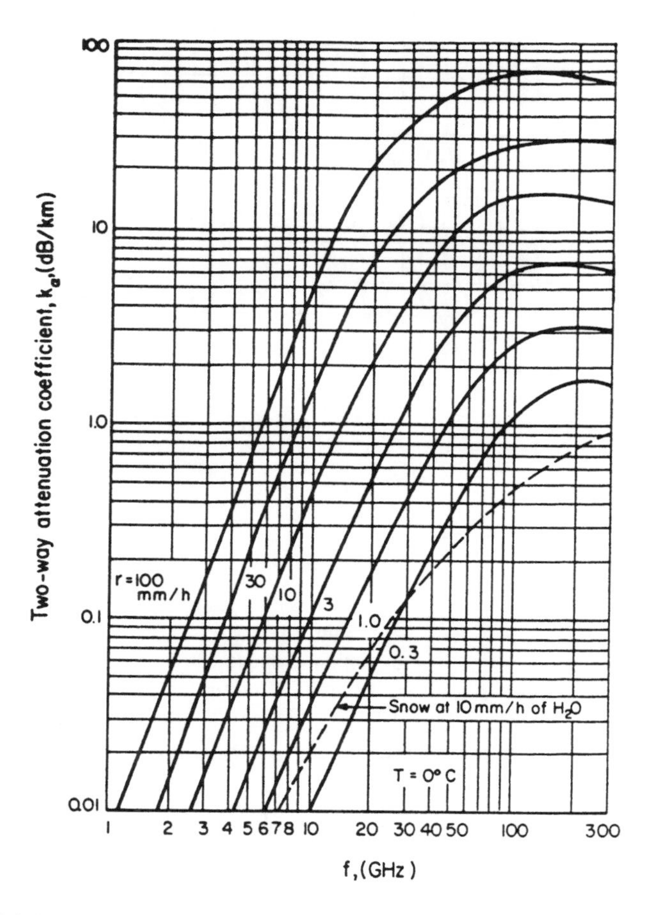

Figure 3.11 (a) Attenuation by rain; (b) attenuation by clouds and fog. (Reproduced from [13], Figures 6.1.5 and 6.1.6, with permission of Artech House.)

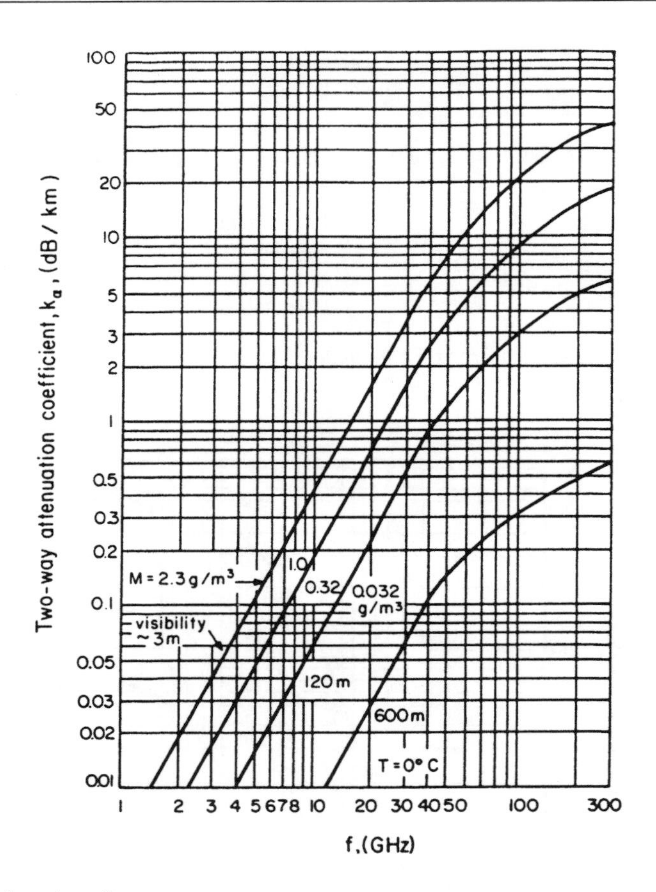

Figure 3.11 (continued).

Table 3.5
Attenuation Example
R(one-way) = 5 km, rainfall rate = 4 mm/hr

f (GHz)	Band	*a* (dB) (two-way)
3	S	0.016
5	C	0.073
10	X	0.61
16	Ku	1.9
30	Ka	7.4
100	W	31

3.2.5 Reflection of Radar Waves from the Earth's Surface

When a ray of EM radiation strikes a perfectly smooth surface, the angle of reflection is equal to the angle of incidence, a fact noted by Snell around 1600 that can be easily deduced from Huygens' principle. We refer to such reflection as specular, from the Latin *speculum*, meaning *mirror*. We also can consider a "perfectly rough" surface, where the amplitude of roughness is large compared to a wavelength. In that case, the scattering from the surface is isotropic or diffuse, and the intensity versus θ, the angle from the perpendicular to the surface, is $\sim\cos\theta$, due to the projection of the surface as seen from the direction of reflection. The expression is known as Lambert's law ([15, p. 182]; the adjective is *Lambertian*). It results simply from the projected surface area along the direction of the radiation.

If σ_h, the standard deviation of surface-height distribution (assumed Gaussian), is comparable to λ, the forward-reflected power ratio can be approximated by Ament's formula [7]:

$$\rho_s = \exp\left[-2\left(\frac{2\pi\sigma_h \sin\psi}{\lambda}\right)^2\right] \tag{3.34}$$

where ψ is the grazing angle for both incident and reflected rays. Blake [7] states that Ament's formula gives too low an estimate when the calculated value of ρ_s is less than 0.4. Ament's formula is useful in analyses involving either multipath or bistatic radar with scattering from the Earth's surface.

3.2.6 Multipath

The term *multipath* characterizes the situation in which a radar ray strikes not only a target of interest but also some other object, such as the Earth's surface, before returning to the radar. That can produce anomalous returns. As an example, consider a target observed by an airborne radar such as a corner reflector on a post. In Figure 3.12, B is denoted as the target location, C as the base of the (vertical) post, B′ as the image point of the target under the ground, and ψ as the grazing angle. D is the specular point, that is, the ground-bounce point such that the angle of incidence equals the angle of reflection (ψ). The line segment EF is perpendicular to the incoming ray FB, and the distance from EF to the radar is R_∞. There are three possible reflection paths, each with a length $2R_\infty + l$, as follows:

- Direct path, FBF: $l = 2b$
- Single-bounce path, FBDE or EDBF: $l = b + (a + b) + a = 2b + 2a$
- Double-bounce path, EDBDE: $l = a + 2\,(a + b) + a = 2b + 4a$.

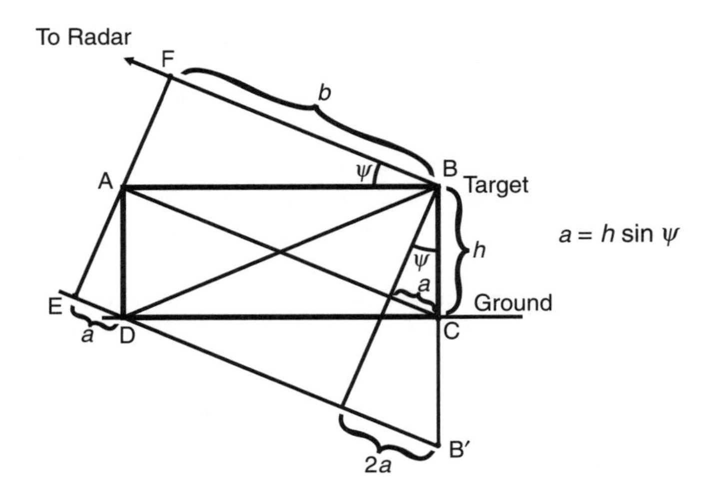

Figure 3.12 Multipath from corner reflector.

If the radar emits a short pulse, the echo of which is sampled, the echo will indicate, in addition to the true target, additional "targets" at the ranges corresponding to the pole base and to the target image below the ground.

To provide an absolute reference for the RCS measurements, corner reflectors on posts are often used as calibration targets for airborne radars. The possibility of multipath must be taken into account in the design of the calibration procedure.

As a second example of multipath, consider an outdoor range. Several such ranges exist in the United States [1] for obtaining precise measurements of the RCS of a complex target of interest, such as an aircraft. As shown in Figure 3.13, the target is placed on top of a pole of height h_t. A radar is located at a distance of 1 to several kilometers, with antenna at height h_a. Clearly, there will be one ray with ψ (incidence) = ψ (reflection); the point O at which it strikes the Earth is at distance d_1 from the radar and d_2 from the target, and $R = d_1 + d_2$. Recalling that at very low grazing angles the reflected ray

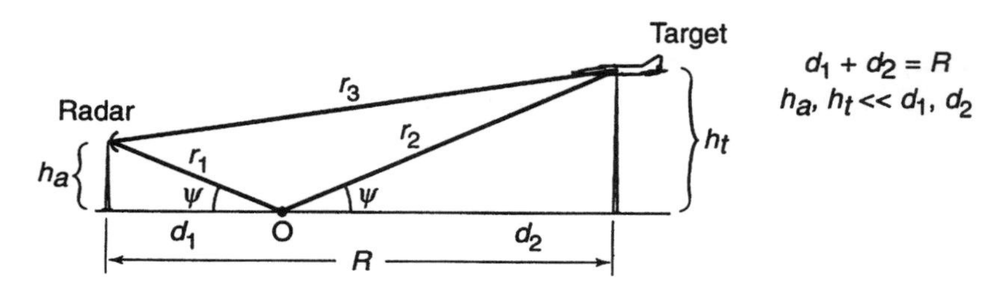

Figure 3.13 Outdoor range.

undergoes a phase change of 180 degrees (Problem 3.9) [16, 17], we design the range so that the multipath distance is $\lambda/2$ greater than the direct path:

$$r_1 + r_2 = r_3 + \lambda/2 \tag{3.35}$$

The E field at the target is then ideally twice its free-space value (constructive interference, assuming a reflection coefficient of unity), and the received power at the target is four times its free-space value. The same effect occurs on return, and the received power at the radar is ideally 16 times its free-space value. Thus, the range ideally provides a bounce gain of 16, or 12 dB. It can readily be shown (Problem 3.10) that [13, p. 290]

$$h_a h_t \cong R\lambda/4 \tag{3.36}$$

As a final example of the effects of multipath, consider a surface-based radar looking out over a smooth sea. Its effective antenna pattern will exhibit a series of strong lobes in elevation, due to the apparent interference between the radar and its image under the water [7].

3.2.7 Reflectivity of Surface Clutter

The term *clutter* describes undesired radar returns from objects that are "in the way" of targets of interest. Evidently, one person's clutter may be another person's target. Nevertheless, the admittedly pejorative term is widely used for radar returns from the Earth's surface (surface clutter) or atmosphere (volume clutter).

Consider radiation from an airborne radar striking a portion of the Earth's surface of area A. We represent the RCS of that region as $\sigma_c = \sigma^0 A$. σ^0, the parameter that characterizes the illuminated surface, is dimensionless (m^2/m^2) and is typically measured in decibels. For a diffuse (Lambertian) surface (see Section 3.2.5), $\sigma^0 = \gamma \sin\psi$, where ψ is the grazing angle. If all the incident radiation is diffusely scattered and none is absorbed, then γ attains its maximum theoretical value, which is $1/\pi$ (Problem 3.12).

If we denote that

- θ = antenna elevation beamwidth
- ϕ = antenna azimuth beamwidth
- τ = pulse width

then (assuming no pulse compression; see Section 4.2.6) for high grazing angle where $(c\tau/2) >> R\theta/\sin\psi$,

$$A \cong \left(\frac{\pi}{4} R^2 \frac{\theta\phi}{\sin\psi}\right) \tag{3.37}$$

For low grazing angle where $(c\tau/2) << R\theta/\sin\psi$,

$$A \cong R\phi \cdot \frac{c\tau}{2\cos\psi} \tag{3.38}$$

A high grazing angle is shown in Figure 3.14(a), a low grazing angle in Figure 3.14(b).

Given the surface type and the grazing angle, we can estimate (based on measurements) values of σ^0. Figure 3.15 summarizes the variation of σ^0 with ψ [13]. At $\psi \sim 90$ degrees, a strong specular component occurs.

3.2.8 Radar Backscatter from Precipitation

For volume clutter, $\sigma_c = \eta V$, where the units of η are $[\eta] = \text{m}^2/\text{m}^3$. From Rayleigh scattering theory (see Section 3.1.1.1), $\eta \sim 1/\lambda^4$. For a radar beam intersecting a region containing rain, the volume of a range gate is

$$V \cong \frac{\pi}{4} R^2 \theta\phi \frac{c\tau}{2} \tag{3.39}$$

Figure 3.16 illustrates η versus frequency band and rainfall rate.

Consider a radar observing a distant target, and assume a region of rain in the area. If the rain occurs between the radar and the target, attenuation occurs. If the rain is located at the target position, both attenuation and backscatter interfere with the measurement of the target (see Problem 3.13).

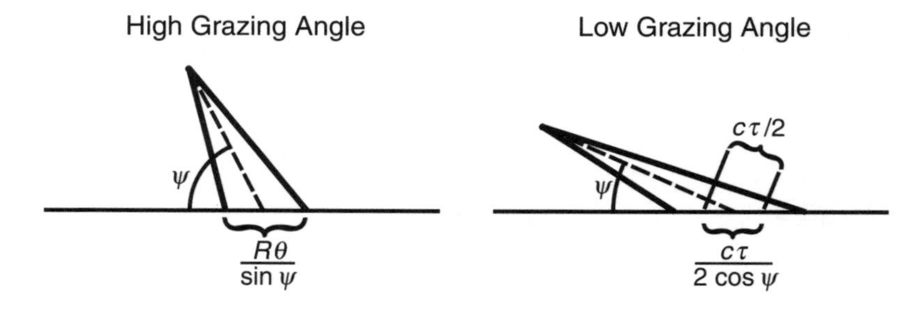

Figure 3.14 Radar return from Earth surface: (a) high grazing angle and (b) low grazing angle.

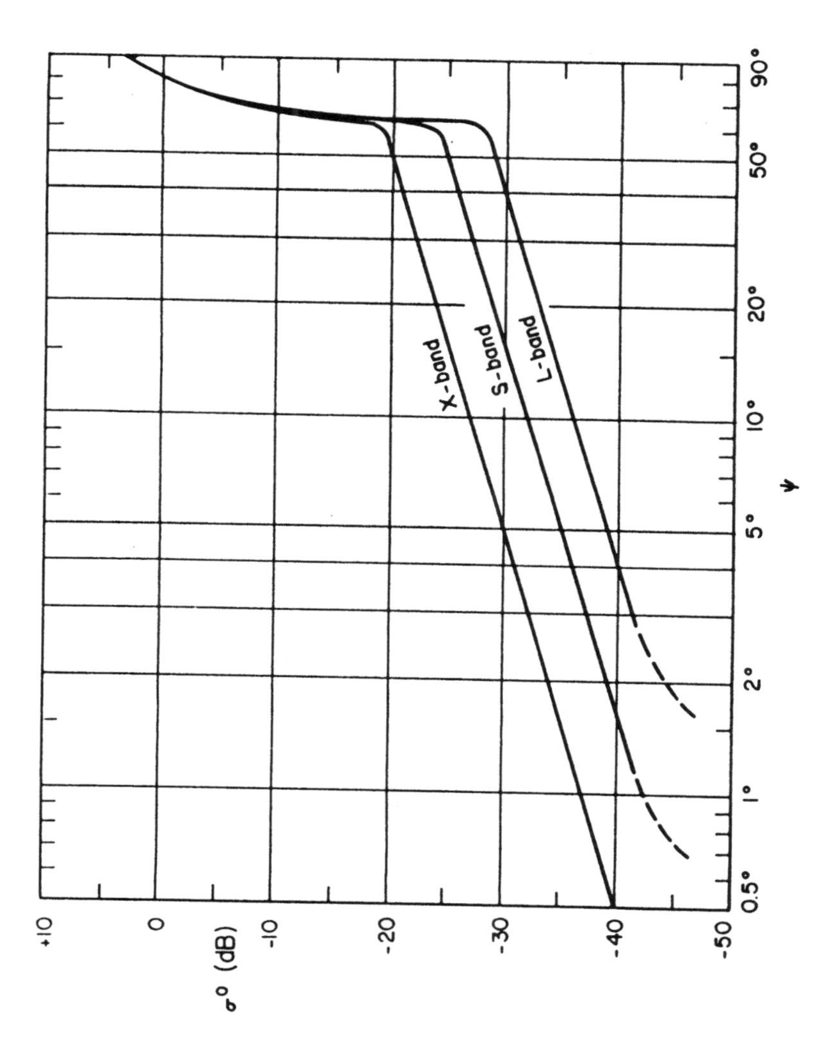

Figure 3.15 Surface clutter: (a) reflectivity of sea clutter; (b) reflectivity of land clutter. (Reproduced from [13], Figures 3.6.1 and 3.6.2, with permission of Artech House.)

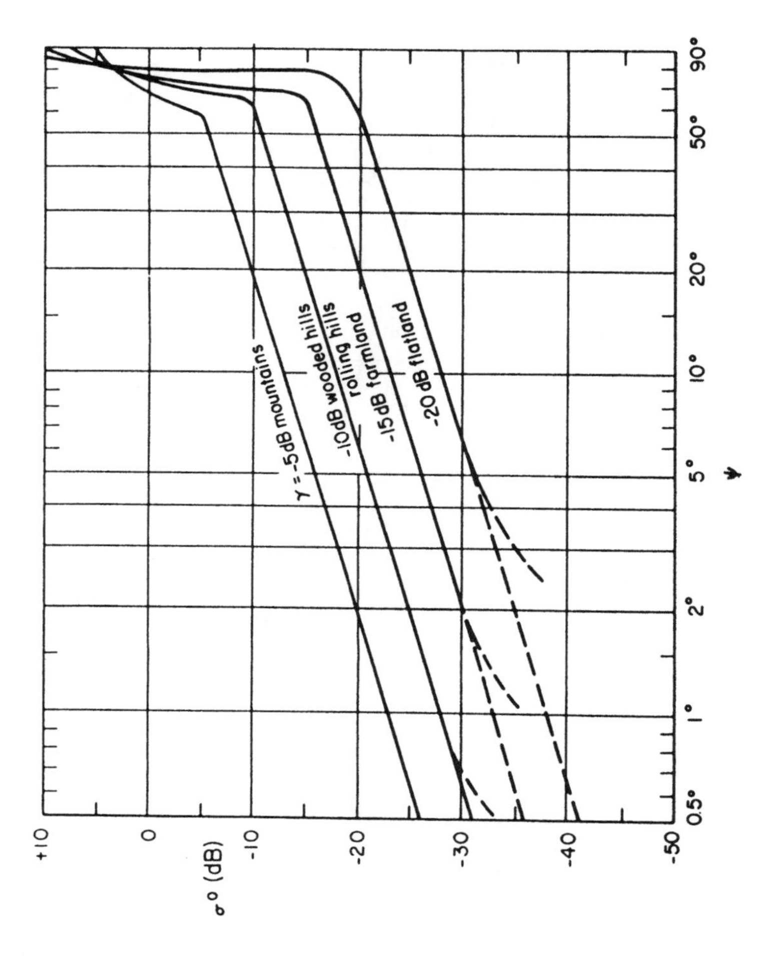

Figure 3.15 (continued).

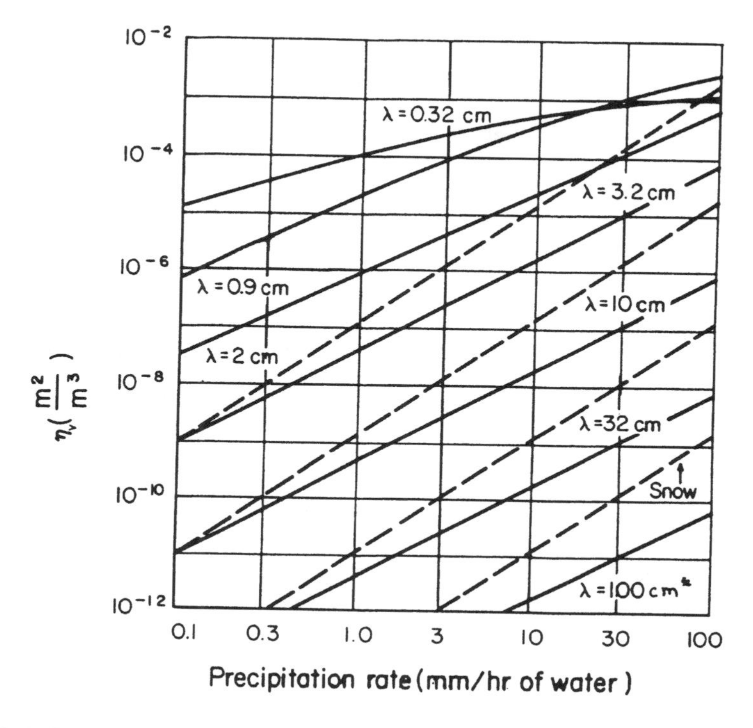

Figure 3.16 Backscatter by rain and snow. (Reproduced from [13], Figure 3.6.6, with permission of Artech House.)

3.3 External Noise

As discussed in Section 2.2.11, the noise in a radar is due to both the internal noise, resulting from the temperature of the receiver, and the external noise, resulting from the temperature of the scene. This section considers the external scene to be the sky and follows the discussion of radio astronomy by Kraus [18, Chap. 3].

The brightness distribution of a scene, measured in W m^{-2} Hz^{-1} sr^{-1}, is denoted as $B(\theta, \phi)$. Using the normalized antenna power pattern $|f(\theta, \phi)|^2$ (see Section 1.6), the power density (W Hz^{-1}) received at an antenna observing the scene is

$$w_{\text{ext}} = \frac{1}{2}A_{\text{eff}}\iint B(\theta, \phi)|f(\theta, \phi)|^2 d\Omega \qquad (3.40)$$

The factor 1/2 is introduced because it is assumed that the scene radiation is unpolarized and the antenna receives only one polarization (half the power).

The beam solid angle is defined as

$$\Omega_A = \iint |f(\theta, \phi)|^2 d\Omega \tag{3.41}$$

It can be shown [18] that the gain for a lossless antenna is

$$G = 4\pi/\Omega_A \tag{3.42}$$

A thick surface at temperature T will emit thermal radiation with a frequency distribution (Planck distribution):

$$B_f(\text{W m}^{-2}\ \text{Hz}^{-1}\ \text{sr}^{-1}) = \frac{2hf^3}{c^2} \cdot \frac{1}{e^{hf/kT} - 1} \cdot \epsilon \tag{3.43}$$

where $h \equiv$ Planck's constant = 6.623×10^{-34} J-sec, and ϵ = emissivity of the surface ($0 \leq \epsilon \leq 1$). Kirchhoff's law states that the emissivity of a surface is equal to its absorptivity [19]. If we integrate (3.43) over all frequencies and over 2π sr (Problem 3.14), we obtain the Stefan-Boltzmann law:

$$B'(\text{W m}^{-2}) = \sigma T^4 \epsilon \tag{3.44}$$

where the Stefan-Boltzmann constant [20, p. 101] is

$$\sigma = \frac{2\pi^5 k^4}{15c^2 h^3} = 5.672 \times 10^{-8}\ \text{W m}^{-2}\ K^{-4} \tag{3.45}$$

The Planck distribution has a maximum ($dB_f/df = 0$) at $f_{\max}$ such that $\lambda_{\max}(f) \equiv c/f_{\max}$ is given by Wien's displacement law:

$$\lambda_{\max}(f)\text{T} = 0.0051 \text{ m-K} \tag{3.46}$$

The Planck distribution can be expressed in terms of either frequency or wavelength. When wavelength is the independent variable, the maximum ($dB_\lambda/d\lambda = 0$) occurs at [18 p. 84]

$$\lambda_{\max}(\lambda)\text{T} = 0.0029 \text{ m-K} \tag{3.47}$$

Table 3.6 lists some values of $\lambda_{\max}(f)$ and $f_{\max}$.

Table 3.6
f_{max} for Various Temperatures

Emitting Surface	Approximate Temperature (K)	$\lambda_{max}(f)$ (μm)	f_{max} (Hz)	Spectral Region of f_{max}
Sun	6,000	0.85	3.53×10^{14}	Near infrared
Earth	300	17	1.76×10^{13}	Far infrared
Space	3	1,700	1.76×10^{11} = 176 GHz	Millimeter wave

From Table 3.6, it can be seen that for microwave radars with typical frequencies 0.1 GHz $\leq f_{RF} \leq$ 100 GHz, $f_{RF} << f_{max}$, even if the radar is above the atmosphere and pointing toward deep space. Thus, $h_f << kT$ and

$$e^{hf/kT} \cong 1 + \frac{hf}{kT}$$
$$B_f \cong \frac{2hf^3}{c^2} \cdot \frac{kT}{hf} = \frac{2f^2kT}{c^2} = \frac{2kT}{\lambda^2} \tag{3.48}$$

That is the Rayleigh-Jeans (R-J) law, and the region $f << f_{max}$ is known as the R-J region. The region $hf >> kT$ is the Wien region, where

$$B_f \sim e^{-hf/kT} \tag{3.49}$$

For an isothermal R-J scene, from (3.40) we have

$$w_{ext} = \frac{1}{2}A_{eff} \cdot \frac{2kT}{\lambda^2} \cdot \Omega_A \epsilon = kT\frac{A_{eff}\Omega_A}{\lambda^2}\epsilon \tag{3.50}$$

Furthermore, from (3.42), for a lossless antenna

$$\Omega_A = \frac{4\pi}{G} \tag{3.51}$$

and from [8]

$$A_{eff} = \frac{G\lambda^2}{4\pi} = \frac{\lambda^2}{\Omega_A} \tag{3.52}$$

Thus,

$$A_{\text{eff}}\Omega_A = \lambda^2 \tag{3.53}$$

and

$$w_{\text{ext}} = kT\epsilon \tag{3.54}$$

independent of the antenna radiation pattern.

Thus, given T_{scene}, we can immediately calculate w_{ext}. We now consider the sky as the "scene." Although most radiation from astronomical sources is nonthermal, that is, it does not follow the Planck distribution, nevertheless, we typically refer to T_{sky} as the temperature that a blackbody sky would require to radiate the signal observed at a particular frequency.

Narrowbeam radars that typically observe faint targets against the sky sometimes use cooled receivers. Because T_{sky} is low, it is useful to reduce the internal radar temperature also, to minimize the noise power input to the receiver (see Section 2.2.11).

Figure 3.17 illustrates T_{sky} at 408 MHz versus right ascension (α) and declination (δ), the spherical coordinate system appropriate for the sky. Obviously, what is sky noise to the radar engineer is a highly interesting signal to the radio astronomer. If the noise is important, the radar boresight direction in (α, δ) should be calculated and T_{sky} incorporated into the estimate of the noise power (Problem 3.17).

The sun is also a significant radio source; it experiences times of "quiet sun" and "active sun." If the sun is expected to be near the radar mainlobe, its effects also should be included [7, 18].

References

[1] Knott, E. F., J. F. Shaeffer, and M. T. Tuley, *Radar Cross Section*, 2nd ed., Norwood, MA: Artech House, 1993.

[2] Ruck, G. T., et al., *Radar Cross Section Handbook*, New York: Plenum, 1970 (2 Volumes).

[3] Crispin, J. W., and K. M. Siegel, *Methods of Radar Cross-Section Analysis*, New York: Academic, 1968.

[4] Mie, G., "A Contribution to the Optics of Turbid Media, Especially Colloidal Metallic Suspensions," *Ann. Physik*, Vol. 25, 1908, pp. 377–445.

[5] Stratton, J. A., *Electromagnetic Theory*, New York: McGraw-Hill, 1941.

[6] Kerr, D. E., et al., *Propagation of Short Radio Waves*, New York: Dover Publications, 1965.

[7] Blake, L. V., *Radar Range-Performance Analysis*, Norwood, MA: Artech House, 1986.

[8] Stutzman, W. L., and G. A. Thiele, *Antenna Theory and Design*, New York: Wiley, 1981.

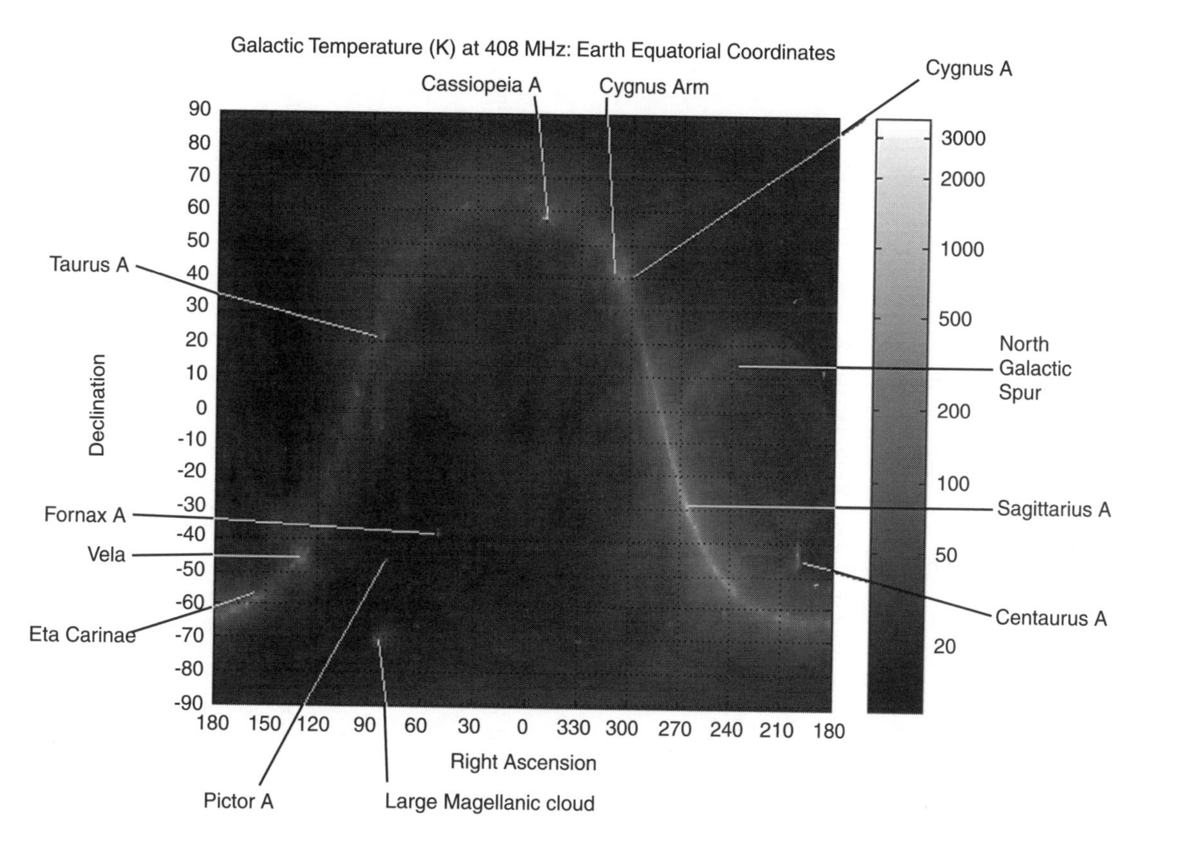

Figure 3.17 Galactic temperature (K) at 408 MHz: Earth equatorial coordinates (provided courtesy of Dr. James Heagy, Mr. Joel Iams, and Dr. James Ralston of the Institute for Defense Analyses, based on [21]).

[9] Spencer, R. C., "Optical Theory of the Corner Reflector," MIT Radiation Laboratory Report 435, 1944.

[10] Mott, H., *Polarization in Antennas and Radar*, New York: Wiley, 1986.

[11] Skolnik, M. I., *Introduction to Radar Systems*, 2nd edition, New York: McGraw-Hill, 1980.

[12] Townes, C. H., and A. L. Schawlow, *Microwave Spectroscopy*, New York: Dover, 1975.

[13] Barton, D. K., *Modern Radar System Analysis*, Norwood, MA: Artech House, 1988.

[14] Crane, R. K., *Electromagnetic Wave Propagation Through Rain*, New York: Wiley, 1996.

[15] Born, M., and E. Wolf, *Principles of Optics*, 4th ed., New York: Pergamon, 1969.

[16] Jackson, J. D., *Classical Electrodynamics*, 2nd ed., New York: Wiley, 1975.

[17] Rossi, B., *Optics*, Reading, MA: Addison-Wesley, 1957.

[18] Kraus, J. D., *Radio Astronomy*, 2nd ed., Powell, OH: Cygnus-Quasar Books, 1986.

[19] Leighton, R. B., *Principles of Modern Physics*, New York: McGraw-Hill, 1959.

[20] Allen, C. W., *Astrophysical Quantities*, 2nd ed., New York: Oxford University Press, 1963.

[21] Haslam, C. G. T., et al. "A 408 MHz All-Sky Continuum Survey. II. The Atlas of Contour Maps," *Astron. Astrophys. Suppl. Ser.*, Vol. 47, 1982, p. 1.

Problems

Problem 3.1

Show that, assuming geometrical optics ($\lambda << a$) and a perfectly reflecting sphere, the incident radiation is scattered uniformly into all 4π steradians (isotropic scattering) and the RCS is

$$\sigma = \pi a^2$$

Problem 3.2

Calculate the RCS (in dBsm) for a square, flat plate with side = 0.1m, 1m, and 3m for mid-L, -S, -C, -X, and -Ku bands.

Problem 3.3

The broadside RCS of a flat plate can be speciously derived as follows:

$$Q = \frac{SA}{\Omega_A} = \frac{SAG}{4\pi} = \frac{SA^2}{\lambda^2}$$

$$\sigma = \frac{4\pi Q}{S} = \frac{4\pi A^2}{\lambda^2}$$

Does that derivation apply off broadside? Comment on its validity.

Problem 3.4

Show that for a corner reflector the direction of the trihedral angle boresight makes an angle of 35.26 degrees with each planar side.

Problem 3.5

Show that the RCS of a triangular trihedral at boresight is

$$\sigma = \frac{4\pi A_{\text{eff}}^2}{\lambda^2} = \frac{4\pi a^4}{3\lambda^2}$$

Problem 3.6

From (3.12), use a computer to plot the triangular trihedral RCS = $4\pi A^2/\lambda^2$ as a function of azimuth and elevation (each coordinate axis contains an edge). Verify that the peak RCS value agrees with that in Problem 3.5 and that its elevation angle agrees with that in Problem 3.4.

Problem 3.7

a. Compute the PSM for circular polarization for the flat plate, trihedral, and dihedral at arbitrary ψ. Show that for a dihedral, it is independent of ψ, except for phase.
b. Show that the normalized PSM is unitary (see Table 8.1).

Problem 3.8

(Note: Do not use 4/3 Earth approximation in this problem.) For a satellite at altitude 600 km, calculate the slant range R_s and ground range R_g (arc) to the horizon. If R_g is half this value find

$$\frac{R_s(\textit{flat Earth})}{R_s(\textit{curved Earth})} = \frac{(h^2 + R_g^2)^{1/2}}{R_s}$$

Show that

$$\lim_{R_g \to 0}\left[\frac{R_s(\textit{flat Earth})}{R_s(\textit{curved Earth})}\right] = 1$$

Problem 3.9

Verify that, for grazing angle ≅ 0, an EM wave undergoes a 180-degree phase shift on reflection from a flat, dielectric surface. (Hint: See [5, 16, or 17]).

Problem 3.10

Show that, for a flat RCS measurement range,

$$h_a h_t \cong \frac{R\lambda}{4}$$

Problem 3.11

For a lossless isotropic antenna transmitting at wavelength λ a height H above a flat plane of reflectivity γ, show that

$$\Gamma = \frac{P_{\text{Rx-peak}}}{P_{\text{Tx-peak}}} = \frac{P_{\text{Rx-avg}}}{P_{\text{Tx-avg}}} = \frac{\gamma\lambda^2}{96\pi^2 H^2}$$

Problem 3.12

Show that the maximum theoretical value of γ is $1/\pi$.

Problem 3.13

An airborne radar has h_{radar} = 20 km, $\theta = 2^0$, $\phi = 6^0$, $f \cong$ 10 GHz, and τ = 100 ms. While observing terrain with $\sigma^0 = -25$ dB at R = 200 km, it encounters a large rainstorm located between R_{storm} and R_{storm} + 25 km, and height 5 km. Estimate the attenuation (decibels) and ratio of returns from the ground and from the rain, for rainfall rates of 1, 4, 16, and 64 mm/hr; for f = 10 and 16 GHz; and for R_{storm} = 150 km and 180 km. Would you, the designer, have chosen X band or Ku band for the radar?

Problem 3.14

Integrate (3.43) over frequency and solid angle to show that the total radiation emitted from a surface is σT^4, where

$$\sigma = \text{Stefan-Boltzmann constant} = \frac{2\pi^5 k^4}{15 c^2 h^3} = 5.672 \times 10^{-8}\ \text{W m}^{-2}\ \text{K}^{-4}$$

Problem 3.15

From Kirchhoff's law and Figure 3.10, estimate P_{noise} for a 10-GHz antenna at sea level pointing at the zenith on a clear night, with receiver cooled by liquid helium (T = 4.2K). (See Section 2.2.11 and [18, Chap. 8].)

Problem 3.16

Sidereal time is defined as the right ascension of the zenith. For a ground-based, narrowbeam 408-MHz radar near Los Angeles observing a distant aerostat at azimuth 120 degrees (clockwise from north) and elevation 40 degrees, at what sidereal time will the noise figure be at its maximum? Approximately what local time is this in January? (Use Figure 3.17 and assume quiet sun.)

4

Elementary Radar Signal Processing

Having examined how a radar signal is generated, is emitted from the antenna, scatters from external objects, reenters the radar, and is recorded as a voltage by an A/D converter, we now begin consideration of how to interpret the resulting measurement. Throughout this chapter, the returned echo is considered as a signal voltage plus a noise voltage, both analog (continuous). Effects of A/D quantization noise can be considered part of the noise voltage.

4.1 Detection of Radar Signals in Noise and Clutter

Consider this simple question: Suppose we transmit monochromatic pulses (by which we mean pulses with a single frequency, except for the inevitable bandwidth resulting from the finite pulse width), sample the return from a particular range gate, and ask, "Is there a point target in the range gate?"

4.1.1 Detector Characteristics

The term *detector* refers to "that portion of the radar receiver from the output of the IF amplifier to the input of the indicator or data processor" [1, p. 382]. *Detection* refers to the process of deciding whether to declare the presence of a target [2]. The relationship between the signal input and the detector output may be complicated. For simplicity, this chapter considers the following detector types:

- Coherent detector, which utilizes both amplitude and phase information (phase of signal is known);

- Envelope detector (noncoherent detector), which samples the envelope (magnitude) of the output voltage (phase of signal is unknown).

The following relationships between envelope detector output and input may exist:

- Linear detector: Detector output voltage is proportional to the signal input voltage.
- Square-law detector: Detector output voltage is proportional to the square of the signal input voltage.

4.1.2 Fundamentals of Detection

As illustrated in Figure 4.1, two types of errors can be made in the declaration of the presence or absence of a target:

- Target is absent, detector says target is present (false alarm).
- Target is present, detector says target is absent (missed detection).

Assume that we know the probability distribution of the measurement of received voltage $V \geq 0$ for the cases of (1) target absent and (2) target present. The former is noise only, the latter is a signal voltage plus the noise. Figure 4.2 represents the two probability density functions (PDFs), $p_0(V)$ and

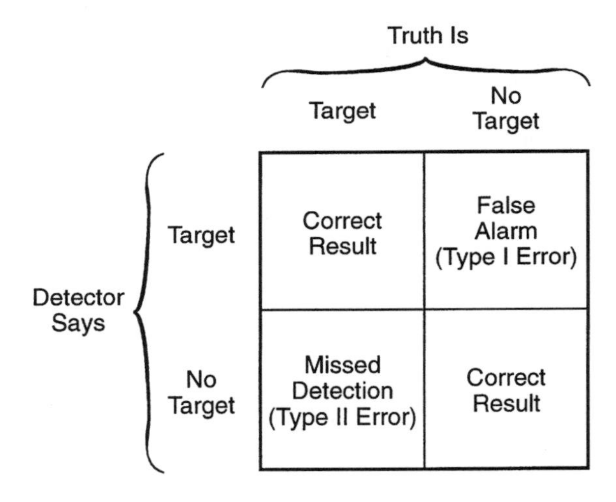

Figure 4.1 Types of target-declaration errors.

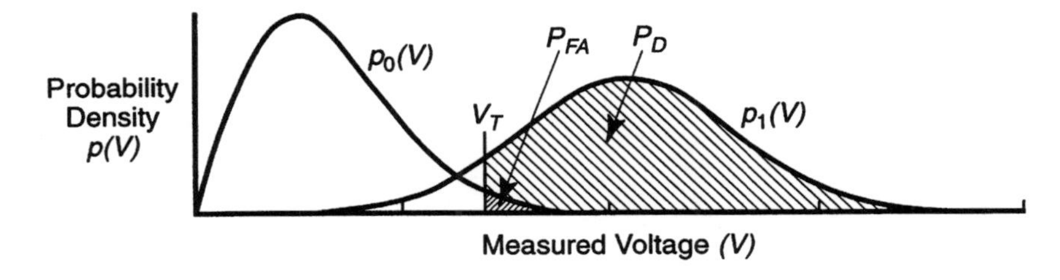

Figure 4.2 Calculation of P_D and P_{FA}.

$p_1(V)$. The probability that the measured voltage is between V and $V + dV$ is $p_i(V)dV$, where $i = 0$ (target absent) or 1 (target present). The integral of each p_i from 0 to infinity is unity. Each p_i is characterized by a mean voltage μ_i.

We make the detection decision by choosing a threshold voltage, V_T, and declaring

- $V > V_T \Rightarrow$ target present.
- $V < V_T \Rightarrow$ target absent.

The probability of false alarm (P_{FA}) and the probability of correct detection (P_D) are the integrals of p_0 and p_1, respectively, from V_T to infinity.

With noise present, it is not possible to achieve the ideal case of $P_D = 1$, $P_{FA} = 0$. As V_T increases, P_{FA} decreases, but so does P_D. A curve of P_D versus P_{FA} (as V_T varies) is called a receiver operating characteristic (ROC) curve (Figure 4.3). The closer the curve is to the upper left corner ($P_D = 1$, $P_{FA} = 0$), the better is the detection procedure [3].

The threshold V_T can be determined if we know the costs of false alarms (Type I errors) and missed detections (Type II errors) [3, p. 24]. We define

- C_{MD} = cost of a missed detection = cost of deciding target is absent (*hypothesis* H_0) if target is present (*hypothesis* H_1).
- C_{FA} = cost of a false alarm = cost of deciding H_1 if H_0 is true.
- Risk = $C_{MD}(1 - P_D)N_T + C_{FA}P_{FA}N_A$, where N_T and N_A are the a priori numbers of opportunities for detections and false alarms, respectively (i.e., expected numbers of targets and dwells, respectively).

The Bayes criterion states that we should minimize the risk. We define the ratio between $p_1(V)$ and $p_0(V)$ (from Figure 4.2) as the likelihood ratio $\Lambda(V)$. The Bayes criterion then leads to the condition that

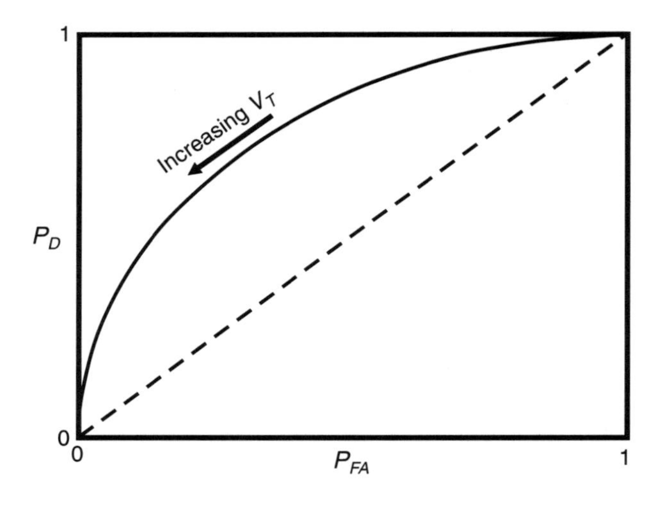

Figure 4.3 ROC curve (schematic).

$$\Lambda(V) = \frac{p_1(V)}{p_0(V)} \underset{H_0}{\overset{H_1}{\gtrless}} \frac{C_{FA}N_A}{C_{MD}N_T} \tag{4.1}$$

The likelihood ratio is a useful test statistic, that is, a positive scalar quantity to be compared with a threshold, with the result determining whether the target is declared to be present or absent. In actual situations, C_{FA} and C_{MD} may be impossible to quantify; in such cases it is useful to obtain a ROC curve and allow the user to select the operating point, that is, the values of P_D and P_{FA} to be chosen.

If $p_0(V)$ and $p_1(V)$ can be calculated, then the detection problem can be solved. For given assumptions about target and noise PDFs, there are four parameters of interest: SNR (ratio of μ_1^2 to μ_0^2), V_T, P_D, and P_{FA}. We also may have many (n) pulses; integration of those pulses can lead to a higher P_D than in the single-pulse case. Integration may be noncoherent or coherent. Results of all these cases have been calculated and published in many references (e.g., [1, 4–9]).

Unless indicated otherwise, this book makes the usual assumption that, for the noise, the PDFs of I and Q are zero-mean Gaussian (Problem 4.1):

$$p_{\substack{I\\Q}}(V_{\substack{I\\Q}}) = \frac{1}{\sigma_v(2\pi)^{1/2}} \exp\left(\frac{-V_{\substack{I\\Q}}^{\,2}}{2\sigma_v^2}\right) \tag{4.2}$$

where σ_v is the standard deviation of the noise voltage.

Precise calculation of P_D and P_{FA} is tedious. However, good approximations are well summarized by Barton [4, Chap. 2] and may be readily programmed on a personal computer. The detectability factor, D, is "in pulsed radar, the ratio of single-pulse signal energy to noise power per unit bandwidth that provides stated probabilities of detection and false alarm" [2]. Following [4], we define four types of detection. "The methods are listed in order of declining efficiency of integration, and also declining complexity of implementation" [4, p. 69].

- *Coherent integration,* in which pulses are added prior to envelope detection;
- *Noncoherent (or video) integration,* in which each pulse is envelope detected, and the resulting video pulses are added together prior to application of thresholding;
- *Binary integration,* in which each pulse is applied to a threshold, and the number M of threshold crossings is used as the criterion for an output alarm;
- *Cumulative detection,* in which $M = 1$ is the alarm criterion.

A brief summary of detection results follows; further details are provided in [4, Chap. 2].

4.1.3 Detection of Nonfluctuating Target in Noise

We first consider detection of a point target that is nonfluctuating.

4.1.3.1 Coherent Detection of a Single Pulse of Known Phase

From [4, Eq. 2.2.8],

$$D_C(1) = [erfc^{-1}(2P_{FA}) - erfc^{-1}(2P_D)]^2 \tag{4.3}$$

where we use a function derived from the error function erf(V) [4, p. 66]:

$$\begin{aligned} \mathrm{erf}(V) &\equiv \frac{2}{\sqrt{\pi}}\int_0^V \exp(-v^2)dv = X, \ \mathrm{erf}^{-1}(X) = V \\ \mathrm{erfc}(V) &\equiv 1 - \mathrm{erf}(V) = 1 - X, \ \mathrm{erfc}^{-1}(1 - X) = V \end{aligned} \tag{4.4}$$

4.1.3.2 Envelope Detection of a Single Pulse

From [4, Eq. 2.2.11] (North's approximation),

$$D_0(1) \approx [\sqrt{\ln(1/P_{FA})} - \text{erfc}^{-1}(2P_D)]^2 - 1/2 \tag{4.5}$$

It is instructive to examine this important case in more detail. For a single pulse, we denote the following outputs from envelope detector:

- r = magnitude of noise voltage (variable);
- A = magnitude of signal (target) voltage (constant).

As originally shown by Rice for a square-law detector and summarized by many others (e.g., [4, 6, 7, 10]; notation is from [10]), the PDF of r, the received voltage, for the target-present case is

$$p_1(r) = \frac{r}{\sigma_v^2} \exp\left[-\frac{(r^2 + A^2)}{2\sigma_v^2}\right] \cdot I_0\left(\frac{rA}{\sigma_v^2}\right) \tag{4.6}$$

where A is the received signal (without noise) voltage, $I_0(z) = J_0(jz)$, a modified Bessel function [11, Chap. 9]. This is a Rician PDF. For the target-absent case, we set $A = 0$ and obtain the PDF for noise only:

$$p_0(r) = \frac{r}{\sigma_v^2} \exp\left(-\frac{r^2}{2\sigma_v^2}\right) \tag{4.7}$$

which is a Rayleigh PDF. Then,

$$P_{FA} = \int_{V_T}^{\infty} \frac{r}{\sigma_v^2} e^{-r^2/2\sigma_v^2} dr = e^{-V_T^2/2\sigma_v^2} \tag{4.8}$$

A good approximation to P_D is [4, Eq. 2.2.10]:

$$P_D \cong \frac{1}{2}\left[1 - \text{erfc}\left(\sqrt{\ln\left(\frac{1}{P_{FA}}\right)} - \sqrt{SNR + \frac{1}{2}}\right)\right] \tag{4.9}$$

4.1.3.3 Coherent Integration of n Pulses

When n pulses are coherently integrated, the resulting SNR is n times the SNR for a single coherently integrated pulse (Figure 4.4). Thus [4, p. 69],

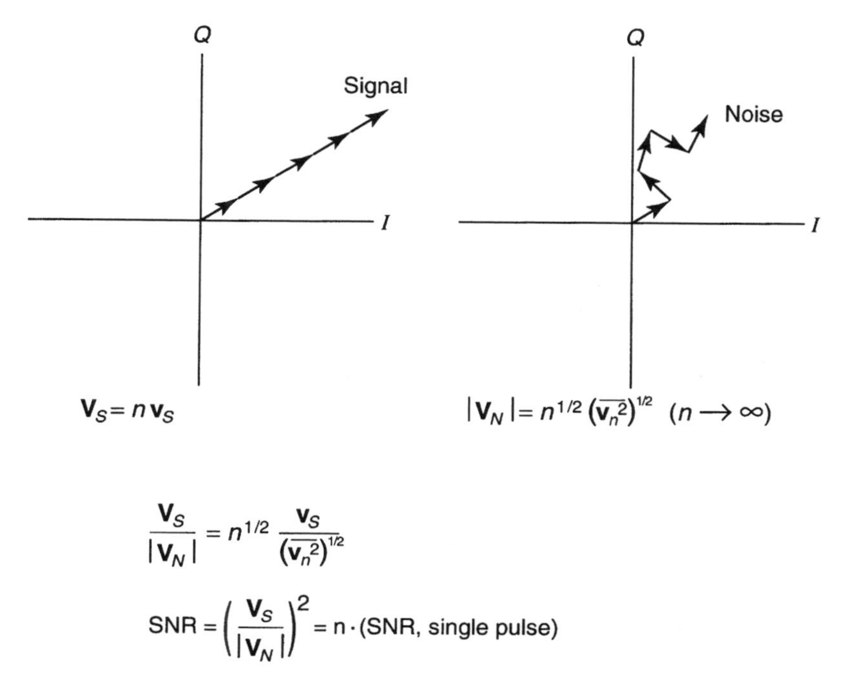

Figure 4.4 Coherent integration.

$$D_c(n) = \frac{D_c(1)}{n} \tag{4.10}$$

4.1.3.4 Integration Loss for n Noncoherent Pulses

Integration loss[1] is the required increase in total collected energy for n pulse compared with the value for single-pulse detection. From [4, Eq. 2.3.8],

$$L_i(n) \equiv \frac{nD_0(n)}{D_0(1)} \sim \frac{1 + \sqrt{1 + 9.2n/D_c(1)}}{1 + \sqrt{9.2/D_c(1)}} \tag{4.11}$$

4.1.4 Detection of Fluctuating Target in Noise

For fluctuating targets, we assume a probability distribution for the target voltage as well as the noise voltage.

4.1.4.1 Swerling Cases

As discussed in [4–6, 10] and other sources, Swerling defined four interesting cases (where σ is radar cross-section [1, p. 47]).

1. Detector loss is defined as $D_0(1)/D_c(1)$.

Swerling Case 1

$p(\sigma) = (1/\overline{\sigma})e^{-\sigma/\overline{\sigma}}$, dwell-to-dwell ("scan-to-scan") decorrelation (exponential PDF). Within a single dwell (a set of pulses to be integrated), all pulses are assumed to have the same magnitude. The value is drawn randomly from the distribution. The overall assumption is that the target consists of many scatterers of about the same size and fluctuates slowly compared with the dwell time. (This dwell sometimes corresponds to a scan over the target by an antenna rotating uniformly about a vertical axis. In such a case, a correction should also be made for the beamshape loss resulting from the varying gain as the antenna scans over the target.)

Swerling Case 2

The distribution is the same as in case 1, except for pulse-to-pulse decorrelation. The signal from each pulse is independently drawn from the distribution (fast fluctuations), and the target fluctuates rapidly compared with the dwell time.

Swerling Case 3

$p(\sigma) = (4\sigma/\overline{\sigma}^2)e^{-2\sigma/\overline{\sigma}}$. In case 3, there is dwell-to-dwell decorrelation; the distribution corresponds to a target with one large scatterer and many smaller ones.

Swerling Case 4

In case 4 , there is the same distribution as case 3, with pulse-to-pulse decorrelation.

These distributions are special cases of the chi-square PDF [1]:

$$p(\sigma) = \frac{k}{(k-1)\overline{\sigma}}\left(\frac{k\sigma}{\overline{\sigma}}\right)^{k-1} e^{-k\sigma/\overline{\sigma}} \tag{4.12}$$

Cases 1 and 2 correspond to $k = 1$; cases 3 and 4 correspond to $k = 2$.

For the case of a single pulse from a target that is Swerling-1 or Swerling-2 (for a single pulse they are the same), the result is surprisingly simple [4, Eq. 2.4.3]:

$$D_1(1) = SNR = \frac{\ln(P_{FA})}{\ln(P_D)} - 1 \tag{4.13}$$

$$P_D = P_{FA}^{1/(1+SNR)}$$

4.1.4.2 Fluctuation Loss

Fluctuation loss is the required increase in signal energy for detection of a fluctuating target compared with a nonfluctuating target. For a Swerling-1 target, single pulse [4, Eq. 2.4.4],

$$L_f(1) \equiv \frac{D_1(1)}{D_0(1)} \tag{4.14}$$

For a more general representation of a target, we can use a chi-square target model. From [4, Eq. 2.4.6],

$$L_f(n_e) = [L_f(1)]^{1/n_e} \tag{4.15}$$

where $n_e = 1$ for Swerling-1, $n_e = n$ for Swerling-2, $n_e = 2$ for Swerling-3, and $n_e = 2n$ for Swerling-4, but any value of $n_e > 1$ may be used. For the nonfluctuating target, we can take $n_e \to \infty$ and $L_f(n_e) = 1$.

4.1.5 Summary of Detection of Target in Noise

The overall result is [4, Eq. 2.4.8]

$$SNR_{\text{required}} = D_e(n, n_e) = \frac{D_0(1)L_i(n)L_f(n_e)}{n} \tag{4.16}$$

Calculation of the required SNR for a particular P_D, P_{FA}, n, and n_e can be made by hand-using Figures 2.2.2, 2.3.2, and 2.4.2 from [4]. Results are accurate to about ±0.3 dB (see Problem 4.2).

Blake [5, 6] presents useful graphs of those results in the following format: abscissa = n; ordinate = SNR (dB); parameter = P_{FA}; constant for graph = P_D. Several other authors [7–9] have prepared extensive graphical summaries using variations of "false-alarm number" instead of P_{FA}. (Caution: the graphs in [9] define SNR as the ratio of the peak of the received sinusoidal signal voltage to the rms noise voltage; thus, SNR [9] = 2 · *SNR* [1, 4–8].)

4.1.6 Binary and Cumulative Integration: Nonfluctuating Target

The term *binary integration* refers to the case for which we gather information from a number of dwells, N, declare for each that the target is present or absent (a binary decision), then combine the results of the binary decisions. The following discussion applies to a nonfluctuating target, where the probabilities for successive dwells are independent; the case for a fluctuating target is more complicated [4, p. 87]. Thus,

- The probability of detecting the target in one dwell is P_D.
- The probability of not detecting the target in one dwell is $1 - P_D$.

- The probability of not detecting the target in N dwells is $(1 - P_D)^N$.
- The probability of at least one detection in N dwells is $1 - (1 - P_D)^N$.

For M detections in N dwells,

- The probability of exactly M detections in N dwells is

$$P_D'(M, N) = \binom{N}{M} P_D^M (1 - P_D)^{N-M}, \text{ where } \binom{N}{M} = \frac{N!}{(N-M)!M!} \tag{4.17}$$

- The probability of at least M detections in N dwells (M-out-of-N detection) is

$$P_D(M, N) = \sum_{k=M}^{N} \binom{N}{k} P_D^k (1 - P_D)^{N-k} \tag{4.18}$$

(See Problem 4.3.) Similar reasoning holds for P_{FA}.

Binary integration is often performed for convenience, for example, for a relatively slow-scanning radar searching for a target in nonstationary clutter. However, if the clutter distribution is stationary, the overall probability of detection is less for binary integration than for noncoherent integration. Cumulative detection is equivalent to binary detection with $M = 1$ [4, p. 74].

4.1.7 Targets in Clutter

If the interference is clutter (land, sea, or air) rather than random noise, the detection calculations follow the same principle, but the interference distribution changes. It has been shown that for clutter, the following distributions may be more accurate representations of reality, depending on the details [12]:

$$\text{Log-normal: } p(V) = \frac{1}{Vs\sqrt{2\pi}} \exp\left\{-\frac{1}{2s^2}\left[\ln\left(\frac{V}{\mu}\right)\right]^2\right\} \tag{4.19}$$

Note the typographical error in [12, Eq. 1], where μ (voltage) and s (dimensionless) are the log-normal distribution parameters:

$$\text{Weibull: } p(V) = \frac{\eta}{\nu}\left(\frac{V}{\nu}\right)^{\eta-1} \exp\left[-\left(\frac{V}{\nu}\right)^{\eta}\right] \tag{4.20}$$

where η (dimensionless) and ν (voltage) are the Weibull distribution parameters. For $m = 2$, the distribution is Rayleigh.

4.2 Radar Waveforms

So far, we have considered only the monochromatic pulse (i.e., the product of a sinusoid and a rectangular function of duration t_p, which still has a nonzero bandwidth $\sim 1/t_p$). Let us now expand our horizons and consider more general radar signals, the properties of which can be effectively utilized for determining information about a target or a scene. The discussion here follows Levanon [10, Chaps. 5–7], who presents considerably more detail.

4.2.1 Generalized Radar Signals

This section utilizes the form of FTs involving ω (= $2\pi f$) rather than f [13, p. 22]:

$$\begin{aligned} H(\omega) &= \int_{-\infty}^{\infty} h(t)e^{-j\omega t}dt \\ h(t) &= \frac{1}{2\pi}\int_{-\infty}^{\infty} H(\omega)e^{j\omega t}d\omega \end{aligned} \tag{4.21}$$

The following denotations are made:

$$\begin{aligned} \text{voltage} &= s(t) \\ \text{power} &= |s(t)|^2 \\ \text{energy } (E) &= \int_{-\infty}^{\infty} |s(t)|^2 dt \end{aligned} \tag{4.22}$$

Proportionality constants are omitted because they disappear in normalization.

This section considers only bandpass signals, that is, those with narrow bandwidth $\Delta\omega << \omega_c$, centered about a carrier $\pm\omega_c$. There are at least four ways to represent such a signal [10]:

$$s(t) = g(t)\cos[\omega_c t + \phi(t)],\ g(t) = \textit{natural envelope} \geq 0 \tag{4.23a}$$

$$s(t) = g_c(t)\cos\omega_c t - g_s(t)\sin\omega_c t \tag{4.23b}$$

where $g_c(t) = g(t)\cos\phi(t) = I$ (the in-phase component), and $g_s(t) = g(t)\sin\phi(t) = Q$ (the quadrature component).

$$u(t) \equiv g_c(t) + jg_s(t) = \textit{complex envelope} \tag{4.23c}$$

$$s(t) = \mathrm{Re}[u(t)e^{j\omega_c t}]$$

Note: $g(t) = |u(t)|$.

$$s(t) = \frac{1}{2}[u(t)e^{j\omega_c t} + u*(t)e^{-j\omega_c t}] \text{ (Problem 4.4)} \tag{4.23d}$$

where * represents the complex conjugate. According to representation (4.23c), modulating $|u(t)|$ is amplitude modulation (AM); modulating arg $[u(t)]$ [the phase of $u(t)$] is phase modulation (PM), or frequency modulation (FM).

We also refer to positive and negative frequencies, which are defined simply as follows (Figure 4.5):

- Positive frequency occurs when the signal vector moves counter-clockwise in *I, Q* diagram.
- Negative frequency occurs when the signal vector moves clockwise in *I, Q* diagram.

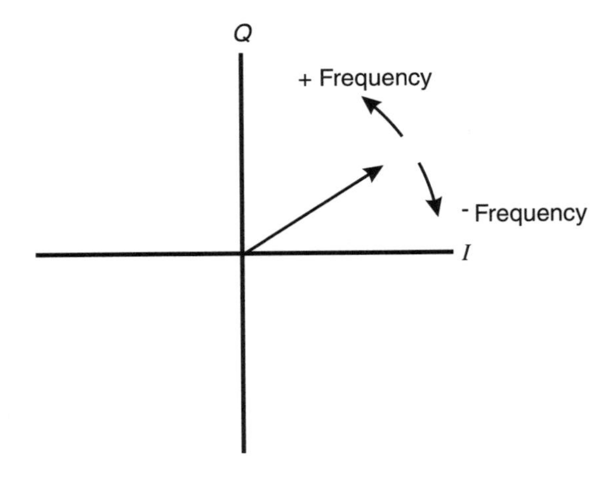

Figure 4.5 Positive and negative frequencies.

4.2.2 Matched Filter

Still following [10], we consider a baseband signal $s(t)$ with zero-mean Gaussian noise of spectral density (watts per hertz = joules) $N_0/2$. The noise density is double-sided, that is, it involves positive and negative frequencies. (Alternatively, we could use a single-sided representation, for which the noise density would be N_0.) The signal-plus noise is passed through a linear filter with transfer function $H(\omega)$. Let $u(t) \Leftrightarrow U(\omega)$ (the FT relationship); then at the measurement time, t_m,

$$u_{\text{out}}(t_m) = \frac{1}{2\pi}\int_{-\infty}^{\infty} H(\omega)U(\omega)e^{j\omega t_m}d\omega \tag{4.24}$$

The mean square value of the noise (independent of t) is

$$\langle n_{\text{out}}^2 \rangle = \frac{N_0}{4\pi}\int_{-\infty}^{\infty} |H(\omega)|^2 d\omega \tag{4.25}$$

The SNR at the filter output is then

$$SNR = \frac{\left| \int_{-\infty}^{\infty} H(\omega)U(\omega)e^{j\omega t_m}d\omega \right|^2}{\pi N_0 \int_{-\infty}^{\infty} |H(\omega)|^2 d\omega} \tag{4.26}$$

We use the Schwarz inequality: For any two complex signals $A(\omega)$, $B(\omega)$,

$$\left| \int_{-\infty}^{\infty} A(\omega)B(\omega)d\omega \right|^2 \leq \int_{-\infty}^{\infty} |A(\omega)|^2 d\omega \cdot \int_{-\infty}^{\infty} |B(\omega)|^2 d\omega \tag{4.27}$$

The equality holds if and only if

$$A(\omega) = \alpha B^*(\omega) \tag{4.28}$$

where α is a constant. Therefore,

$$SNR \leq \frac{1}{\pi N_0}\int_{-\infty}^{\infty} |U(\omega)|^2 d\omega = \left(\frac{1}{2\pi}\int_{-\infty}^{\infty} |U(\omega)|^2 d\omega\right) \cdot \frac{1}{N_0/2} = \frac{E}{(N_0/2)} = \frac{2E}{N_0} \tag{4.29}$$

Equality (maximum SNR) holds when

$$H(\omega) = KU^*(\omega)e^{-j\omega t_m} \tag{4.30}$$

where K is a constant;

$$H(\omega) \Leftrightarrow h(t) = Ku^*(t_m - t) \tag{4.31}$$

The filter impulse response, $h(t)$, is a delayed, mirror-image conjugate of the signal.

In summary, for any received radar signal, where $u(t) \Leftrightarrow U(\omega)$, we can use a matched filter processing method:

$$h(t) = Ku^*(t_m - t) \Leftrightarrow H(\omega) = KU^*(\omega)e^{-j\omega t_m} \tag{4.32}$$

with output

$$\begin{aligned} u_{\text{out}}(t) &= \frac{1}{2\pi}\int_{-\infty}^{\infty} H(\omega)U(\omega)e^{j\omega t}d\omega = \frac{K}{2\pi}\int_{-\infty}^{\infty} U(\omega)U^*(\omega)e^{j\omega(t-t_m)}d\omega \\ &= \int_{-\infty}^{\infty} u(\tau)h(t-\tau)d\tau = K\int_{-\infty}^{\infty} u(\tau)u^*[\tau - (t - t_m)]d\tau \qquad (4.33) \\ u_{\text{out}}(t_m) &= \frac{E}{N_0/2} \end{aligned}$$

N_0 is the noise power density that is identified with Boltzmann's constant multiplied by system temperature, kT_s [10, p. 11].

As explained by Nathanson [14], $u_{\text{out}}(t_m)$ is the ratio of the peak instantaneous signal power to the mean noise power. "If the output signal is a pulse of sine wave, it is conventional to define the output signal-to-noise ratio as the mean signal power . . . divided by the mean noise power, so that for the optimum filter . . . we have $S/N = E/N_0$" [14, p. 357].

The salient characteristics of a matched filter are the following:

- Its output SNR is the collected energy divided by the (single-sided) noise power density, kT_s [4, p. 67]: $SNR = E/N_0$.
- It is the optimum processor; no other processor (filter) can produce a higher SNR.

Thus, we want the radar signal processor to approximate a matched filter as closely as possible.

When we transmit a radar signal, we usually have some idea about what the target characteristics will be. Qualitatively, the idea behind a matched filter is this: We postulate a target with specific characteristics, and then we match the processor (filter) to that target. If our guess is exactly correct, the filter output is E/N_0, which we generally normalize to unity. If our guess is incorrect, the filter output is between 0 and 1.

Yet another way to represent the matched filter output is [10, pp. 112–113]

$$s_{\text{out}}(t) \cong \text{Re}[u_{\text{out}}(t)e^{j\omega_c t}] \tag{4.34}$$

where

$$u_{\text{out}}(t) = \frac{1}{2}Ke^{-j\omega_c t_m}\int_{-\infty}^{\infty} u(\tau)u*(\tau - t + t_m)d\tau \tag{4.35}$$

4.2.3 Matched-Filter Response to Its Delayed, Doppler-Shifted Signal

This section considers a radar processor that is testing for the existence of a point target, possibly moving. The received pulse will then be a delayed, doppler-shifted replica of the transmitted pulse. The response of the matched filter to its own signal (delayed by t and doppler-shifted by a frequency ν) is

$$u_{\text{out}}(t, \nu) = \frac{1}{2}Ke^{-j\omega_c t_m}\int_{-\infty}^{\infty} u(\tau)u*(\tau - t + t_m)e^{j2\pi\nu\tau}d\tau \tag{4.36}$$

We choose $t_m = 0$ and $K = 2$ for normalization; then

$$u_{\text{out}}(t, \nu) = \int_{-\infty}^{\infty} u(\tau)e^{j2\pi\nu\tau}u*(\tau - t)d\tau \tag{4.37}$$

We change notation by reversing the roles of t and τ denote u_{out} as χ:

$$\chi(\tau, \nu) = \int_{-\infty}^{\infty} u(t)u*(t - \tau)e^{j2\pi\nu t}dt \tag{4.38}$$

Of course, one could consider nonpoint targets, which are of great importance in actual radar operation. However, the mathematics become highly complicated.

4.2.4 Radar Ambiguity Function

We now have a function $\chi(\tau, \nu)$ that conveniently summarizes the response of a matched filter to a point target that is delayed and/or doppler-shifted with respect to the expected target for which the matched filter was "tuned." We shall also frequently refer to $|\chi(\tau, \nu)|$ and $|\chi(\tau, \nu)|^2$, as well as $\chi(\tau, \nu)$. χ or $|\chi|$ is analogous to voltage, and $|\chi|^2$ is analogous to power or energy. $|\chi(\tau, \nu)|^2$ is the radar ambiguity function (AF) [2].

If the radar transmits a signal and receives an echo $s(t)$ from a target, the processor (matched filter) is matched to a target at a particular expected delay (range) and doppler shift (LOS velocity). With respect to the expected delay and doppler shift, the echo exhibits an additional delay τ and doppler-shift ν; then the normalized (power) output of the matched-filter processor is the value of $|\chi(\tau, \nu)|^2$. Therefore, if we have a radar/waveform/processor matched to a target at a particular range and velocity (LOS), examination of the ambiguity function $|\chi(\tau, \nu)|^2$ gives information about the extent to which the radar can distinguish between that target and another target at the respective relative range and velocity delays, $c\tau/2$ and $\lambda\nu/2$, that is, the resolution and possible ambiguity in range and velocity. Hence, the term *ambiguity function.*

The normalization is such that

- The matched filter output when the target is just as expected ($\tau = 0$, $\nu = 0$) is

$$|\chi(0, 0)| = 1 \tag{4.39}$$

- When the target is not just as expected, the filter output cannot be greater than $|\chi(0, 0)|$ and generally is less:

$$0 \leq |\chi(\tau, \nu)| \leq 1 \tag{4.40}$$

- the integral of $|\chi(\tau, \nu)|^2 = 1$:

$$\iint_{-\infty}^{\infty} |\chi(\tau, \nu)|^2 d\tau d\nu = 1 \tag{4.41}$$

- $|\chi|$ exhibits symmetry across the origin:

$$|\chi(-\tau, -\nu)| = |\chi(\tau, \nu)| \tag{4.42}$$

The proofs of these statements are straightforward and are given as Problem 4.5.

Because $\chi(\tau, \nu)$ is a function of two variables, it is useful to consider "cuts" through χ, where we set either τ or ν equal to a constant and observe the resulting function of one variable. To consider the behavior in delay (therefore range) only, we examine the delay cut ($\nu = 0$):

$$\chi(\tau, 0) = \int_{-\infty}^{\infty} u(t)u*(t - \tau)dt \equiv R(\tau) \tag{4.43}$$

defined as the autocorrelation of the time-domain signal. By the Einstein-Wiener-Khintchine relations [15, p. 254], the autocorrelation of a signal is equal to the inverse FT of the power spectral density (PSD) in the frequency domain, $|S(\omega)|^2$. Therefore, the range behavior of χ is determined by (only) the PSD of the signal.

To examine the doppler behavior of χ, we consider the doppler cut ($\tau = 0$):

$$\chi(0, \nu) = \int_{-\infty}^{\infty} |u(t)|^2 e^{j2\pi\nu t} dt \tag{4.44}$$

Therefore, $\chi(0, \nu)$ is determined by (only) $|u(t)|$, the AM of the signal.

To clarify the meaning and application of the AF, we next consider some relatively simple examples.

4.2.5 Example 1: One Monochromatic Pulse; Range and Velocity Resolution

First we define the rectangular function:

$$\text{rect}(x) = \begin{cases} 1, & |x| < 1/2 \\ 0, & |x| \geq 1/2 \end{cases} \tag{4.45}$$

Then we consider the signal:

$$s(t) = Re\left[u(t)e^{j\omega_c t}\right] \tag{4.46}$$

where

$$u(t) = \frac{1}{\sqrt{t_p}}\text{rect}\left(\frac{t}{t_p}\right)$$
$$\int_{-\infty}^{\infty} |u(t)|^2 dt = 1 \tag{4.47}$$

Then (Problem 4.6),

$$|\chi(\tau, \nu)| = \left(1 - \frac{|\tau|}{t_p}\right)\text{sinc}\left[\pi\nu t_p\left(1 - \frac{|\tau|}{t_p}\right)\right], \ |\tau| \le t_p; \ 0 \text{ otherwise} \tag{4.48}$$

$$\text{Delay cut: } |\chi(\tau, 0)| = 1 - \frac{|\tau|}{t_p}, \ |\tau| \le t_p; \ 0 \text{ otherwise} \tag{4.49}$$

$$\text{Doppler cut: } |\chi(0, \nu)| = \text{sinc}(\pi\nu t_p) \tag{4.50}$$

$|\chi(\tau, 0)|$ is a triangular function and $|\chi(0, \nu)|$ is a sinc function.

With this example, the AF leads to the subject of resolution. Consider a radar observing two zero-velocity point targets of equal RCS at different ranges. The range resolution of the radar is the minimum distance between them such that they can be resolved as two distinct targets. The velocity resolution is the analogous parameter in the velocity dimension.

As will be made clearer in Section 4.3, the resolution (range or velocity) of a radar depends on the SNR; the higher the SNR, the easier it is to resolve the two targets. We seek an SNR-independent way to qualitatively indicate the radar resolution. For that, as with an antenna pattern, we usually choose either (a) the peak-to-first-null value of $|\chi(\tau, \nu)|^2$, the power version of the AF, which we designate by the subscript *pn*, or (b) the half-power width (or 3-dB width) of the peak, which we designate by the subscript *3dB*. With those definitions, for the single monochromatic pulse,

$$\begin{aligned}\delta\tau_{pn} &= t_p \\ \delta f_{d\text{-}pn} &= (1/t_p) \\ \delta\tau_{3dB} &= t_p \cdot 2(1 - 1/\sqrt{2}) = (0.586)t_p \\ \delta f_{d\text{-}3dB} &= (0.886)(1/t_p) \text{ (Section 2.2)}\end{aligned} \tag{4.51}$$

4.2.6 Example 2: One Linear Frequency Modulated Pulse; Pulse Compression

We next consider a single pulse, the frequency of which changes linearly with time:

$$s(t) = \frac{1}{\sqrt{t_p}}\text{rect}\left(\frac{t}{t_p}\right)e^{jt(\omega_c + \pi\gamma t)}, \; u(t) = \frac{1}{\sqrt{t_p}}\text{rect}\left(\frac{t}{t_p}\right)e^{j\pi\gamma t^2} \tag{4.52}$$

with variable frequency

$$f(t) = \frac{1}{2\pi}\frac{d}{dt}(\omega_c t + \pi\gamma t^2) = \frac{\omega_c}{2\pi} + \gamma t \tag{4.53}$$

The units of γ are (second)$^{-2}$. Via the FT, it can be verified that the RF bandwidth $\cong \gamma t_p = B$, for $B >> 1/t_p$ ($\gamma >> 1/t_p^2$) (Problem 4.7).

The AF for the single linear frequency modulated (LFM) pulse is

$$|\chi(\tau, \nu)| = \left(1 - \frac{|\tau|}{t_p}\right)\text{sinc}\left[\pi t_p\left(1 - \frac{|\tau|}{t_p}\right)(\nu + \gamma\tau)\right], \; |\tau| \le t_p; \; 0 \text{ otherwise} \tag{4.54}$$

(Problem 4.8). The doppler cut is quite simple:

$$|\chi(0, \nu)| = \text{sinc}(\pi\nu t_p) \tag{4.55}$$

which is the same as the single monochromatic pulse [recall that $\chi(0, \nu)$ is determined only by the amplitude modulation]. The delay cut is

$$|\chi(\tau, 0)| = \left(1 - \frac{|\tau|}{t_p}\right)\text{sinc}\left[\pi t_p\left(1 - \frac{|\tau|}{t_p}\right)\gamma\tau\right] \tag{4.56}$$

$$= \left(1 - \frac{|\tau|}{t_p}\right)\text{sinc}[\pi\gamma\tau(t_p - |\tau|)] \tag{4.57}$$

$$= \left(1 - \frac{|\tau|}{t_p}\right)\text{sinc}[\pi(\gamma\tau t_p - \gamma\tau/\tau)], \; |\tau| \le t_p, \; 0 \text{ otherwise} \tag{4.58}$$

Let $\tau \equiv \alpha t_p$, $0 < |\alpha| \tilde{<} 0.01$, and let $\beta = \gamma t_p^2$, $\beta >> 1$; $\beta\alpha = \gamma t_p \tau = B\tau$. Then,

$$
\begin{aligned}
|\chi(\tau, 0)| &= (1 - |\alpha|)\operatorname{sinc}[\pi(\gamma\alpha t_p^2 - \gamma\alpha|\alpha|t_p^2)] \\
&= (1 - |\alpha|)\operatorname{sinc}[\pi\beta\alpha(1 - |\alpha|)]
\end{aligned} \tag{4.59}
$$

Because $|\alpha| << 1$, $|\chi(\tau, 0)| \cong \operatorname{sinc}(\pi\beta\alpha) = \operatorname{sinc}(\pi B\tau)$.
Thus, the peak-to-first-null values are

$$\delta\tau_{pn} \cong 1/B \tag{4.60a}$$

$$\delta r_{pn} \cong c/2B \tag{4.60b}$$

and the 3-dB width in range is

$$\delta r_{3dB} \cong (0.886)(c/2B) \text{ (Section 2.2)} \tag{4.60c}$$

That is an extremely important result. If we modulate (LFM) a pulse of width t_p so that it has a bandwidth $B >> 1/t_p$, then δr_{pn} is reduced from the monochromatic-pulse value $ct_p/2$ to a "compressed" value $c/2B$, where

$$\frac{\delta r_{pn} \text{ (monochromatic)}}{\delta r_{pn} \ (LFM)} = \frac{ct_p/2}{c/2B} = Bt_p \tag{4.61}$$

the time-bandwidth product of the LFM pulse. It is as if we had compressed the pulse to a shorter width by a factor of Bt_p, and reduced δr_{pn} by approximately that factor. Therefore, this technique for attaining good range resolution is called *pulse compression* and is widely used in modern radar. Although it is beyond the scope of this book, it turns out that if the pulse bandwidth is increased from $1/t_p$ to B by virtually any type of modulation (e.g., nonlinear FM or phase coding), pulse compression by a factor of Bt_p generally can be achieved [10, Chap. 8] (see also Section 7.2.1).

4.2.7 Example 3: Coherent Pulse Train; Resolution and Ambiguity in Range and Velocity

We now consider a set of N coherent pulses

$$s(t) = \operatorname{Re}[u(t)e^{j\omega_c t}] \tag{4.62}$$

$$u(t) = \frac{1}{\sqrt{N}} \sum_{n=0}^{N-1} u_n(t - nT_R) \tag{4.63}$$

where T_R = pulse-repetition interval = 1/(PRF) = $1/f_R$. The factor of $1/\sqrt{N}$ is for normalization. Note that the carrier frequency is coherent (referenced to the same zero of phase) throughout the pulse train. That corresponds to each pulse being formed by mixing the pulse modulation function with the STALO output. This pulse-to-pulse coherence is essential to modern radar operation.

Also, assume that $u_n(t) = u_m(t)$ for all n, m, that is, all pulses have the same complex envelope: the same amplitude, pulse width, and, if LFM, the same γ. Furthermore, it is assumed that $T_R >> t_p$, which normally is the case for radars. Defining

$$\text{sind}(N, x) \equiv \frac{\sin(Nx)}{\sin(x)}, \quad \text{sind}(N, 0) = N \tag{4.64}$$

we have [10, p. 139],

$$|\chi(\tau, \nu)| = \frac{1}{N} \sum_{p=-(N-1)}^{N-1} |\chi_c(\tau - pT_R, \nu)| \cdot \text{sind}[(N - |p|), \pi\nu T_R] \tag{4.65}$$

where $\chi_c(\tau, \nu)$ is the ambiguity function of an individual pulse.

4.2.7.1 Train of Monochromatic Pulses

For a train of monochromatic pulses, from (4.48) and (4.65),

$$|\chi(\tau, \nu)| = \frac{1}{N}\sum_p \text{sind}[(N - |p|), \pi\nu T_R] \cdot \left(1 - \frac{|\tau - pT_R|}{t_p}\right)\text{sinc}\left[\pi\nu t_p\left(1 - \frac{|\tau - pT_R|}{t_p}\right)\right] \tag{4.66}$$

$|\tau - pT_R| \leq t_p$; 0 otherwise (this condition, denoted here as the "truncation condition," is understood throughout the remainder of Section 4.2.7).

The delay cut is

$$|\chi(\tau, 0)| = \frac{1}{N}\sum_p (N - |p|)\left(1 - \frac{|\tau - pT_R|}{t_p}\right) \tag{4.67}$$

The term for $p = 0$ is $\chi_0(\tau, 0) = 1 - |\tau|/\tau_p$. Thus, $\delta\tau_{pn} = t_p$, $\delta r_{pn} = ct_p/2$, as found in Section 4.2.5. For $N >> 1$, the first few ambiguities (small p values) occur at the peaks of

$$\chi_p(\tau, 0) = 1 - \frac{|\tau - pT_R|}{t_p} \tag{4.68}$$

which peaks at $\tau = \pm pT_R$. The corresponding range ambiguities are

$$\Delta r = p \cdot \frac{cT_R}{2}, \; p = 0, \pm 1, \pm 2, \ldots \tag{4.69}$$

The first ambiguity in delay (and range) is

$$\tau_{\text{first}} = T_R \text{ (delay)}, \quad r_{\text{first}} = \frac{cT_R}{2} \text{ (range)} \tag{4.70}$$

For the doppler cut, because of the truncation condition, only the $p = 0$ term of the series survives,

$$|\chi(0, \nu)| = \frac{1}{N} \operatorname{sinc}(\pi \nu t_p) \cdot \operatorname{sind}(N, \pi \nu T_R) \tag{4.71}$$

That formula is completely analogous to the radiation pattern for an array or diffraction grating (Section 2.3). Because $T_R >> t_p$, the sinc factor varies very slowly relative to the sind factor. For ν near 0,

$$|\chi(0, \nu)| \cong \frac{1}{N} \cdot \operatorname{sind}(N, \pi \nu T_R) \tag{4.72}$$

Then,

$$\delta f_{d\text{-pn}} = \frac{1}{NT_R}, \; v_{\text{first}} = \frac{\lambda}{2} \cdot \frac{1}{NT_R} \tag{4.73}$$

and, consistent with Section 1.15,

$$f_{d\text{-first}} = \frac{1}{T_R}, \; v_{\text{first}} = \frac{\lambda}{2T_R} \tag{4.74}$$

4.2.7.2 Train of LFM Pulses

For a train of LFM pulses ($T_R >> t_p$, $\gamma t_p >> 1/t_p$), we have from (4.54) and (4.65),

$$|\chi(\tau, \nu)| = \frac{1}{N} \sum_{p=-(N-1)}^{N-1} |\chi_c(\tau - pT_R, \nu)| \operatorname{sind}[(N - |p|), \pi\nu T_R] \tag{4.75}$$

$$|\chi_c(\tau - pT_R, \nu)| = \left(1 - \frac{|\tau - pT_R|}{t_p}\right) \operatorname{sinc}\left[\pi t_p \left(1 - \frac{|\tau - pT_R|}{t_p}\right)(\nu + \gamma\tau)\right] \tag{4.76}$$

Figure 4.6 is a schematic illustration of this rather complicated function. Four distinct times are involved: $t_{\text{compressed}}$, t_{pulse}, $t_{\text{repetition}}$, and t_{dwell}. For a synthetic aperture radar (Chapter 7), typical values might be

- $t_{\text{compressed}} \sim 2$ ns;
- $t_{\text{pulse}} \sim 100$ μs;
- $t_{\text{repetition}} \sim 1$ ms;
- $t_{\text{dwell}} \sim 20$ s.

The ratio $t_{\text{dwell}}/t_{\text{compressed}}$ is $\sim 10^{10}$.

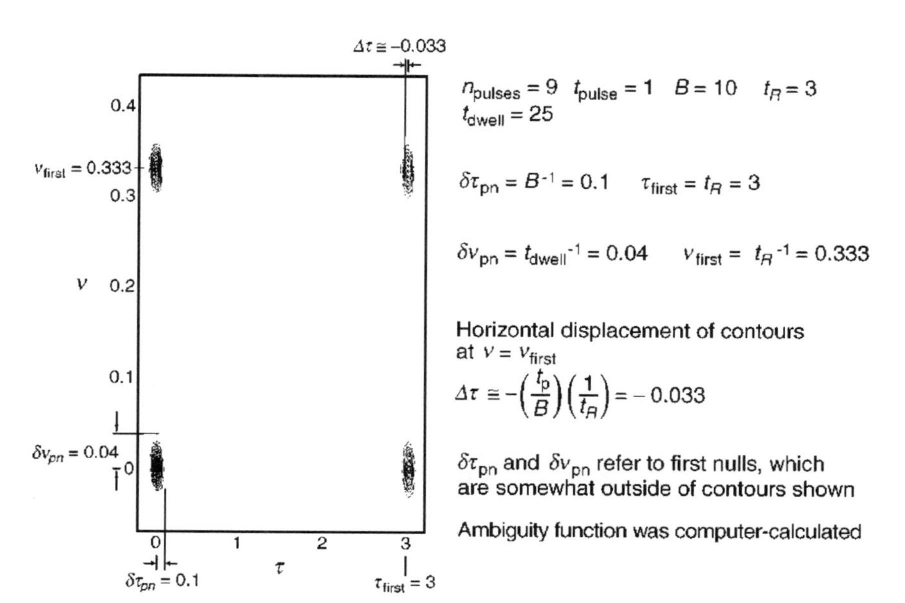

Figure 4.6 Ambiguity function for train of LFM pulses.

It is interesting to compare the relative values of the four times to relative distances within the solar system. The ratio of the semimajor axis of the orbit of the planet Pluto to the radius of Pluto itself is [16, p. 142]

$$\frac{R_{\text{Pluto}}}{r_{\text{Pluto}}} \cong \frac{6 \times 10^9 \text{ km}}{3 \times 10^3 \text{ km}} \sim 2 \times 10^6 \tag{4.77}$$

Thus, it is much harder to draw the LFM AF to scale than to draw the solar system to scale, and Figure 4.6 must remain a schematic drawing.

The delay cut is

$$|\chi(\tau, 0)| = \frac{1}{N}\sum_p (N - |p|)\left(1 - \frac{|\tau - pT_R|}{t_p}\right) \text{sinc}\left[\pi\gamma t_p \tau\left(1 - \frac{|\tau - pT_R|}{t_p}\right)\right] \tag{4.78}$$

To find $\delta\tau_{pn}$, we need consider only the $p = 0$ term:

$$|\chi_0(\tau, 0)| = \left(1 - \frac{|\tau|}{t_p}\right) \text{sinc}\left[\pi\gamma t_p \tau\left(1 - \frac{|\tau|}{t_p}\right)\right] \tag{4.79}$$

That is the same as for the single LFM pulse (4.58):

$$\delta\tau_{pn} = 1/B, \ \delta r_{pn} = c/2B \tag{4.80}$$

From Section 4.2.7.1, ambiguities will clearly occur at $t = pt$, $p = \pm 1, \pm 2, \ldots$ Thus,

$$\tau_{\text{first}} = T_R, \ r_{\text{first}} = cT_R/2 \tag{4.81}$$

as with the train of monochromatic pulses, consistent with Section 1.11.

The doppler cut is

$$|\chi(0, \nu)| = \frac{1}{N}\sum_p \left(1 - \frac{|pT_R|}{t_p}\right) \text{sinc}\left(\pi\nu t_p\left[1 - \frac{|pT_R|}{t_p}\right]\right) \cdot \text{sind}(N - |p|, \pi\nu T_R) \tag{4.82}$$

Because $T_R >> t_p$, and because of the truncation condition only the $p = 0$ term survives; thus,

$$|\chi(0,\ \nu)| = \frac{1}{N}\operatorname{sinc}(\pi\nu t_p)\operatorname{sind}(N,\ \pi\nu T_R) \cong \frac{1}{N}\operatorname{sind}(N,\ \pi\nu T_R) \tag{4.83}$$

As with the train of monochromatic pulses,

$$\delta f_{d\text{-}pn} = \frac{1}{NT_R},\ \delta v_{pn} = \frac{\lambda}{2NT_R} \tag{4.84}$$

$$f_{d\text{-first}} = \frac{1}{T_R},\ v_{\text{first}} = \frac{\lambda}{2T_R} \tag{4.85}$$

Table 4.1 summarizes peak-to-first-null and first-ambiguity values for the four cases considered.

4.2.7.3 Other Pulse Trains

Rihaczek [17] presents AFs for many functions. Vannicola et al. [18] describe AFs for chirp-diverse waveforms, where the chirp slope γ varies from one pulse to the next.

As shown in Figure 4.6, the AF for the train of LFM pulses consists of "ridges" with a slope of B/t_p. The ideal AF would equal unity near (0, 0) and zero elsewhere; such an AF is referred to as a *thumbtack AF* (recall that the integral must be unity). Costas [19] developed a class of waveforms that nearly have that property. A Costas pulse of order N consists of N contiguous chips (subpulses), each of duration T and each at a different frequency, selected from N frequencies spaced $1/T$ apart. The order of the frequencies determines the AF. The order may be represented by an *N-by-N* matrix of 1s and 0s. Each frequency is used once, and only one frequency is used at a given time; thus the matrix has one 1 in each row and one 1 in each column. If the correct Costas sequence is chosen, the AF will approximate a thumbtack, with sidelobes no greater than $1/N$ [20].

Consider a pulse train of M Costas pulses. If the sidelobes of the AFs of successive pulses did not overlap, the overall sidelobe level would be $1/MN$. "Regrettably, it was proven [21] that there do not exist two Costas [pulses] of order $N > 3$ with completely disjointed ambiguity sidelobe patterns" [20, p. 606]. Nevertheless, Freedman and Levanon [20] have developed a class of staggered Costas signals, that is, trains of pulses, each with a different Costas sequence, that closely approximate the thumbtack AF. [10] provides further details.

Table 4.1
Summary of Peak-to-First-Null and First-Ambiguity Values

	Delay (Range) Cut				Doppler (Velocity) Cut			
	Peak-to-First-Null		First Ambiguity		Peak-to-First-Null		First Ambiguity	
Waveforms	**Delay**	**Range**	**Delay**	**Range**	**Doppler**	**Velocity**	**Doppler**	**Velocity**
Single monochromatic pulse (t_p)	t_p	$\frac{ct_p}{2}$	—	—	$\frac{1}{t_p}$	$\frac{\lambda}{2t_p}$	—	—
Single LFM pulse (t_p, B)	$\frac{1}{B}$	$\frac{c}{2B}$	—	—	$\frac{1}{t_p}$	$\frac{\lambda}{2t_p}$	—	—
Train of monochromatic pulses (t_p, N, T_R)	t_p	$\frac{ct_p}{2}$	T_R	$\frac{cT_R}{2}$	$\frac{1}{NT_R}$	$\frac{\lambda}{2NT_R}$	$\frac{1}{T_R}$	$\frac{\lambda}{2T_R}$
Train of LFM pulses (t_p, B, N, T_R)	$\frac{1}{B}$	$\frac{c}{2B}$	T_R	$\frac{cT_R}{2}$	$\frac{1}{NT_R}$	$\frac{\lambda}{2NT_R}$	$\frac{1}{T_R}$	$\frac{\lambda}{2T_R}$

4.2.8 Coherent Processing Interval

As already discussed, for matched-filter processing (optimum processing), the SNR is equal to the collected energy E divided by kT_s. A necessary condition is that the waveform be coherent; otherwise, information is lost, and the processing cannot be optimum. Accordingly, when a radar is observing a target coherently during a specified time—known as a coherent dwell, or coherent processing interval (CPI)—and the return is predictable enough that processing essentially can be via a matched filter, then the SNR is essentially equal to E/kT_s. That justifies the form of the radar equation discussed in Section 1.11:

$$SNR \equiv \frac{E}{kT_s C_B} = \frac{P_{\text{avg}} G^2 \lambda^2 \sigma t_{\text{dwell}}}{(4\pi)^3 R^4 k T_s C_B L} = \frac{P_{\text{avg}} A^2 \eta^2 \sigma t_{\text{dwell}}}{4\pi \lambda^2 R^4 k T_s C_B L} \tag{4.86}$$

where C_B is the processing loss relative to the matched filter.

4.2.9 Example of CPI: The Search Radar Equation

We frequently consider a radar that scans a particular region of space to search for targets of interest. That is called a volume-search radar and may or may not use an ESA. A common example is a ground-based or airborne radar searching for aircraft. Consider a radar with a mechanically scanned antenna of dimensions L_x by L_y. We denote the beam solid angle of the radar by the equation $\Omega_b = \beta^2\lambda^2/L_xL_y = \beta^2\lambda^2/A$, where β is of the order of unity and depends on the aperture taper (Section 2.3.4). The solid angle to be searched is Ω_s. We also denote t_b as the beam dwell time between 3-dB beamwidths (assumed to be one CPI) and t_s as the time to search the region. A beamshape loss L_{bm} is included, approximately 3 dB [22]. (Fielding [22] discusses appropriate factors to use when the beam packing is more or less dense than that which achieves beam intersection at the 3-dB points, or when beam packing is not in a rectangular lattice.) Then,

$$\begin{aligned} \frac{\Omega_s}{t_s} &= \frac{\Omega_b}{t_b} \\ t_b &= \frac{\Omega_b t_s}{\Omega_s} = \frac{t_s}{\Omega_s} \cdot \frac{\beta^2\lambda^2}{A} \end{aligned} \tag{4.87}$$

For one beam dwell t_b, the collected energy is

$$E = \frac{P_{\text{avg}} A^2 \eta^2 \sigma}{4\pi R^4 \lambda^2 L L_{\text{bm}}} \cdot t_b = \frac{P_{\text{avg}} A^2 \eta^2 \sigma}{4\pi R^4 \lambda^2 L L_{\text{bm}}} \cdot \frac{t_s}{\Omega_s} \cdot \beta^2 \frac{\lambda^2}{A} = \frac{P_{\text{avg}} A \eta^2 \sigma \beta^2 t_s}{4\pi R^4 L L_{\text{bm}} \Omega_s} \tag{4.88}$$

Then,

$$\text{SNR} = \frac{P_{\text{avg}} A \eta^2 \sigma \beta^2 t_s}{4\pi R^4 L L_{\text{bm}} \Omega_s k T_{sys}^{C_B L}} \tag{4.89}$$

Thus, for a volume-search radar, the SNR is, to first order, proportional to the power-aperture product and independent of frequency band [1, p. 64; 4, p. 26].

That interesting fact is responsible for what may be the only case of a technical equation influencing an international treaty. Article III of the 1972 Anti-Ballistic Missile (ABM) Treaty between the United States and the Soviet Union states, in part [23, p. 257]:

> Each party undertakes not to deploy ABM systems or their components except . . . [for] no more than 18 ABM radars each having a potential less than the potential of the smaller of [aforementioned] two large phased-array ABM radars.

"Agreed statement B" then states,

> The parties understand that the potential (*the product of mean emitted power in watts and antenna area in square meters*) [emphasis added] of the smaller of the two large phased-array ABM radars . . . is considered for purposes of the Treaty to be three million.

4.3 Accuracy of Radar Measurements

When a radar is used to measure a parameter of a target, such as its range, velocity, or angular position, there evidently is some theoretical limit on the minimum error associated with the measurement. Intuitively we would expect that such error would decrease as the SNR increases. This section, following [10, Chap. 9] and [24, Chap. 2], briefly summarizes the methods for determining theoretical minimum measurement errors.

4.3.1 Monochromatic Pulse

A linearly polarized monochromatic (single-frequency) pulse is characterized by the following (assumed constant during pulse): amplitude = E_0, frequency = f_c, phase = ϕ_0, delay = τ, pulse width = t_p, and polarization = $\hat{\mathbf{E}}$. The corresponding equation for the pulse is

$$\mathbf{E}(t) = \text{rect}\left(\frac{t-\tau}{t_p}\right) \cdot E_0 \cos[2\pi f_c t + \phi_0] \cdot \hat{\mathbf{E}} \tag{4.90}$$

We consider the reception of an echo pulse (not necessarily monochromatic) from a point target, and we wish to consider the potential accuracy of measurement of the delay (→ range), phase, and frequency (→ velocity).

4.3.2 Cramer-Rao Bounds

Following [10, Chap. 9], assume the following:

- Signal $s(t) = Re[q(t)] = Re[u(t)\exp(j2\pi f_c t)]$.
- FT of $q(t)$ is band-limited and equals

$$Q(f) = \begin{cases} 2S(f) & \left(f_c - \frac{B}{2}\right) \le f \le \left(f_c + \frac{B}{2}\right) \\ 0 & \text{Otherwise} \end{cases} \tag{4.91}$$

- $q(t) \cong 0$ for $|t| > T/2$. (Note: $q(t)$ cannot exactly be both band limited and time limited.)
- Noise is additive zero-mean Gaussian.
- SNR is high.
- Processing is by means of a matched filter.
- $q(t)$ has only one unknown parameter μ.
- μ can be amplitude, frequency f, phase ϕ, or delay τ.

The Cramer-Rao bound is a theoretical lower bound on the variance in the measurement error for $\mu(\mu_0 = \text{true value})$, given by

$$CRB_\mu \equiv \langle(\mu - \mu_0)^2\rangle = \frac{N_0}{\int_{-T/2}^{T/2} dt \left(\frac{\partial q}{\partial \mu}\right)\left(\frac{\partial q}{\partial \mu}\right)^*} \equiv \frac{N_0}{I},\ \sigma_\mu^2 \ge CRB_\mu \tag{4.92}$$

4.3.2.1 Cramer-Rao Bound for Frequency

For $\mu = f =$ frequency,

$$\frac{\partial q}{\partial \mu} = \frac{\partial q}{\partial f} = j2\pi t q(t) \tag{4.93}$$

$$\sigma_f^2 \geq \frac{N_0}{I} \tag{4.94}$$

$$\begin{aligned} I &= \int_{-\infty}^{\infty} dt [j2\pi t q(t)][-j2\pi t q^*(t)] \\ &= (2\pi)^2 \int_{-\infty}^{\infty} dt t^2 |q(t)|^2 = (2\pi)^2 \int_{-\infty}^{\infty} dt t^2 |u(t)|^2 \end{aligned} \tag{4.95}$$

We define an effective time duration α, such that

$$\alpha^2 \equiv \frac{(2\pi)^2 \int_{-\infty}^{\infty} dt t^2 |u(t)|^2}{\int_{-\infty}^{\infty} dt |u(t)|^2} \tag{4.96}$$

and note that

$$2E = \int_{-\infty}^{\infty} dt |q(t)|^2 = \int_{-\infty}^{\infty} dt |u(t)|^2 \tag{4.97}$$

Then

$$\begin{aligned} \sigma_f^2 &\geq \frac{N_0}{I} = \frac{N_0}{2E\alpha^2} \\ \sigma_f &\geq \frac{1/\alpha}{(2E/N_0)^{1/2}} \end{aligned} \tag{4.98}$$

4.3.2.2 Cramer-Rao Bound for Delay

For $\mu = \tau =$ delay, using a similar but more cumbersome argument [10, pp. 177–180], we can show (Problem 4.9) that, for measurement at baseband ($f_c = 0$),

$$\sigma_\tau^2 \geq \frac{N_0}{2E\beta^2} \tag{4.99}$$

where

$$\beta^2 \equiv \frac{(2\pi)^2 \int_{-\infty}^{\infty} df f^2 |S(f)|^2}{\int_{-\infty}^{\infty} df\, |S(f)|^2}, \; \beta = \text{effective bandwidth} \tag{4.100}$$

$$\sigma_\tau \geq \frac{1/\beta}{(2E/N_0)^{1/2}} \tag{4.101}$$

It can be shown that $\beta\alpha \geq \pi$, which is the radar uncertainty relation [1, p. 408].

4.3.2.3 Cramer-Rao Bound for Angular Position

Furthermore, we can consider the accuracy of measuring the angular position θ of a point target [1, Sec. 11.3], using a linear slit aperture antenna (Section 2.2). The aperture is infinite in the y-direction, is of width D in the x-direction, and has an illumination function $A(x, y) = A(x)$. Then, from Section 2.3.2, expressing the normalized radiation pattern as $g(\theta/\lambda)$ instead of $f(\theta)$,

$$\begin{aligned} g(\theta/\lambda) &= \int_{-D/2}^{D/2} dx A(x) e^{j2\pi x(\sin\theta)/\lambda} \\ &\cong \int_{-D/2}^{D/2} dx A(x) e^{j2\pi x\theta/\lambda} \end{aligned} \tag{4.102}$$

As shown in Table 4.2, if we consider $A(x) = \text{rect}(x/D)$, then there is a complete mathematical analogy between the measurement of angular position and the measurement of time delay.

Table 4.2
Analogy Between Time Delay and Angular Accuracy
$f \leftrightarrow x,\ t \leftrightarrow \theta/\lambda,\ S(f) \leftrightarrow A(x)$

Parameter	Time Delay	Angular Accuracy
Independent parameter	$S(f)$	$A(x)$
Fourier Transform	$s(t) = \int_{-\infty}^{\infty} df S(f) e^{j2\pi ft}$	$g(\theta/\lambda) = \int_{-\infty}^{\infty} dx A(x) e^{j2\pi x(\theta/\lambda)}$
β^2	$\beta^2 \equiv \dfrac{(2\pi)^2 \int_{-\infty}^{\infty} df f^2 \lvert S(f)\rvert^2}{\int_{-\infty}^{\infty} df \lvert S(f)\rvert^2}$	$\beta'^2 = \dfrac{(2\pi)^2 \int_{-\infty}^{\infty} dx x^2 \lvert A(x)\rvert^2}{\int_{-\infty}^{\infty} dx \lvert A(x)\rvert^2}$
σ	$\sigma_t = \dfrac{1/\beta}{(2E/N_0)^{1/2}}$	$\sigma_{\theta/\lambda} = \dfrac{1/\beta'}{(2E/N_0)^{1/2}}$

4.3.2.4 Examples of Cramer-Rao Bounds

As examples of Cramer-Rao bounds, we consider the theoretical minimum measurement errors on (1) the frequency of a monochromatic pulse, (2) delay of an LFM pulse, and (3) the angular position of a point target as measured by a uniform-aperture antenna.

Frequency of a monochromatic pulse

We set

$$u(t) = \mathrm{rect}(t/T) \tag{4.103}$$

Then,

$$\alpha^2 = \frac{1}{T}(2\pi)^2 \int_{-T/2}^{T/2} t^2 dt = \frac{1}{T}(2\pi)^2 \frac{T^3}{12} = \frac{\pi^2}{3} T^2 \tag{4.104}$$

$$\alpha = T\frac{\pi}{\sqrt{3}}$$

$$\sigma_f \geq \frac{1/\alpha}{(2E/N_0)^{1/2}} = \frac{\sqrt{3}}{\pi} \cdot \frac{1/T}{(2E/N_0)^{1/2}} \tag{4.105}$$

Delay of an LFM Pulse

Similarly, we can estimate the delay of an LFM pulse of bandwidth B reflected from a stationary target (measured at baseband) by setting

$$S(f) = \text{rect}(f/B) \tag{4.106}$$

Then,

$$\beta^2 = \frac{1}{B}(2\pi)^2 \int_{-B/2}^{B/2} f^2 dt = \frac{1}{B}(2\pi)^2 \cdot \frac{B^3}{12} = \frac{\pi^2}{3}B^2$$

$$\beta = B \cdot \frac{\pi}{\sqrt{3}} \tag{4.107}$$

$$\sigma_\tau \geq \frac{1/\beta}{(2E/N_0)^{1/2}} = \frac{\sqrt{3}}{\pi} \cdot \frac{1/B}{(2E/N_0)^{1/2}}$$

Angular Position of Point Target Using a Uniform-Aperture Antenna

From Table 4.2, if $A(x) = \text{rect}(x/D)$, then

$$\beta' = D \cdot \frac{\pi}{\sqrt{3}} \tag{4.108}$$

$$\sigma(\theta/\lambda) = \frac{\sqrt{3}}{\pi} \cdot \frac{1/D}{(2E/N_0)^{1/2}}, \quad \sigma_\theta = \frac{\sqrt{3}}{\pi} \cdot \frac{\lambda/D}{(2E/N_0)^{1/2}} \tag{4.109}$$

For a moving target (such as an aircraft), a tracking radar may obtain a series of range-angle coordinates and produce a track for the target. Because of the multiple measurements, the overall uncertainty of track location will be considerably less than the uncertainty associated with a single measurement of range or angle [25, 26].

4.3.2.5 Summary

In general, if in one domain we have a rectangular function, then its FT in the conjugate domain is a sinc function with peak-to-first-null value δ_{pn}. The Cramer-Rao bound on a measurement of the peak of the sinc is

$$\sigma = (\delta_{pn}) \cdot \frac{\sqrt{3}}{\pi} \cdot \frac{1}{(2E/N_0)^{1/2}} \tag{4.110}$$

Table 4.3 summarizes the results for the three examples.

Table 4.3
Summary of Cramer-Rao Bounds

Measurement	**Domain #1**	**Domain #2**	**Equation***
Frequency of monochromatic pulse, duration t_p	Time	Frequency	$\sigma_f = \frac{\sqrt{3}}{\pi} \cdot \frac{1/t_p}{(2E/N_0)^{1/2}}$
Delay of LFM pulse, bandwidth B	Frequency	Time (delay)	$\sigma_\tau = \frac{\sqrt{3}}{\pi} \cdot \frac{1/B}{(2E/N_0)^{1/2}}$
Angular position of target (slit aperture, uniform illumination, width D, wavelength λ)	Aperture	Beam	$\sigma_\theta = \frac{\sqrt{3}}{\pi} \cdot \frac{\lambda/D}{(2E/N_0)^{1/2}}$

*The equations in this table correspond to Skolnik's [1] equations (11.29), (11.26), and (11.42a), respectively.

References

[1] Skolnik, M., *Introduction to Radar Systems*, 2nd ed., New York: McGraw-Hill, 1980.

[2] Kurpis, G. P., and C. J. Booth (eds.), *The New IEEE Standard Dictionary of Electrical and Electronics Terms*, 5th ed., New York: Institute of Electrical and Electronics Engineers, 1993.

[3] Van Trees, H. L., *Detection, Estimation, and Modulation Theory*, 3 vols., New York: Wiley, 1968.

[4] Barton, D. K., *Modern Radar System Analysis*, Norwood, MA: Artech House, 1988. [See also D. K. Barton, C. E. Cook, and P. Hamilton, (eds.), *Radar Evaluation Handbook*, Norwood, MA: Artech House, 1991.]

[5] Blake, L., *Radar Range-Performance Analysis*, Norwood, MA: Artech House, 1986.

[6] Blake, L. V., "Prediction of Radar Range," Chap. 2 in M. Skolnik (ed.), *Radar Handbook*, 2nd ed., New York: McGraw-Hill, 1990.

[7] Fehlner, L. F., *Target Detection by a Pulsed Radar*, Baltimore: Johns Hopkins University, Applied Physics Laboratory, July 1962.

[8] Meyer, D. P., and H. A. Mayer, *Radar Target Detection*, New York: Academic, 1973.

[9] DiFranco, J. V., and W. L. Rubin, *Radar Detection*, Norwood, MA: Artech House, 1980.

[10] Levanon, N., *Radar Principles*, New York: Wiley-Interscience, 1988.

[11] Abramowitz, M., and Stegun, I. A., *Handbook of Mathematical Functions*, Washington, DC: U.S. Government Printing Office, 1972.

[12] Goldstein, G. B., "False-Alarm Regulation in Log-Normal and Weibull Clutter," *IEEE Trans. Aerospace and Electronic Systems*, Vol. AES-9, No. 1, Jan. 1973, pp. 84–92.

[13] Brigham, E. O., *The Fast Fourier Transform and Its Applications*, Englewood Cliffs, NJ: Prentice Hall, 1988.

[14] Nathanson, F. E., J. P. Reilly, and M. N. Cohen, *Radar Design Principles*, 2nd ed., New York: McGraw-Hill, 1991.

[15] Haykin, S., *Communication Systems*, 3rd ed., New York: Wiley, 1994.

[16] Allen, C. W., *Astrophysical Quantities*, 2nd ed., New York: Oxford University Press, 1963.

[17] Rihaczek, A., *Principles of High-Resolution Radar*, Los Altos, CA: Peninsula Publishing, 1985.

[18] Vannicola, V., et al., "Ambiguity Function Analysis for Chirp-Diverse Waveforms," *Proc. 2000 IEEE National Radar Conference*.

[19] Costas, J. P., "A Study of a Class of Detection Waveforms Having Nearly Ideal Range-Doppler Ambiguity Functions," *Proc. IEEE*, Vol. 72, No. 8, Aug. 1984, pp. 996–1009.

[20] Freedman, A., and N. Levanon, "Staggered Costas Signals," *IEEE Trans. Aerospace and Electronic Systems*, Vol. AES-22, No. 6, Nov. 1986, pp. 695–701.

[21] Freedman, A., and N. Levanon, "Any Two $N \times N$ Costas Signals Must Have at Least One Common Ambiguity Sidelobe if $N > 3$—A Proof," *Proc. IEEE*, Vol. 73, No. 10, Oct. 1985, pp. 1530–1531.

[22] Fielding, J. E., "Beam Overlap Impact on Phased-Array Target Detection," *IEEE Trans. Aerospace and Electronic Systems*, Vol. 29, No. 2, April 1993, pp. 404–411.

[23] Boffey, P. M., et al., *Claiming the Heavens: Complete Guide to the Star Wars Debate*, New York: New York Times, 1988.

[24] Barton, D. K., and Ward, H. R., *Handbook of Radar Measurement*, Norwood, MA: Artech House, 1984.

[25] Brookner, E., *Tracking and Kalman Filtering Made Easy*, New York: Wiley, 1998.

[26] Blackman, S., and R. Popoli, *Design and Analysis of Modern Tracking Systems*, Norwood, MA: Artech House, 1999.

Problems

Problem 4.1

Using tables or a computer, for a zero-mean Gaussian distribution, determine the probability that x (the independent variable) exceeds 1σ, 2σ, 3σ, 4σ.

Problem 4.2

Choose a set of reasonable values of P_D, P_{FA}, n, and n_e, and compute D_e = required *SNR* for detection, using the methodology outlined in Section 4.1.

Problem 4.3

For a binary integrator, assume $P_D\ (1/1) = 0.7$. Find the probability of *exactly* five out of eight detections and the probability of *at least* 5 out of 8 detections.

Problem 4.4

In (4.23), show that $g(t) = |u(t)|$ and

$$s(t) = \frac{1}{2}[u(t)e^{j\omega_c t} + u*(t)e^{-j\omega_c t}]$$

Problem 4.5

Prove the four properties of the ambiguity function given in Section 4.2.4. Hint: see [10], pp. 121–123.

Problem 4.6

For a single monochromatic pulse, show that

$$|\chi(\tau, \nu)| = \left(1 - \frac{|\tau|}{t_p}\right)\text{sinc}\left[\pi\nu t_p\left(1 - \frac{|\tau|}{t_p}\right)\right],\ |\tau| \le t_p;\ 0 \text{ otherwise}$$

Problem 4.7

For a single LFM pulse, show that if $B >> 1/t_p$, then $B \cong \gamma t_p$.

Problem 4.8

For a single LFM pulse, show that

$$|\chi(\tau, \nu)| = \left(1 - \frac{|\tau|}{t_p}\right)\text{sinc}\left[\pi t_p\left(1 - \frac{|\tau|}{t_p}\right)(\nu + \gamma\tau)\right],\ |\tau| < t_p;\ 0 \text{ otherwise}$$

(Hint: See Levanon [10], pp. 132–135.)

Problem 4.9

Show that the Cramer-Rao bound for delay, measured at baseband, is

$$\sigma_\tau^2 \geq \frac{N_0}{2E\beta^2}$$

where

$$\beta^2 \equiv \frac{(2\pi)^2 \int_{-\infty}^{\infty} f^2 |S(f)|^2 df}{\int_{-\infty}^{\infty} |S(f)|^2 df}$$

(Hint: See Levanon [10], pp. 177–180.)

5

Angle Measurement

Section 4.3 examined the potential accuracy of target angle measurement using a single-aperture radar. This chapter investigates techniques for measuring the angular location of a target by the use of multiple subapertures and by comparing the signals received from them.

5.1 History and Terminology [1]

Suppose an operator is using a stationary radar with a parabolic dish antenna to measure the angular position (e.g., elevation θ) of a stationary point target. The operator probably would first find the antenna position that maximizes the returned signal; the target is now on the antenna boresight, which corresponds to maximum gain G. If the antenna is moved in elevation so the target is slightly off boresight, the change in returned signal is not great because the slope of the beam pattern, $dG/d\theta$ is small (in fact, zero on boresight; see Problem 5.1). However, if the antenna is moved in elevation so the target is farther off boresight, $dG/d\theta$ is larger, but G is smaller. There is some optimum angle off boresight, θ_0, such that a small change in θ produces the maximum observable difference in the overall returned signal. The operator might choose to set the antenna so the target is at that position in the beam, thereby creating the maximum sensitivity of the returned signal to a small change in angle and allowing the most precise measurement of θ.

The lobe switching technique was developed to take advantage of that approach (Figure 5.1). Two feeds are placed in the dish; they are offset, to make the maximum-gain directions off boresight by $\pm\theta_0$. Successive pulses are then transmitted and received alternately by the feeds. When the target

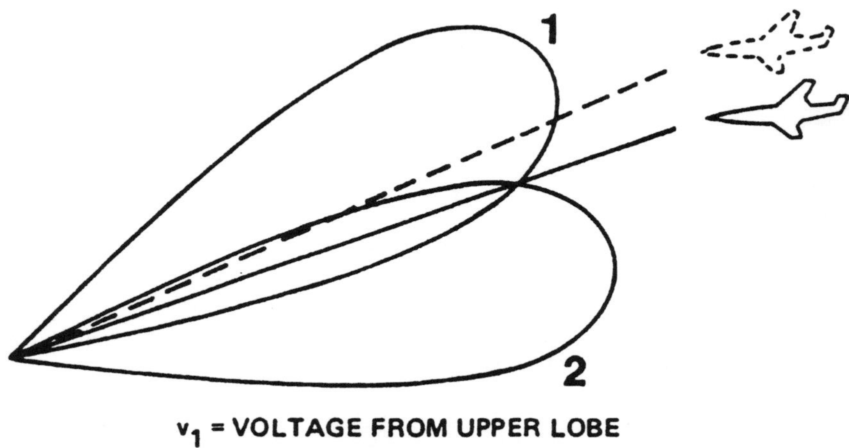

Figure 5.1 Lobe switching (reproduced from [1], Figure 1.2-1, with permission of Artech House).

has constant RCS and is on boresight, the difference between the returned signals from the two feeds is zero; when it is off boresight, the angle can be determined from the difference between the two signals. Lobe switching can be performed in both azimuth and elevation, using four feeds (Figure 5.2).

Figure 5.2 Two-dimensional lobe switching (reproduced from [1], Figure 1.2-2, with permission of Artech House).

Historically, the technique evolved into a conical scan (Figure 5.3); in that case, the feed revolves around the dish axis, and the off-boresight angle of the target can be determined in two dimensions. The error signal for a conical scan is a sine wave, the amplitude of which is proportional to the target angle off boresight, and the phase of which indicates the angular direction off boresight.

Lobe switching and conical scan possess two significant disadvantages: (1) target RCS fluctuations cause errors in the angle measurement, and (2) it is relatively straightforward for an adversary to interpret the distinctive transmitted signal of either and counter it with appropriate jamming.

The next step in the evolution of angle measurement was to generalize the lobe switching technique to the case for which the pulse is transmitted from the two (or four) feeds simultaneously and received simultaneously on each feed. Because the angular measurement can be accomplished with only one pulse instead of two (or four), this technique was termed *monopulse*.

5.2 Monopulse

Many modern monopulse radars involve a planar aperture divided into two (or four) subapertures. We refer to the output signals from the two (or four) subapertures as the *primary outputs*. They typically are combined into sum (Σ) and difference (Δ) channels (or, as shown in Figure 5.4, Σ, Δ_{az}, and Δ_{el}

Figure 5.3 Conical scan (reproduced from [1], Figure 1.2-2, with permission of Artech House).

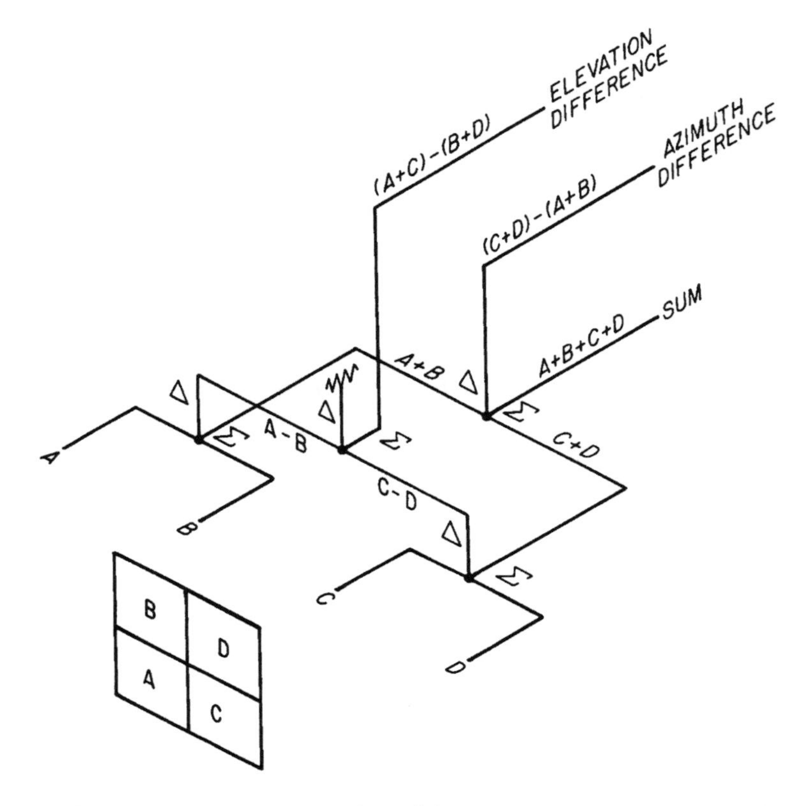

Figure 5.4 Two-dimensional monopulse (after [1], Figure 4.4-6, with permission of Artech House).

channels; the fourth channel, Δ-diagonal, is not necessary). This approach therefore requires two (or three) identical receiver channels.

5.2.1 Monopulse: A General Case

A thorough discussion of monopulse is given by Barton and Ward [2] and the references therein. A brief summary, based on [2, pp. 1–25], follows; notation is essentially the same as in the reference. Other discussions of monopulse can be found in [1, 3–9].

We consider a lossless planar-aperture antenna of arbitrary shape with area A and overall dimensions w (along the x axis) and h (along the y axis). We use the θ–ϕ coordinate system (see Figure 2.7) and consider the radiation pattern in the xz plane ($\phi = 0$). We assume bandwidth, B, is $<< c/\lambda$, $\lambda/w << 1$, and $\theta << 1$. Noise power density (watts per hertz) is $N_0 = kT_{sys}$ (see Sections 2.2.11 and 4.2.2). Collected energy is E, and we define the commonly used energy ratio $2E/N_0$ [2, p. 9]. The aperture produces both sum and difference beams.

For the sum beam, the aperture illumination function is $g(x, y)$, and the voltage and power gain patterns are, respectively, $F(\theta)$ and $G(\theta) = F^2(\theta)$. The one-way power-pattern 3-dB width is θ_{3dB}. Maximum gain is $G(0) = G_m$; that defines the beam axis. If the aperture were uniformly illuminated, its gain would be G_0. The aperture efficiency is $\eta_i = G_m/G_0$, effective aperture area is $A_e = \eta_i A$, and $G_m = 4\pi A_e/\lambda^2$ (see Section 2.3.3). We define the following:

E_0 = energy collected by the full aperture A (i.e., by both receiver channels) for uniform illumination ($G_{max} = G_0$)
E_m = energy collected by the full aperture for arbitrary illumination ($G_{max} = G_m$)

$$\text{Effective aperture width} = L_s = \left[\frac{\int dA(2\pi x)^2|g(x, y)|^2}{\int dA|g(x, y)|^2}\right]^{1/2} \tag{5.1}$$

$$\text{Effective (voltage) aperture width} = L_\theta = \left[\frac{\int dA(2\pi x)^2 g(x, y)}{\int dA g(x, y)}\right]^{1/2} \tag{5.2}$$

For a tapered aperture, $L_\theta > L_s$.

For the difference beam, the aperture illumination function is $g_d(x, y)$, and the voltage and power gain patterns are, respectively, $F_d(\theta)$ and $G_d(\theta) = F_d^2(\theta)$. Angular error sensitivity is related to the slope of the difference beam at $\theta = 0$. To represent that, we define

$$K = \frac{1}{\sqrt{G_0}}\left(\frac{\partial F_d}{\partial \theta}\right)_{\theta=0} \tag{5.3}$$

K_0 = maximum value of K for a given aperture shape

$$K_r = K/K_0 \tag{5.4}$$

For the theoretically minimum error (standard deviation) in θ,

$$g_d(x, y) = \frac{2\pi x}{L_s} g(x, y) \tag{5.5}$$

and

$$F_d(\theta) = \frac{\lambda}{L_s} \frac{\partial F}{\partial \theta} \tag{5.6}$$

Note that if $g(x, y)$ is a continuous even function in x (the usual case for maximum gain at broadside), then $g_d(x, y)$ is a continuous odd function in x, and $g_d(0, y) = 0$.

The minimum angular error (standard deviation) obtainable from monopulse is then

$$\sigma_\theta = \frac{1}{K\sqrt{2E_0/N_0}} \tag{5.7}$$

That can also be expressed in terms of $\theta_{3\text{dB}}$. We define the normalized monopulse slope as

$$k_m = \frac{\theta_{3\text{dB}}}{\sqrt{G_m}} \left(\frac{\partial F_d}{\partial \theta} \right)_{\theta=0} = \theta_{3\text{dB}} K \sqrt{\frac{G_0}{G_m}} \tag{5.8}$$

Then

$$\sigma_\theta = \frac{\theta_{3\text{dB}}}{k_m \sqrt{2E_m/N_0}} \tag{5.9}$$

If the aperture (sum beam) is uniformly illuminated, then $g(x, y) = 1$, $\eta_i = 1$, $G_m = G_0$, $L_s = L_\theta \equiv L_0$, $K_0 = L_0/\lambda$, and

$$g_d(x, y) = 2\pi x/L_0 \tag{5.10}$$

$$\sigma_\theta = \frac{1}{K_0\sqrt{2E_0/N_0}} = \frac{\lambda/L_0}{\sqrt{2E_0/N_0}} \text{ (uniform illumination)} \tag{5.11}$$

Furthermore, if the aperture is rectangular (w by h), then

$$L_0 = \frac{\pi w}{\sqrt{3}} \text{ (uniform illumination, rectangular aperture)} \tag{5.12}$$

$$\sigma_\theta = \frac{\sqrt{3}}{\pi} \frac{\lambda/w}{\sqrt{2E_0/N_0}} = \frac{\theta_{3\text{dB}}}{(1.607)\sqrt{2E_0/N_0}} \tag{5.13}$$

That agrees with the Cramer-Rao bound for angle measurement error (see Section 4.3). The corresponding linear-odd [3, p. 57] difference-beam illumination function is

$$g_d = \frac{2\pi x}{L_0} = \sqrt{3}\frac{x}{(w/2)} \tag{5.14}$$

Tables 5.1 and 5.2, from [2], summarize monopulse parameters for various apertures. Table 5.1 concerns "ideal" monopulse with a uniform sum illumination function and a linear-odd difference illumination function. Table 5.2 summarizes, for several possible sum illumination functions and the corresponding odd difference illumination functions, θ_{3dB}, η_i, L_θ, L_s, K/K_0, k_m, and G_{sr}, the sum-channel one-way ratio of the first-sidelobe peak to the mainlobe peak.

For the horn-fed dishes and phased arrays used in the majority of monopulse radars, careful antenna design allows even and odd sum and difference functions closely approximating the products of a selected taper function and uniform and linear-odd illuminations (see [9]). The sum beam possesses a gain associated with the desired sidelobe level, and the difference beam has a slope reduced from that of the linear-odd function by a factor dependent on the sidelobe level.

Table 5.3 summarizes results obtained for the cases considered in the rest of this section and in Section 5.3.

5.2.2 Monopulse with Uniform-Odd Illumination on Difference Aperture

In preparation for a discussion of interferometric radar in Section 5.3, we reconsider monopulse from a slightly different point of view. The antenna

Table 5.1
Parameters of Ideal Monopulse
Uniform Sum Illumination Function
Linear-Odd Difference Illumination Function
$g(x, y) = 1$, $g_d(x, y) = 2\pi x/L_0$

Aperture Shape	$\theta_0 w/\lambda$	L_0/w	$k_m = L_0\theta_0/\lambda = K_0\theta_0$
Interferometer	0.500	π	$\pi/2$
Rectangular	0.886	$\pi/\sqrt{3}$	1.607
Circular	1.028	$\pi/2$	1.617
Triangular	1.276	$\pi/\sqrt{6}$	1.636

Source: After [2], p. 24, with permission of Artech House.

Table 5.2
Parameters of Monopulse with Various Sum Illumination Functions
$g_d(x) = (2\pi x/L_s)g(x)$

Sum Illumination Function	$\frac{\theta_{3dB} w}{\lambda}$	η_i	$\frac{L_\theta}{w}$	$\frac{L_s}{w}$	$\frac{L_\theta \theta_{3dB}}{\lambda}$	$\frac{L_s \theta_{3dB}}{\lambda}$	$\frac{K}{K_0}$	k_m	G_{sr} (dB)
Circular, $g(x) = \sqrt{1 - (2x/w)^2}$	1.028	0.924	1.571	1.406	1.617	1.446	0.932	1.81	17.6
Parabolic,[a] $g(x) = 1 - 2x^2/w^2$	0.972	0.969	1.66	1.53	1.613	1.481	0.983	1.75	17.1
Parabolic,[a] $g(x) = 1 - 4x^2/w^2$	1.155	0.833	1.407	1.188	1.624	1.372	0.836	1.92	20.6
Triangular, $g(x) = 1 - \|2x/w\|$	1.276	0.750	1.28	0.994	1.636	1.268	0.79	2.08	26.4
Cosine, $g(x) = \cos(\pi x/w)$	1.189	0.812	1.37	1.136	1.629	1.350	0.812	1.96	23.0
Cosine^2, $g(x) = \cos^2(\pi x/w)$	1.441	0.667	1.134	0.89	1.636	1.283	0.653	2.08	32.0
Cosine^4, $g(x) = \cos^4(\pi x/w)$	1.853	0.515	0.886	0.669	1.645	1.240	0.466	2.18	48.0
Gaussian, $g(x) = \exp(-x^2/2\sigma_x^2)$	[b]	0.0	[b]	[b]	1.662	1.177	0.0	2.35	∞
Gaussian, assume $w = 6\sigma_x$	1.59	0.591	1.045	0.74	1.662	1.177	0.628	2.35	∞

[a]The first parabolic distribution shown has a pedestal level $\Delta = 0.5$ at the edge of the aperture; the second one has $\Delta = 0$ (no pedestal).
[b]For Gaussian illumination, $\theta_{3db}\sigma_x/\lambda = 0.265$, $L_\theta/\sigma_x = 2\pi$, $L_s/\sigma_x = \pi\sqrt{2}$.

Source: After [2], p. 25, with permission of Artech House.

Table 5.3
Summary of Angle Measurement Error—Point Target

	ζ	ξ	$\chi = \zeta\xi$	$\mu = \chi/(0.8859)$
	$\sigma_\theta = \zeta\left(\frac{\lambda}{w}\right)\sigma_{\Delta\phi}$	$\sigma_{\Delta\phi} = \frac{\xi}{\sqrt{2E_0/N_0}}$	$\sigma_\theta = \chi\frac{(\lambda/w)}{\sqrt{2E_0/N_0}}$	$\sigma_\theta = \mu\frac{\theta_{3dB}}{\sqrt{2E_0/N_0}}$
Monopulse Sum: uniform Difference: linear-odd (Cramer-Rao bound)	*	*	$\frac{\sqrt{3}}{\pi} = \frac{1}{1.8138}$	$\frac{1}{1.6068}$
Monopulse Sum: uniform Difference: uniform-odd	$\frac{1}{\pi}$	2	$\frac{2}{\pi} = \frac{1}{1.5708}$	$\frac{1}{1.3916}$
Interferometer, case 1: $S = w/2$, $y = 0$	$\frac{1}{2\pi}$	4	$\frac{2}{\pi} = \frac{1}{1.5708}$	$\frac{1}{1.3916}$
Interferometer, case 2: $S = w/2$, $y = \gamma w \neq 0$	$\frac{1}{2\pi(1+\gamma)}$	4	$\frac{2}{\pi(1+2\gamma)} = \frac{1}{(1.5708)(1+2\gamma)}$	—
Interferometer, case 3: Total aperture = w, $y = \gamma w$	$\frac{1}{2\pi(1+2\gamma)}$	$\frac{4}{(1-\gamma)}$	$\frac{2}{\pi(1-\gamma^2)} = \frac{1}{(1.5708)(1-\gamma^2)}$	—

*See Problem 5.5.

geometry is given in Figure 5.5. We transmit on the full aperture and receive simultaneously on each subaperture, observing the difference between the phases in the two subapertures. We assume that the difference-aperture illumination function is uniform-odd [3, p. 57], that is, each subaperture has uniform illumination, and the difference operation causes the phase of the weighting function for one subaperture to be opposite to that for the other. We assume that all SNRs are high and, therefore, that all phase differences can be measured accurately; alternatively, each phase can be measured with respect to a noise-free local reference. The phase measured by each subaperture is the same as the phase received at the subaperture center (Problem 5.4).

From Figure 5.5, the phases measured in the two subapertures (A and B) are

$$\begin{aligned} \phi_A &= \frac{2\pi}{\lambda}\frac{w}{4}\sin\theta \\ \phi_B &= -\frac{2\pi}{\lambda}\frac{w}{4}\sin\theta \\ \Delta\phi \equiv \phi_A - \phi_B &= \frac{\pi w}{\lambda}\sin\theta \approx \frac{\pi w}{\lambda}\theta \end{aligned} \tag{5.15}$$

Thus,

$$\sigma_\theta = \frac{\lambda}{\pi w}\sigma_{\Delta\phi} \tag{5.16}$$

We define,

$$\sigma_\theta \equiv \zeta\left(\frac{\lambda}{w}\right)\sigma_{\Delta\phi} \tag{5.17}$$

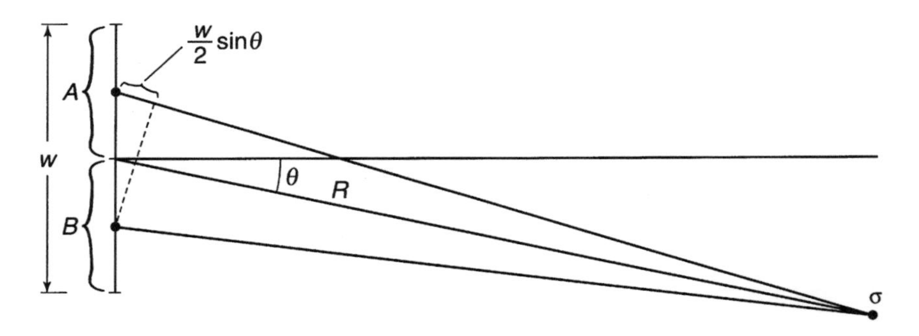

Figure 5.5 Flat-plate antenna with two subapertures.

Thus, $\zeta = 1/\pi$.

From Section 5.2, E_0 is the energy collected when transmission and reception each occur on the full aperture. Because each primary output represents the signal received by half the aperture, the energy collected in each receiver channel is $E = E_0/2$.

From [6, pp. 170–172], the standard deviation of the phase uncertainty for each channel is

$$\sigma_{\phi 1} = \frac{1}{\sqrt{2E/N_0}} = \frac{\sqrt{2}}{\sqrt{2E_0/N_0}} \tag{5.18}$$

The standard deviation of the phase difference between the two channels is

$$\sigma_{\Delta\phi} = \sigma_{\phi 1}\sqrt{2} = \frac{2}{\sqrt{2E_0/N_0}} \tag{5.19}$$

We define

$$\sigma_{\Delta\phi} \equiv \frac{\xi}{\sqrt{2E_0/N_0}} \tag{5.20}$$

Thus, $\xi = 2$, and

$$\sigma_\theta = \zeta\xi\frac{\lambda/w}{\sqrt{2E_0/N_0}} = \frac{2}{\pi}\frac{\lambda/w}{\sqrt{2E_0/N_0}} = \frac{1}{1.5708}\frac{\lambda/w}{\sqrt{2E_0/N_0}} \tag{5.21}$$

We also define

$$\sigma_\theta \equiv \chi\frac{\lambda/w}{\sqrt{2E_0/N_0}},\ \chi = \zeta\xi \tag{5.22}$$

In that case, $\chi = 2/\pi$.

For the Cramer-Rao bound, $\chi = (\sqrt{3})/\pi$. Thus, given a uniformly weighted sum aperture, monopulse with uniform-odd weighting on the difference aperture gives an angular error of $2/\sqrt{3} = 1.155$ times the Cramer-Rao bound, achievable with linear-odd weighting on the difference aperture. That agrees with [3, pp. 39 and 57], and formulas in [2, p. 260].[1]

1. In [2], p. 260, Table A.11, column 2, which describes a "uniform-odd" difference illumination function, the values for F, G, H, K_r, and 20 log(K_r) are slightly incorrect, which one can verify by directly computing the integrals given in column 1.

5.3 Interferometric Radar

We now consider the case of an interferometric radar, consisting of two subapertures spaced some distance apart, to achieve more precise angular measurement accuracy. The apertures each have diameter S, the space between their centers is $Y > S$, and the spacing between their edges is $y = Y - S$ (Figure 5.6). For simplicity, we assume uniform illumination on each subaperture, recognizing that for practical cases, especially for small y, the sidelobe level typically would be high and tapering would be necessary.

We assume that the radar alternately transmits and receives on one subaperture, then transmits and receives on the other. In that case, a change in θ produces a greater $\Delta\phi$ than is produced for monopulse. On the other hand, over the total collection time, each subaperture transmits and receives only half the total average power.

We have

$$\Delta\phi = \frac{4\pi}{\lambda} Y \sin\theta \approx \frac{4\pi}{\lambda}(S + y)\theta \tag{5.23}$$

$$\sigma_\theta = \frac{\lambda}{4\pi(S + y)}\sigma_{\Delta\phi} \tag{5.24}$$

We next consider three cases.

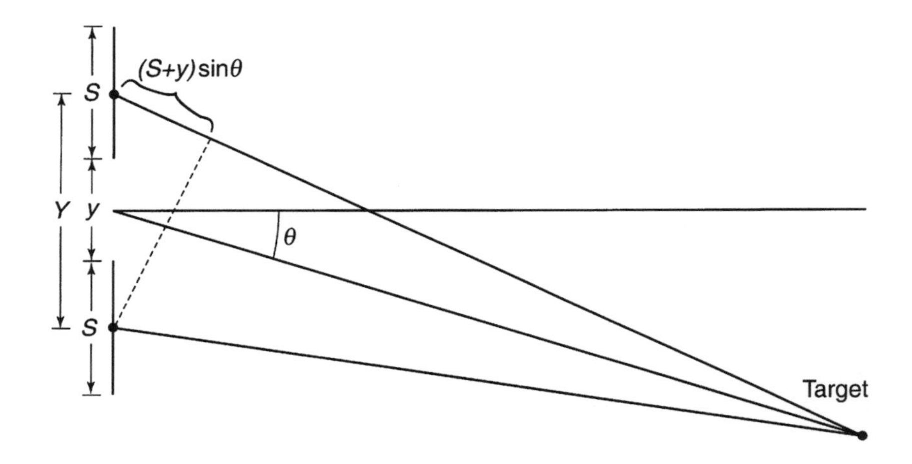

Figure 5.6 Interferometric radar.

5.3.1 Case 1: Zero Spacing Between Subaperture Edges ($w = 2S$)

This case is the same as the monopulse case except that we have interferometric radar with Tx-Rx on subaperture A, then Tx-Rx on subaperture B. Therefore

$$\sigma_\theta = \frac{\lambda}{2\pi w}\sigma_{\Delta\phi},\ \zeta = \frac{1}{2\pi} \tag{5.25}$$

Relative to E_0, the collected energy in each channel, E, is reduced by three factors of 2 (an overall factor of 8), because (1) the transmit aperture is halved, (2) the receive aperture is halved, and (3) the average power is halved. Thus,

$$\begin{aligned} E &= \frac{E_0}{8} \\ \sigma_{\phi 1} &= \left(\frac{N_0}{2E}\right)^{1/2} = \frac{\sqrt{8}}{\sqrt{2E_0/N_0}} \\ \sigma_{\Delta\phi} &= \sigma_{\phi 1}\sqrt{2} = \frac{4}{\sqrt{2E_0/N_0}},\ \xi = 4 \end{aligned} \tag{5.26}$$

Therefore,

$$\sigma_\theta = \frac{2}{\pi}\frac{\lambda/w}{\sqrt{2E_0/N_0}},\ \chi = \zeta\xi = \frac{2}{\pi} \tag{5.27}$$

That is the same result as for monopulse with uniform-odd weighting on the difference aperture. The advantageous reduction in ζ is negated by the increase in ξ.

5.3.2 Case 2: Nonzero Spacing, Constant Subaperture Width ($w = 2S$)

We define $\gamma = y/w$. Then, from (5.24),

$$\sigma_\theta = \frac{1}{2\pi(1+2\gamma)}\frac{\lambda}{w}\sigma_{\Delta\phi},\ \zeta = \frac{1}{2\pi(1+2\gamma)} \tag{5.28}$$

As in case 1, $\xi = 4$. Thus,

$$\sigma_\theta = \frac{2}{\pi(1+2\gamma)}\frac{(\lambda/w)}{\sqrt{2E_0/N_0}},\ \chi = \frac{2}{\pi(1+2\gamma)} \tag{5.29}$$

which of course reduces to (5.27) for $\gamma = 0$.

As the aperture separation, y, increases, σ_θ decreases, but so does θ_{amb}, the value of θ for which $\Delta\phi = 2\pi$, and the target angle is ambiguous with the boresight direction. Their ratio, however, remains constant:

$$\frac{\sigma_\theta}{\theta_{amb}} = \frac{\sigma_{\Delta\phi}}{2\pi} = \frac{2}{\pi\sqrt{2E_0/N_0}} \tag{5.30}$$

which we assume to be << 1.

5.3.3 Case 3: Nonzero Spacing, Constant Overall Aperture Width

We now consider the case for which the overall aperture is constrained to be equal to w, and as y increases, S decreases. Again, $\gamma = y/w$. From Figure 5.6,

$$\begin{aligned} w &= 2S + y \\ Y &= \frac{w-y}{2} + y = \frac{w+y}{2} = \frac{w(1+\gamma)}{2} \\ \Delta\phi &= \frac{4\pi}{\lambda}Y\theta = \frac{2\pi}{\lambda}w(1+\gamma)\theta \\ \sigma_\theta &= \frac{\lambda/w}{2\pi(1+\gamma)}\sigma_{\Delta\phi},\ \zeta = \frac{1}{2\pi(1+\gamma)} \end{aligned} \tag{5.31}$$

Furthermore,

$$\begin{aligned} S &= \frac{w-y}{2} = \frac{w(1-\gamma)}{2} \\ E &= \frac{E_0}{8}(1-\gamma)^2 \\ \sigma_{\phi 1} &= \frac{1}{\sqrt{2E/N_0}} = \frac{\sqrt{8}}{(1-\gamma)\sqrt{2E_0/N_0}} \\ \sigma_{\Delta\phi} &= \sqrt{2}\sigma_{\phi 1} = \frac{4}{(1-\gamma)\sqrt{2E_0/N_0}},\ \xi = \frac{4}{1-\gamma} \end{aligned} \tag{5.32}$$

Then,

$$\sigma_\theta = \frac{2}{\pi(1-\gamma^2)}\frac{\lambda/w}{\sqrt{2E_0/N_0}},\ \chi = \frac{2}{\pi(1-\gamma^2)} \tag{5.33}$$

Because we prefer that χ be as small as possible, we see that the optimum value of γ is zero, at which point (5.33) reduces to (5.27). Thus, if the overall aperture is constrained, it is best to fill it completely with the two subapertures, a result obtained by Rhodes [4, p. 101–102].

5.4 Glint

Heretofore in this chapter we have assumed that the target is a single point. For complex targets, monopulse can produce anomalous results. When the target has more than one scatterer, the measured angle actually can correspond to a position that is displaced from the target centroid, a phenomenon known as *glint*. Summaries are provided in [2, 5, and 6]. Following [6, Chap. 14], we estimate the glint for a two-scatterer target using monopulse. Figure 5.7 illustrates the geometry.

The sum and difference returns from the target (which consists of scatterers A and B at angles θ_A and θ_B) are

$$\begin{aligned} \Sigma &= \Sigma_A + \Sigma_B \\ \Delta &= \Delta_A + \Delta_B \end{aligned} \tag{5.34}$$

We assume that the return from scatterer B differs from that from scatterer A both in amplitude and phase by

$$\Sigma_B = \Sigma_A \rho e^{j\phi} \tag{5.35}$$

Then, a monopulse radar will estimate θ as [6]

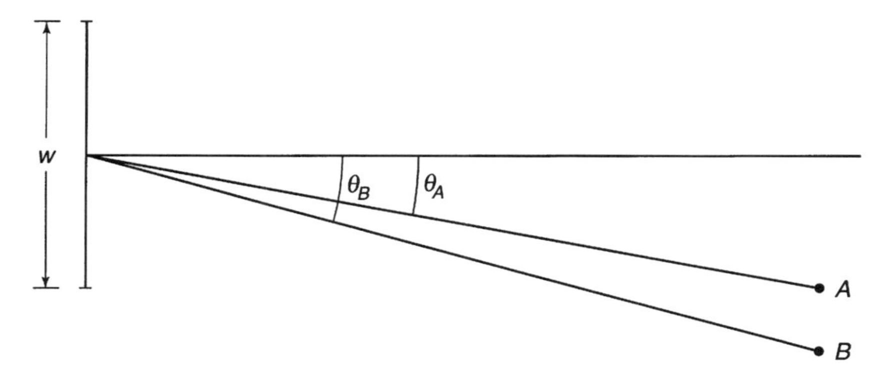

Figure 5.7 Glint for the two-scatterer target.

$$\hat{\theta} = \text{Re}\left(\frac{\theta_A + \rho\theta_B e^{j\phi}}{1 + \rho e^{j\phi}}\right) \tag{5.36}$$

We multiply numerator and denominator of the quantity in parentheses by $1 + \rho e^{-j\phi}$ and find

$$\hat{\theta} = \frac{\theta_A + \rho(\theta_A + \theta_B)\cos\phi + \rho^2\theta_B}{1 + 2\rho\cos\phi + \rho^2} \tag{5.37}$$

It is *not* true that $\theta_A \leq \hat{\theta} \leq \theta_B$. $\hat{\theta}$ can be outside those limits; Barton [5, Fig. 3.4.1] provides a summary. Barton also shows that, given reasonable assumptions,

$$\sigma_{\text{GLINT}} = \frac{L}{3R} \tag{5.38}$$

where L is the apparent target diameter and R is the range. Because σ_{GLINT} is inversely proportional to range, it may dominate σ_θ at short ranges [6, p. 300].

References

[1] Sherman, S. M., *Monopulse Principles and Techniques*, Norwood, MA: Artech House, 1984.

[2] Barton, D. K., and H. R. Ward, *Handbook of Radar Measurements*, Norwood, MA: Artech House, 1984.

[3] Barton, D. K., Ed., *Radars*, Volume 1, "Monopulse Radar," Norwood, MA: Artech House, 1974.

[4] Rhodes, D. R., *Introduction to Monopulse*, New York: McGraw-Hill, 1959. Reprinted by Artech House, Norwood, MA, 1980.

[5] Barton, D. K., *Modern Radar System Analysis*, Norwood, MA: Artech House, 1988.

[6] Levanon, N., *Radar Principles*, New York: Wiley, 1988.

[7] Kirkpatrick, G. M., "Final Engineering Report on Angular Accuracy Improvement," Syracuse, NY: General Electric Company, August 1952; reprinted in [3], pp. 17–103.

[8] Kirkpatrick, G. M., "Aperture Illuminations for Radar Angle-of-Arrival Measurements," *IRE Trans. PGAE-9,* Sept. 1953, pp. 20–27.

[9] Lopez, A. R., "Monopulse Networks for Series Feeding an Array Antenna," *IEEE Transactions on Antennas and Propagation*, Volume AP-16, Number 4, July 1968, pp. 436–440; reprinted in [3], pp. 307–311.

Problems

Problem 5.1

Show that

$$\frac{d}{dx}(\text{sinc}x) \approx -\frac{x}{3},\ x << 1$$

where $\text{sinc}x = (\sin x)/x$. Calculate

$$\frac{d}{dx}(\text{sinc}^n x),\ n = 2 \text{ and } 4$$

representing one-way and two-way power antenna patterns for a uniformly weighted aperture.

Problem 5.2

From Table 5.1, choose an aperture shape (nonrectangular) and verify the numbers shown.

Problem 5.3

From Table 5.2, choose an illumination function and verify the numbers shown.

Problem 5.4

For a radar with a rectangular flat-plate aperture observing a point target in the far field displaced in elevation (but not azimuth) from the boresight direction, use the methods in Section 2.3.2 to show that the phase of the overall returned signal is the same as the phase received at the aperture center.

Problem 5.5

Compute ξ and ζ for ideal rectangular-aperture monopulse (the first two entries in Table 5.3, row 2).

Problem 5.6

Analyze the uniform-odd monopulse case for which all the transmitted power is emitted from one subaperture. Is σ_θ improved or degraded relative to (5.21)?

Problem 5.7

Recompute the last row of Table 5.3 assuming that the antenna is an electronically scanned array (ESA) composed of transmit-receive modules (Section 2.4) and that average power is proportional to aperture area.

Problem 5.8

Using (5.37), find values of ρ and θ such that it is not true that $\theta_A \leq \hat{\theta} \leq \theta_B$. (Hint: See [5], Figure 3.4.1.)

Part II:
Imaging Radar

6

Introduction to Imaging Radar

The preceding chapters have devoted some detail to the fundamental aspects of radar: transmission/reception; antennas; waveforms; propagation; RCS; SNR; detection; and accuracy of measurement of range, velocity, and angular position. This chapter combines those fundamentals and considers the particular advantage of applying those techniques to the case of a rotating target, which leads us to the important domain of imaging radar.

6.1 Range-Velocity Compression

Let us consider a stationary radar with constant beam direction observing a region for which we want to obtain the radar return as a function of both range and velocity versus time. As an example, assume the radar is on the shore facing the ocean. The radar is assumed to have a constant PRF (f_R) and to transmit a step-chirp waveform. That is, the radar transmits groups of N ($N >> 1$) monochromatic pulses; within a group, the frequency of a pulse is Δf greater than that of the previous pulse, and the radar transmits f_R/N groups per second. Each pulse group has bandwidth B, and each pulse has pulse width τ. Within a pulse group, the frequency of individual pulse number n is given by

$$\begin{aligned} f_n &= f_0 + (n - 1)\Delta f,\ n = 1, \ldots, N \\ B &= (N - 1)\Delta f \end{aligned} \tag{6.1}$$

The A/D converter obtains one sample per returned pulse (N samples per pulse group), and M groups ($M >> 1$) are collected. By means of pulse compression, the processor then produces a down-range profile with range resolution $\delta_{rpn} \sim c/2B$ and down-range extent $\Delta_r \sim N\delta_{rpn}$. (This will be discussed further in Section 8.1.5. All resolution variables are peak to first null.) Similarly, by means of doppler processing, each range bin is sorted into velocity bins, each with velocity resolution $\delta_{vpn} \sim f_R\lambda/2NM$ and unambiguous LOS velocity $\Delta_v \sim f_R\lambda/2N$ (see Section 4.2). Alternatively, we could consider a sequence of M linear FM (LFM) pulses transmitted at PRF = f_R/N, each of bandwidth B and pulse width τ, with each pulse sampled N times (see Problem 6.1).

For a given dwell time (perhaps 0.1 to 1 second), the output device could produce a two-dimensional plot of radar return as a function of range and LOS velocity. Such a result would be of great interest to scientists studying the ocean, and, indeed, many such measurements have been made.

6.2 Rotating Target; Inverse Synthetic Aperture Radar

Such processing is of particular interest if the radar is observing an extended target that is rotating. For example, as shown in Figure 6.1, suppose the radar, on the negative x axis in the far field, is observing a turntable, the center of which is at the origin, rotating with angular velocity Ω. Consider a point on the turntable with coordinates r, ϕ relative to the turntable center. The speed

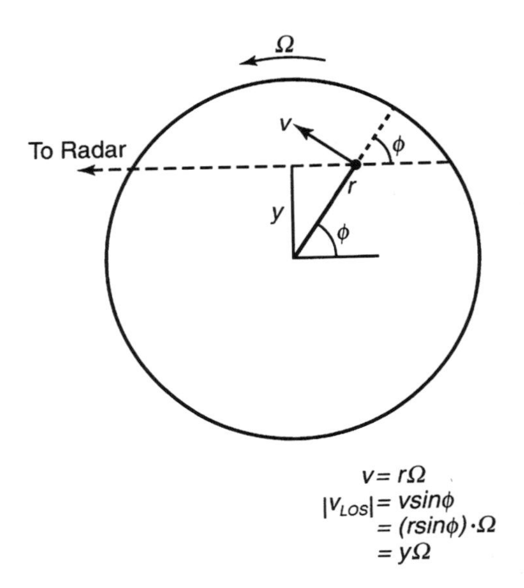

Figure 6.1 Target on turntable.

of the point is $r\Omega$; the LOS velocity is $-r\Omega\sin\phi$. Along a line through the point and parallel to the x axis and at a distance $y = r\sin\phi$ from the origin, the LOS velocity is $-r(\sin\phi)\Omega = -y\Omega$ = constant. Thus, the LOS velocity of a point on the turntable is proportional to its cross-range coordinate y and is independent of its down-range coordinate x.

Therefore, if we perform range-doppler processing on the radar returns from a turntable or other rotating target, the rotation axis of which is perpendicular to the LOS, we can interpret the velocity coordinate as a cross-range coordinate. With appropriate coordinate scaling, we can interpret the resulting two-dimensional plot of radar return versus range and cross-range as an image of the target. The down-range resolution and extent are as discussed in Section 6.1, and, because $r = v/\Omega$, the cross-range resolution and extent are found by dividing the values in Section 6.1 by Ω. Thus,

$$\Delta_{cr} = \frac{f_R\lambda}{2\Omega N} \tag{6.2}$$

$$\delta_{crpn} = \frac{f_R\lambda}{2MN\Omega} = \frac{\lambda}{2\Omega MNt_R} = \frac{\lambda}{2\Omega T_{obs}} = \frac{\lambda}{2\Delta\phi} \tag{6.3}$$

where MN/f_R is the observation time and $\Delta\phi$ is the total angle through which the target rotates during the data collection. (We assume here that $\Delta\phi << \pi/2$, and thus $r\Delta\phi << \delta_{rpn}$, and y is essentially constant during image formation.)

To summarize:

$$\delta_{rpn} = \frac{c}{2B} \tag{6.4}$$

$$\delta_{crpn} = \frac{\lambda}{2\Delta\phi} \tag{6.5}$$

Equations (6.4) and (6.5) are fundamental for the resolution of an imaging radar. It is interesting to note that for an X-band radar (λ = 0.03m) with cross-range resolution of 0.3m (a common design), $\Delta\phi$ = 0.05 radians ~ 3 degrees, a relatively small angle. Figure 6.2 summarizes the derivation of the expression for δ_{crpn}.

So far, we have considered the target to be a simple turntable. The radar waves strike all parts of the turntable, and energy scattered from any part of it returns directly to the radar without interaction with any other matter. For this simple case, an image of the entire target would be produced. Most real

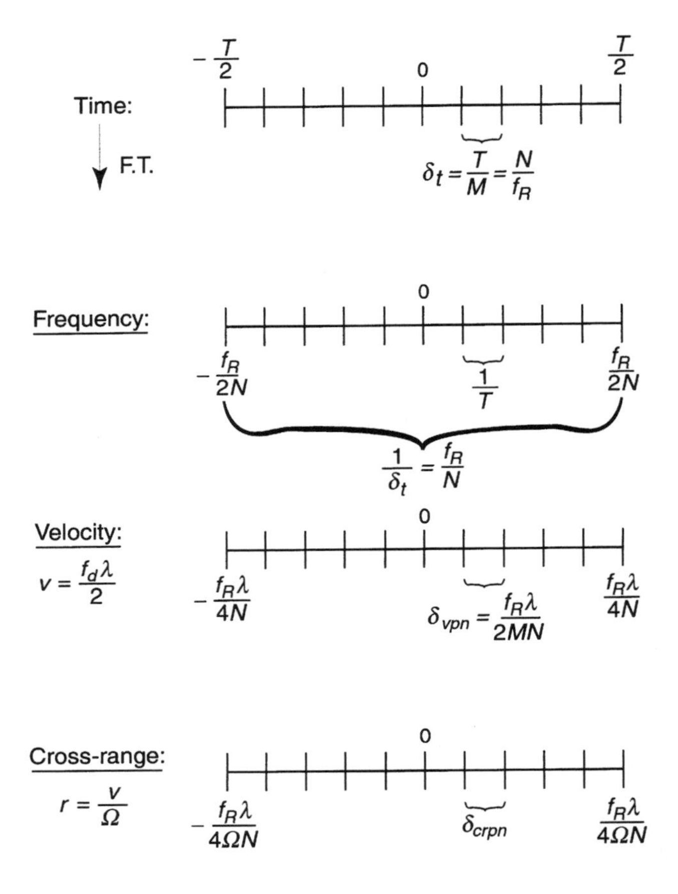

Figure 6.2 Summary of derivation of expression for azimuth resolution.

targets are, of course, more complicated. In many cases, we are interested in a radar image of a vehicle (e.g., a truck or an aircraft). At several locations in the United States, radar ranges [1], both indoor and outdoor, are operated at which vehicles are placed on rotating turntables or poles and their radar images are obtained. For those real targets, the images typically are modified by effects such as the following [1]:

- *Shadowing*. The rear of the target is blocked by the front; thus, it does not appear in the image.
- *Multibounce*. The radiation may bounce first off one scatterer, then another, before returning to the radar, thus producing an apparent return behind and offset in cross-range from the actual scatterer, sometimes appearing to be off the target entirely.

- *Creeping waves*. Delayed returns can produce apparent scatterers behind the physical target.

Suppose the radar is at a distance R from the target. During the data collection, in target-centered coordinates, the radar moves through a distance $R\Delta\phi$ ($\Delta\phi << \pi/2$). If, instead, a radar with a large, rectangular, uniformly illuminated aperture of diameter $R\Delta\phi$ (a real-aperture radar, or RAR) were to observe a stationary target, the angular resolution would be

$$\delta_{crpn} = R\phi_{pn} = \frac{R\lambda}{D} = \frac{R\lambda}{R\Delta\phi} = \frac{\lambda}{\Delta\phi} \tag{6.6}$$

In other words, by using the imaging technique, we have obtained angular resolution somewhat finer than what we would have achieved using a RAR with the large aperture $R\Delta\phi$. The imaging technique thus can be viewed as synthesizing a radar aperture, giving rise to the term *synthetic-aperture radar* (SAR).

Specifically, the 3-dB beamwidths of the two-way RAR and the (perforce two-way) SAR are (see Section 2.3.2)

$$\begin{aligned} \delta_{cr,3\text{dB}}(\text{RAR, 2-way}) &= (0.634)\frac{\lambda}{\Delta\phi} \\ \delta_{cr,3\text{dB}}(\text{SAR}) &= (0.886)\frac{\lambda}{2\Delta\phi} = (0.443)\frac{\lambda}{\Delta\phi} \\ \frac{\delta_{cr,3\text{dB}}(\text{SAR})}{\delta_{cr,3\text{dB}}(\text{RAR, 2-way})} &= 0.699 \end{aligned} \tag{6.7}$$

A stationary radar and a rotating target are mathematically equivalent to a moving radar and a stationary target. By conventional usage, SAR refers to the latter case. For a stationary radar and a rotating target or, more generally, for a case such that most of the radar motion in target-centered coordinates is caused by the target rotation, the term *inverse synthetic aperture radar* (ISAR) is used. The term is a misnomer, because no mathematical inverse is involved. In fact, in many cases (e.g., air-to-air radar imaging), it becomes a matter of semantics as to whether the imaging is considered to be SAR or ISAR. The distinction is clearly a function of the observer's frame of reference. However, the usage is universal and is adopted here.

This book discusses ISAR before SAR. It is the author's contention that that order provides a more logical development.

6.3 ISAR for an Extended Target

This section considers radar imaging of an extended three-dimensional target. We continue to assume that the radar energy strikes each illuminated portion of the target and is returned directly to the radar without further interaction. That corresponds to a target that creates no multibounce conditions. The theory of ISAR strictly holds only under those conditions. However, the approximation is quite good for many actual targets. Furthermore, even when the conditions are violated, the image frequently can still be interpreted and can be quite useful.

We also assume that the scatterers introduce no polarization change. That assumption usually is a good one, although in principle, for a given transmit polarization, an appropriately designed radar can produce two images, one for each received polarization.

As shown in Figure 6.3, the target has its centroid at the origin and is represented by a density function $\rho(\mathbf{r})$, where $\mathbf{r} = (x, y, z)$ is in target-centered coordinates, and ρ represents the square root of the RCS per differential volume element (because we are working with voltage, not power). For portions of the target that the radar waves do not "see" (e.g., shadowed or interior regions), $\rho(\mathbf{r}) = 0$. The radar is assumed to be in the far field near the negative x axis. During the data collection, the radar moves away from the x axis by relatively small angles in both azimuth and elevation. It is also assumed that the waveform is a step-chirp. (For a more complicated waveform, e.g., LFM, the argument is analogous but slightly more cumbersome.) The incident wave for a particular pulse is represented as

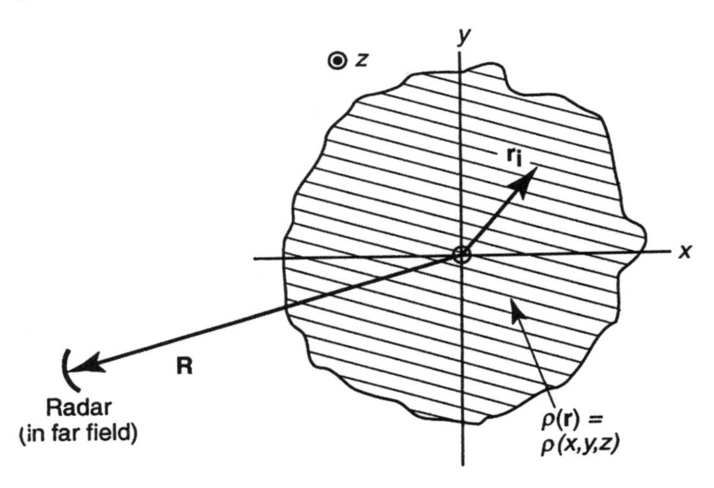

Figure 6.3 Radar return from extended target.

$$\mathbf{E}_{\text{inc}}(\mathbf{R}, \mathbf{r}, \mathbf{k}, \omega, t) = \mathbf{E}_0 \exp[j(\mathbf{k} \cdot [\mathbf{R} + \mathbf{r}] - \omega t)] \tag{6.8}$$

The return from the ith individual scatterer, representing the phase delay relative to the STALO, with voltage reflectivity (square root of RCS) $\sim q_i$, is (dropping the time dependence)

$$\mathbf{E}_{ri}(\mathbf{R}, \mathbf{r}, \mathbf{k}) \sim \mathbf{E}_0 q_i e^{-2j\mathbf{k}\cdot\mathbf{r}_i} e^{-2j\mathbf{k}\cdot\mathbf{R}} \tag{6.9}$$

For the entire target, we sum over the scatterers and let the sum become an integral:

$$\mathbf{E}_r(\mathbf{k}) = \hat{\mathbf{E}} A \int d\mathbf{r} \rho(\mathbf{r}) \mathbf{e}^{-2j\mathbf{k}\cdot\mathbf{r}} \mathbf{e}^{-2j\mathbf{k}\cdot\mathbf{R}} \tag{6.10}$$

where, from Sections 1.3 and 1.8, the proportionality constant A is found from

$$\begin{aligned} P_{Rx\text{-peak}} &= \frac{P_{Tx\text{-peak}} G^2 \lambda^2 \sigma}{(4\pi)^3 R^4 L} = \frac{c\epsilon_0 E_r^2}{2} \\ E_r &= \left(\frac{2}{c\epsilon_0} \frac{P_{Tx\text{-peak}} G^2 \lambda^2}{(4\pi)^3 R^4 L} \right)^{1/2} \sigma^{1/2} = A\sigma^{1/2} \end{aligned} \tag{6.11}$$

We assume here that the range to the radar is constant and known. (If the range is not known, the only effect is a translation of the image along the x axis. Because we usually do not care about the linear position of the image, that is, in general, of no concern. See Problem 6.2.) We correct each returned pulse for the range by multiplying by the known phasor $e^{2j\mathbf{k}\cdot\mathbf{R}}$. Furthermore, in actual processors, the value of the electric field is not of interest. Rather, what is of concern are the phase and the amplitude relative to some constant depending on the radar. We denote that measured complex electric field as $E(\mathbf{k})$, dropping the subscript r:

$$E(\mathbf{k}) = A \int d\mathbf{r} \rho(\mathbf{r}) e^{-2j\mathbf{k}\cdot\mathbf{r}} \sim \int d\mathbf{r} \rho(\mathbf{r}) e^{-2j\mathbf{k}\cdot\mathbf{r}} \tag{6.12}$$

The symbol $\sim$ is used to indicate that the expression should, in general, include the proportionality constant. Equation (6.12) is in the form of a three-dimensional FT. Its inverse is

$$\rho(\mathbf{r}) \sim \int d\mathbf{k}E(\mathbf{k})e^{2j\mathbf{k}\cdot\mathbf{r}} \tag{6.13}$$

To produce an image (in general, a three-dimensional image) of a target, the phases of the returned pulses are determined and processed according to (6.13).

For (6.13), it is assumed that the phases of the returned pulses have been collected over the full 4π steradians and over values of wave number (therefore frequency) from zero to infinity. Obviously, in practice, the collection is over finite intervals of those parameters. In the latter case, instead of reconstructing the exact image of the target, the process produces an estimate of the image $\hat{\rho}(\mathbf{r})$:

$$\text{Pulse: } E(\mathbf{k}) \sim \int d\mathbf{r}\rho(\mathbf{r})e^{-2j\mathbf{k}\cdot\mathbf{r}} \tag{6.14}$$

$$\text{Image: } \hat{\rho}(\mathbf{r}) \sim \int_{\text{limits}} d\mathbf{k}E(\mathbf{k})e^{2j\mathbf{k}\cdot\mathbf{r}} \tag{6.15}$$

Because $E(\mathbf{k})$ is not known over all $\mathbf{k}$ but only over a subset of $\mathbf{k}$, $\hat{\rho}(\mathbf{r})$ will generally not be the same as $\rho(\mathbf{r})$.

6.4 Point-Spread Function

To gain an understanding of the quality of an image, it is particularly convenient to consider the image of a hypothetical single-point scatterer on the target. The image of such a point will, in general, not be a point, but will be spread out. Therefore, the image of a single point is called a point-spread function (PSF). It is also sometimes called an impulse response function (IPR). However, "impulse response" may have other connotations with respect to signal processing; therefore, this chapter uses the term PSF.

If the range to the radar is large, the PSF will not vary within the image. We can, therefore, assume that the point scatterer is at the origin. Then,

$$\rho_0(\mathbf{r}) \sim \delta(\mathbf{r}) \tag{6.16}$$

where $\delta(\mathbf{r})$ is the three-dimensional delta function [2, App. I] and

$$E_0(\mathbf{k}) \sim 1 \tag{6.17}$$

The PSF is

$$\hat{\rho}_0(\mathbf{r}) \sim \int_{\text{limits}} d\mathbf{k} e^{2j\mathbf{k}\cdot\mathbf{r}} \tag{6.18}$$

The PSF is determined entirely by the limits of the integral, which correspond to the set of positions of the radar and of the transmitted frequencies. The integral over $d\mathbf{k}$ can be written as follows:

$$\hat{\rho}_0(\mathbf{r}) \sim \int d\mathbf{k} H(\mathbf{k}) e^{2j\mathbf{k}\cdot\mathbf{r}} \tag{6.19}$$

where $H(\mathbf{k})$ is the *integration limit function*. Here, we consider only cases where H is separable:

$$\text{Cylindrical coordinates: } H(\mathbf{k}) = H_r(k_r)H_z(k_z)H_\phi(\phi) \tag{6.20a}$$

$$\text{Spherical coordinates: } H(\mathbf{k}) = H_k(k)H_\theta(\theta)H_\phi(\phi) \tag{6.20b}$$

The following sections describe several specific ISAR cases. In each case, it is assumed that the frequency is chirped from f_1 to f_2, and, accordingly, k varies from k_1 to k_2. The average value of k is $\overline{k}$; $\Delta k \equiv k_2 - k_1 << \overline{k}$, and $\Delta k/\overline{k} = \Delta f/\overline{f}$ = fractional bandwidth.

Thus,

$$H_k(k) = \text{rect}\left(\frac{k - \overline{k}}{\Delta k}\right) \tag{6.21}$$

6.5 Standard Two-Dimensional ISAR: Small Angles

For this case, in target-centered coordinates, we assume that the radar remains in the xy plane and scans an angular interval $\Delta\phi << \pi$ centered about the $-x$ axis. Because the radar is in the far field, for small values of z, the return from a scatterer at (x, y, z) will not depend on z. Therefore, cylindrical coordinates are most appropriate. We have

$$H_z(k_z) = \delta(k_z),\ H_\phi(\phi) = \text{rect}\left(\frac{\phi}{\Delta\phi}\right),\ H_r(k_r) = H_r(k) = \text{rect}\left[\frac{k - \overline{k}}{\Delta k}\right] \tag{6.22}$$

Then,

$$\hat{\rho}_0(x, y) \sim \int H(\mathbf{k}) d\,\mathbf{k} e^{2j\mathbf{k}\cdot\mathbf{r}} \tag{6.23}$$

$$= \int_{k_1}^{k_2} kdk \int_{\phi_1}^{\phi_2} d\phi e^{2j\mathbf{k}\cdot\mathbf{r}} \tag{6.24}$$

$$= \int_{\bar{k}-\Delta k/2}^{\bar{k}+\Delta k/2} kdk \int_{-\Delta\phi/2}^{\Delta\phi/2} d\phi e^{2j(kx\cos\phi + ky\sin\phi)}$$

Furthermore, because $\Delta\phi << \pi$,

$$\cos\phi \cong 1, \ \sin\phi \cong \phi \tag{6.25}$$

Thus,

$$\hat{\rho}_0(x, y) \cong \int_{\bar{k}-\Delta k/2}^{\bar{k}+\Delta k/2} kdke^{2jkx} \int_{-\Delta\phi/2}^{\Delta\phi/2} d\phi e^{2jky\phi}$$

$$= \int_{\bar{k}-\Delta k/2}^{\bar{k}+\Delta k/2} kdke^{2jkx} \cdot \Delta\phi \frac{\sin(ky\Delta\phi)}{(ky\Delta\phi)} \tag{6.26}$$

$$= \frac{1}{y} \int_{\bar{k}-\Delta k/2}^{\bar{k}+\Delta k/2} dke^{2jkx} \sin(ky\Delta\phi)$$

where we have used the integral

$$\int_{-a/2}^{a/2} dxe^{jbx} = a\frac{\sin(ba/2)}{(ba/2)} = a\,\mathrm{sinc}(ba/2) \tag{6.27}$$

For sufficiently small $\Delta\phi$ and y (to be discussed), as k varies between its limits, the value of $\sin(ky\Delta\phi) \cong ky\Delta\phi$ is essentially constant and equal to $\bar{k}y\Delta\phi$, although e^{2jkx} varies. Using that approximation, we have

$$\hat{\rho}_0(x, y) \sim \frac{\sin(\bar{k}y\Delta\phi)}{y} \int_{\bar{k}-\Delta k/2}^{\bar{k}+\Delta k/2} dk e^{2jkx} \tag{6.28}$$

$$= \bar{k}\Delta\phi \operatorname{sinc}(\bar{k}y\Delta\phi) \cdot e^{2j\bar{k}x} \int_{-\Delta k/2}^{\Delta k/2} dk' e^{2jk'x} \tag{6.29}$$

where $k' = k - \bar{k}$. Thus, the PSF is

$$\hat{\rho}_0(x, y) \sim e^{2j\bar{k}x}\bar{k}\Delta\phi\Delta k \cdot \operatorname{sinc}(\bar{k}y\Delta\phi) \cdot \operatorname{sinc}(x\Delta k) \tag{6.30}$$

The down-range cut through that two-dimensional function is, in absolute value,

$$|\hat{\rho}_0(x, 0)| \sim |\operatorname{sinc}(x\Delta k)| \tag{6.31}$$

The first null occurs at

$$x_{pn}\Delta k = \pi = x_{pn} \cdot \frac{2\pi\Delta f}{c}, \; x_{pn} = \delta_{rpn} = \frac{c}{2\Delta f} = \frac{c}{2B} \tag{6.32}$$

That agrees with the value of range resolution derived in Section 4.2. The cross-range cut is, in absolute value,

$$|\hat{\rho}_0(0, y)| \sim |\operatorname{sinc}(\bar{k}y\Delta\phi)| \tag{6.33}$$

The first null occurs at

$$\bar{k}y_{pn}\Delta\phi = \pi = \frac{2\pi}{\bar{\lambda}} y_{pn}\Delta\phi, \; y_{pn} = \delta_{crpn} = \frac{\bar{\lambda}}{2\Delta\phi} \tag{6.34}$$

in agreement with the value for azimuth resolution derived in Section 6.2.

We now return to discuss the approximation $ky\Delta\phi \cong$ constant. We quantify that statement by requiring $\Delta(ky\Delta\phi) < \pi$ as k varies from

$$\bar{k} - \frac{\Delta k}{2} \text{ to } \bar{k} + \frac{\Delta k}{2} \tag{6.35}$$

that is,

$$y_{\max}\Delta k\Delta\phi < \pi \tag{6.36}$$

$$y_{\max}\Delta\phi < \frac{\pi}{\Delta k} = \frac{\pi c}{2\pi\Delta f} = \frac{c}{2\Delta f} \tag{6.37}$$

$$y_{\max}\Delta\phi < \delta_{rpn}$$

In other words, this condition is equivalent to the condition that the maximum movement through a range cell is less than the down-range resolution element. Indeed, if this condition is violated (a situation called *range migration*; see Section 7.1.6), the small-angle approximation of (6.28) is no longer valid.

If down-range and cross-range resolution are equal, that is, if

$$\delta_{rpn} = \delta_{crpn} \tag{6.38}$$

then

$$\frac{c}{2B} = \frac{\overline{\lambda}}{2\Delta\phi} \tag{6.39}$$

and

$$\Delta\phi = \frac{B\overline{\lambda}}{c} = \frac{B}{\overline{f}} = \text{fractional bandwidth} \tag{6.40}$$

The following set of parameters is typical for this *two-dimensional standard ISAR approximation*:

$$\delta_{rpn} = \delta_{crpn} = 0.3\text{m} \tag{6.41}$$

$$\overline{f} = 10 \text{ GHz}, \ \overline{\lambda} = 0.03\text{m} \tag{6.42}$$

$$\delta_{rpn} = \frac{c}{2B}, \ B = \frac{c}{2\delta_{rpn}} = \frac{300 \text{ m}/\mu\text{s}}{0.6\text{m}} = 500 \text{ MHz} \tag{6.43}$$

$$\delta_{crpn} = \frac{\overline{\lambda}}{2\Delta\phi}, \ \Delta\phi = \frac{\overline{\lambda}}{2\delta_{crpn}} = 0.05 \text{ radians} \cong 3 \text{ degrees} \tag{6.44}$$

$$x_{\max} = y_{\max} = 3\text{m (target extent in } x \text{ and } y \text{ is 6m)} \tag{6.45}$$

$$\Delta k \cdot x_{\max} = \frac{2\pi\Delta f}{c} \cdot x_{\max} = 10\pi > \pi \tag{6.46}$$

$$\Delta k \cdot y_{\max} \cdot \Delta\phi = \frac{2\pi\Delta f}{c} \cdot y_{\max} \cdot \Delta\phi = (0.5)\pi < \pi \tag{6.47}$$

For actual ISAR cases, the assumption of small $\Delta\phi$ is frequently violated. However, focused ISAR images can still be obtained via the polar format algorithm, which is discussed in Section 6.8.

6.6 Two-Dimensional ISAR: Large Angles

For an arbitrary large angle, the two-dimensional ISAR integral equations are not simple. However, for $\Delta\phi = 2\pi$, they become simple again. In this case,

$$H(\mathbf{k}) = H_r(k_r)H_z(k_z)H_\phi(\phi) = \text{rect}\left(\frac{k - \overline{k}}{\Delta k}\right) \cdot \delta(k_z) \cdot 1 \tag{6.48}$$

$$\hat{\rho}_0(r) \sim \int d\mathbf{k} e^{2j\mathbf{k}\cdot\mathbf{r}} H(k_r, k_z, \phi) \tag{6.49}$$

$$\begin{aligned} &= \int_{k_1}^{k_2} kdk \int_0^{2\pi} d\phi e^{2jkr\cos\phi} \\ &= 2\pi \int_{k_1}^{k_2} kdk J_0(2kr) \\ &= \frac{\pi}{r}[k_2 J_1(2k_2 r) - k_1 J_1(2k_1 r)] \end{aligned} \tag{6.50}$$

where $J_n(x)$ is the ordinary Bessel function.

If $\Delta\phi = 2\pi$, then even if k = constant (zero bandwidth signal), imaging can still be performed. For such single-frequency imaging, we have

$$H_r(k_r) = H_r(k) = \delta(k - k_o) \tag{6.51}$$

$$\hat{\rho}_0(r) \sim \int_0^{2\pi} d\phi e^{2jk_0 r\cos\phi} = J_0(2k_0 r) \tag{6.52}$$

That result was previously obtained by Mensa et al. [3] (see Problem 6.4).

6.7 Three-Dimensional ISAR

For three-dimensional ISAR with $\Delta\phi << \pi$, $\Delta\theta << \pi$, $\Delta k << \overline{k}$, the results are completely analogous. It is left as a problem (Problem 6.5) to show that

$$\delta_x = \frac{c}{2B}, \ \delta_y = \frac{\overline{\lambda}}{2\Delta\phi}, \ \delta_z = \frac{\overline{\lambda}}{2\Delta\theta} \tag{6.53}$$

For the case where the radar position varies over all 4π steradians, we use spherical coordinates:

$$H(\mathbf{k}) = \text{rect}\left(\frac{k - \overline{k}}{\Delta k}\right) \tag{6.54}$$

$$\begin{aligned}\hat{\rho}_0(r) &\sim \int_{k_1}^{k_2} k^2 dk \int_0^{\pi} \sin\theta d\theta \int_0^{2\pi} d\phi e^{2jkr\sin\theta\cos\phi} \\ &= 2\pi \int_{k_1}^{k_2} k^2 dk \int_0^{\pi} \sin\theta d\theta J_0(2kr\sin\theta) \\ &= 4\pi \int_{k_1}^{k_2} k^2 dk j_0(2kr) \\ &= \frac{2\pi}{r}[k_2^2 j_1(2k_2 r) - k_1^2 j_1(2k_1 r)]\end{aligned} \tag{6.55}$$

Here, $j_n(x)$ is the spherical Bessel function, and the result is completely analogous to (6.50).

For single-frequency imaging, $H_k(k) = \delta(k - k_0)$, and

$$\hat{\rho}_0(r) \sim j_0(2k_0 r) \tag{6.56}$$

again in complete analogy to (6.52).

6.8 Wavenumber Space and Polar Format Algorithm

We again consider the case of standard two-dimensional ISAR. The vector $\mathbf{r}$ is in "real space," the type of space in which we live. As shown in Figure

6.4(a), in real space (in target-centered coordinates), the radar moves at uniform angular velocity through an angle $\Delta\phi$. During this time, the direction of **k** varies accordingly. We can represent the various values of **k** by expressing them in wavenumber space, or **k**-space, as shown in Figure 6.4(b). The values of **k** lie between radii of k_1 and k_2 and angles of $-\Delta\phi/2$ and $+\Delta\phi/2$. For the case of *standard two-dimensional ISAR,* with uniform sampling in both frequency and angle and using a step-chirp waveform, the sampled values of **k** will lie

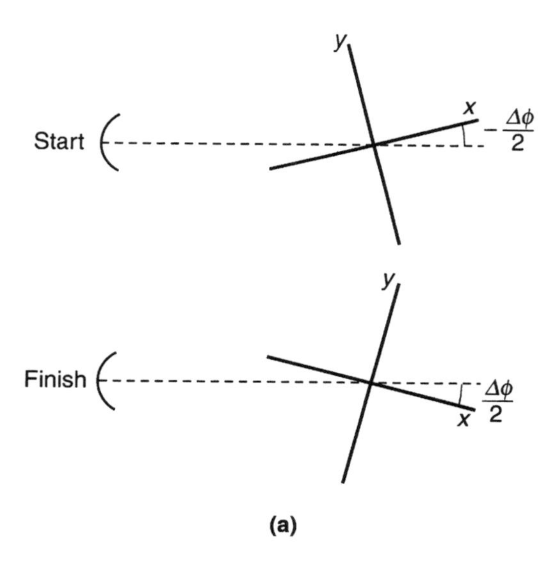

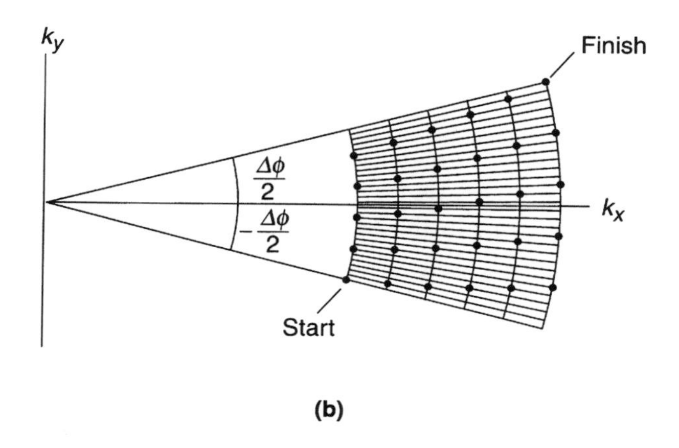

Figure 6.4 Target rotation: (a) in real space; (b) in wavenumber space.

on portions of spirals in **k**-space. For digital processing, it is highly desirable to obtain values of $E(\mathbf{k})$ on a rectangular grid in **k**-space. Doing so permits first performing the Fourier transform in k_y, then in k_x (i.e., it permits separability):

$$\hat{\rho}(\mathbf{r}) \sim \int d\mathbf{k} E(\mathbf{k}) e^{2j\mathbf{k}\cdot\mathbf{r}} \tag{6.57}$$

$$\rho(x, y) \sim \int dk_x e^{2jk_x x} \int dk_y e^{2jk_y y} E(k_x, k_y) \tag{6.58}$$

With the use of a digital processor, each step can be performed via the highly efficient fast Fourier transform (FFT) [4], along with appropriate weighting functions (Section 2.3.4).

We obtain values of $E(k)$ on a square grid in **k**-space by interpolating from the directly measured values on spirals. The overall procedure of interpolation and subsequent image formation is known as the polar format algorithm (PFA). The interpolation procedure (for an LFM waveform) is summarized in Figure 6.5 (reprinted from Carrara et al. [5], who discuss the procedure in some detail). A simple procedure for performing two-dimensional interpolation is the "interp2" command in MATLAB®, a software package manufactured by MathWorks, Inc., of Natick, Massachusetts.

For three-dimensional ISAR, the PFA can be generalized to three dimensions. Data are interpolated to a three-dimensional cubical grid, and a three-dimensional FFT, with weighting, is performed.

6.9 Comments on ISAR

The following comments are germane to the discussion of ISAR.

- In relation to the resolution of a radar image, the preferred terms are *fine* and *coarse*. Better resolution is *finer* (not *greater*). Poorer resolution is *coarser* (not *less*). In that way, ambiguities can be avoided.
- Although step chirp was given as an example, any pulse-compression waveform that has a satisfactory ambiguity function with respect to range/velocity resolution/ambiguity can be used.
- Munson et al. [6] and Jakowatz et al. [7] (with a foreword by Munson) show a strong analogy between ISAR and x-ray computer-aided tomography (CAT). Using an array of x-ray emitters and a parallel array of x-ray detectors on the other side of a target (e.g., part of a human

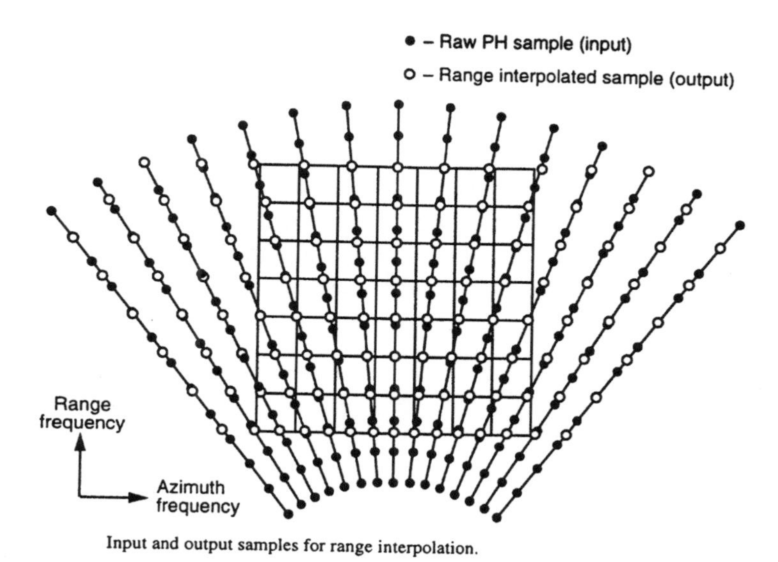

Input and output samples for range interpolation.

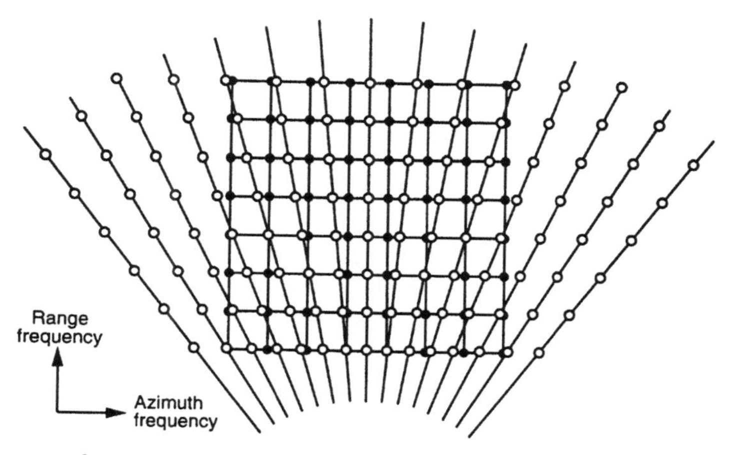

Input and output samples for azimuth interpolation.

Figure 6.5 Polar formatting (LFM chirp assumed): (a) input and output samples for range interpolation; (b) input and output samples for azimuth interpolation. (Reproduced from [5], Figures 4.17 and 4.18, with permission of Artech House.)

body), the CAT technique (noncoherent) produces projections of the density function of the target at a series of angles. By use of the projection-slice theorem and the convolution backprojection (CBP) algorithm, the two-dimensional target density function can be determined. "The CBP algorithm is the one employed in virtually all modern medical CAT scanners" [7, p. 59]. ISAR can be described in a similar manner. At a single target angle, the projection is normal to the LOS in CAT and along the LOS in ISAR. See also [5, App. B; 8, Chap. 7].

6.10 Other ISAR Cases

For complicated radar motions (in target-centered coordinates), the PSF also is more complicated. As pointed out by Ausherman et al. [9], if the radar motion describes a small circle (or latitude line) on a sphere surrounding the target, the IPR is a cone with two nappes.

This discussion gives rise to some questions. What is the IPR for a fixed radar observing a target such as a ship or an aircraft that is pitching, rolling, and yawing in various ways? What about an aircraft-borne radar making several passes over a target at various altitudes?

In discrete formalism, (6.18) becomes

$$\hat{\rho}_0(r) \sim \sum_{k_x, k_y, k_z} e^{2j(k_x x + k_y y + k_z z)} \sim \hat{\rho}_0(x, y, z) \tag{6.59}$$

This section illustrates several examples of $\hat{\rho}_0(x, y, z)$. The value actually plotted is $\sim|\hat{\rho}|^2$ because it represents the energy in various parts of the image. However, phase information is also present in the PSF and is sometimes of value (see Section 8.2).

The following discussion illustrates several specific cases of three-dimensional PSFs. The cases can be visualized most easily by considering an azimuth-elevation (AZ-EL) plot, like the ones shown in Figure 6.6. The horizontal axis is azimuth, covering only the azimuth interval desired (typically a few degrees), and the vertical axis is elevation, again covering only the interval desired. Each dot on the graph signifies the sweep over the bandwidth. For the cases analyzed, AZ (ϕ) varies from −2.5 to +2.5, and EL (ϕ) varies from 0 to +7.

Figure 6.6 illustrates the cases analyzed, as follows:

- Case A: $\theta = 0$. This case is the simplest one, corresponding to a rotating target on a turntable or a pole.

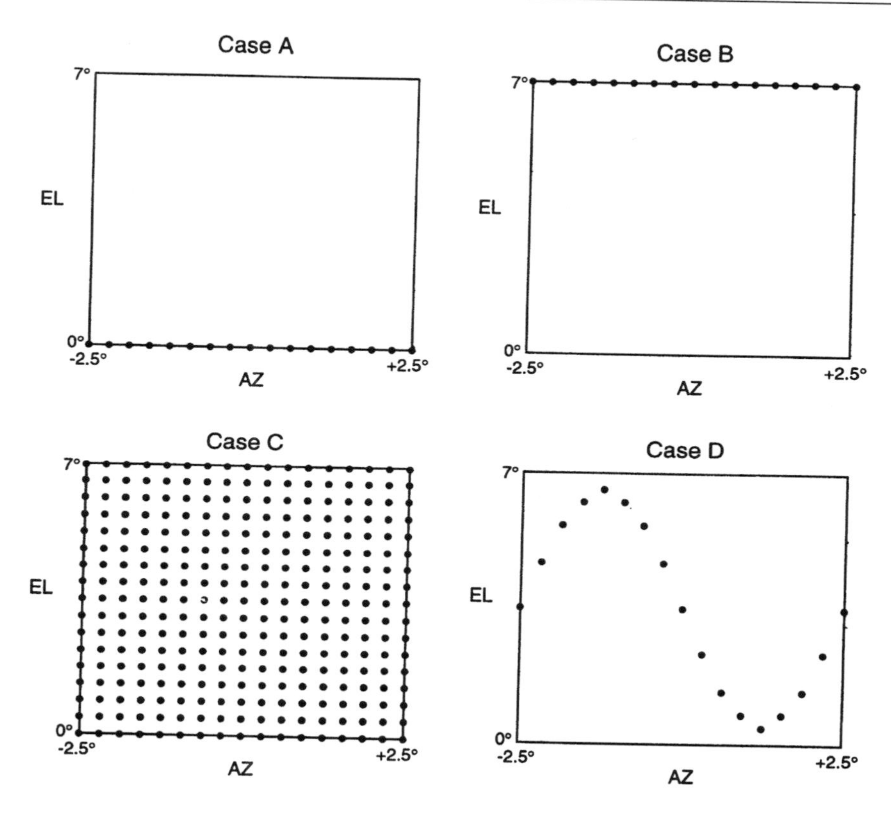

Figure 6.6 Three-dimensional ISAR cases. (Courtesy of Veridian ERIM-International.)

- Case B: $0 < \theta < 90$ degrees. This case corresponds to a nonzero look-down angle. The conical shape of the locus of viewing directions provides some resolution in the third dimension.
- Case C: $\Delta\theta \cong \Delta\phi$. Data are sampled thoroughly throughout the AZ-EL space. This case corresponds to a full three-dimensional image.
- Case D. As radar is transmitting, the target is pitching and rolling, or, equivalently, the radar platform (aircraft) is climbing and diving.

Figure 6.7 through 6.10 show the results[1] for cases A through D for far-field geometry. A summary of those results follows.

1. Results were prepared by R. J. Sullivan and P. J. Gutowski while the authors were at ERIM, Ann Arbor, Michigan, and are provided courtesy of Veridian ERIM-International.

- Case A (Figure 6.7). For this simple turntable case, both the x-axis cut and the y-axis cut are sinc functions. The z-axis cut has no resolution. It is a straight line.
- Case B (Figure 6.8). The x and y axis cuts are the same as in case A, while the slight elevation gives the z-axis cut a slight curve.
- Case C (Figure 6.9). The full three-dimensional image is produced. The x-, y-, and z-axis cuts are all sinc functions.
- Case D (Figure 6.10). The target performs a full-cycle pitch over ±3 degrees while the radar sweeps in azimuth. The x and y resolutions are good. The z resolution is also good but with high sidelobes.

6.11 Near-Field ISAR

If the radar is in the near field (specifically the Fresnel region; see Section 2.3.1), that is, if $R < 2D^2/\lambda$ (D = target cross-range dimension > radar antenna diameter), then the foregoing must be generalized. In that case, (6.12) becomes

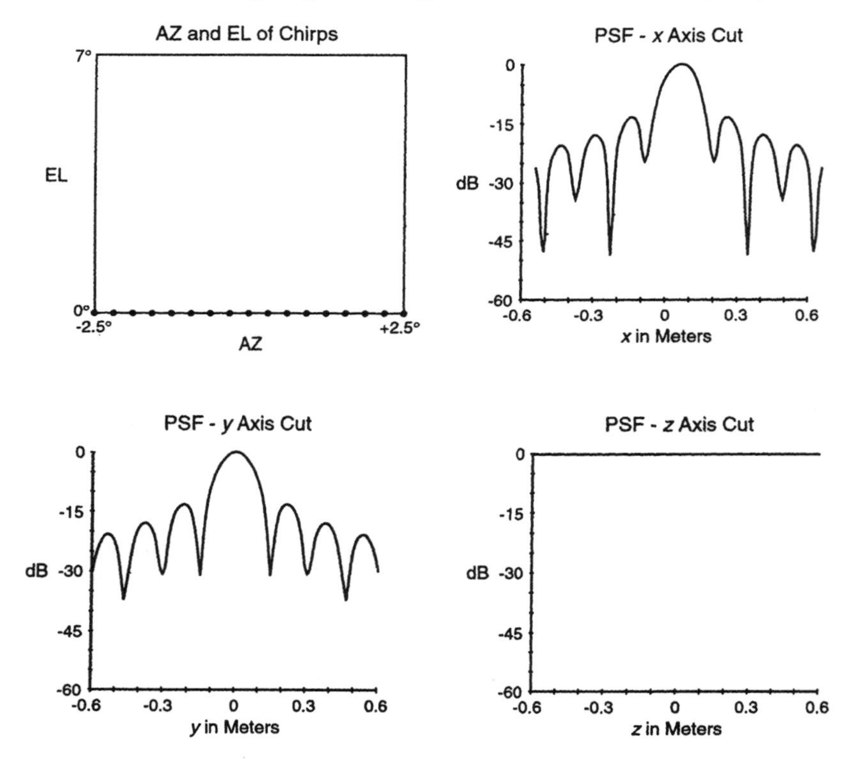

Figure 6.7 Three-dimensional ISAR results for case A. (Courtesy of Veridian ERIM-International.)

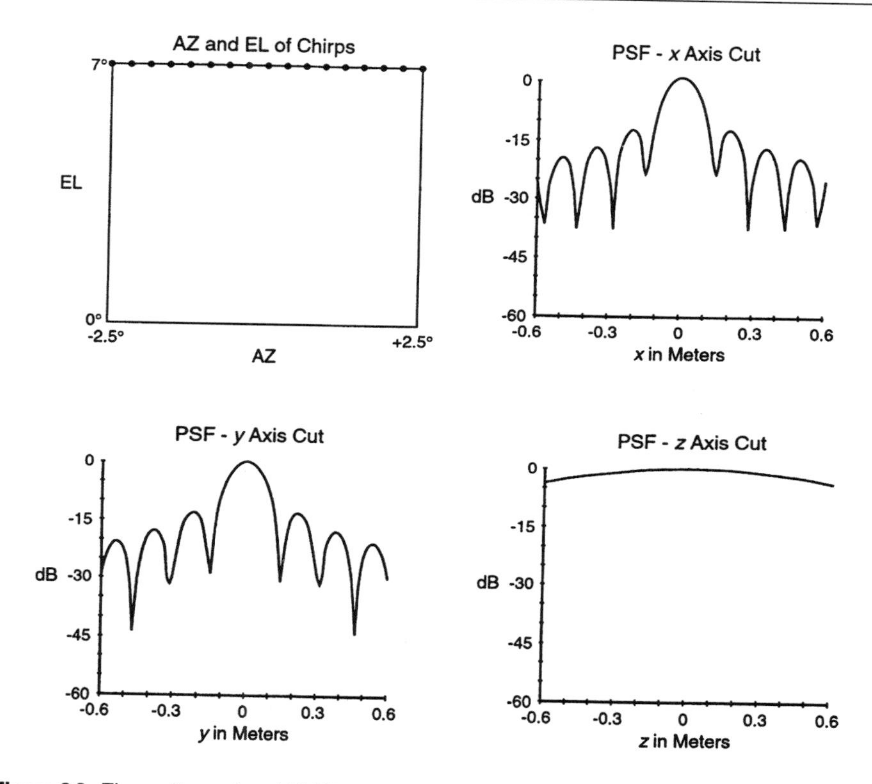

Figure 6.8 Three-dimensional ISAR results for case B. (Courtesy of Veridian ERIM-International.)

$$E(\mathbf{k}) \sim \int d\mathbf{r}\,\rho(\mathbf{r})e^{-2jk|\mathbf{r}-\mathbf{R}|} \tag{6.60}$$

The radar image is represented by the inverse of (6.60), which is given by Norton and Linzer [10] in an unnumbered equation following their Equation (66), as

$$\hat{\rho}(\mathbf{r}) \sim \int d\mathbf{k}\,E(\mathbf{k})e^{2jk|\mathbf{r}-\mathbf{R}|} \tag{6.61}$$

(Notation is somewhat different in [10].) The assumptions are that R is a constant over the imaging process (spherical aperture); that a backprojection approximation, described in [10], is valid; and that the target size is small enough so that the $1/R^4$ variation in the returned power from different parts of the target can be neglected.

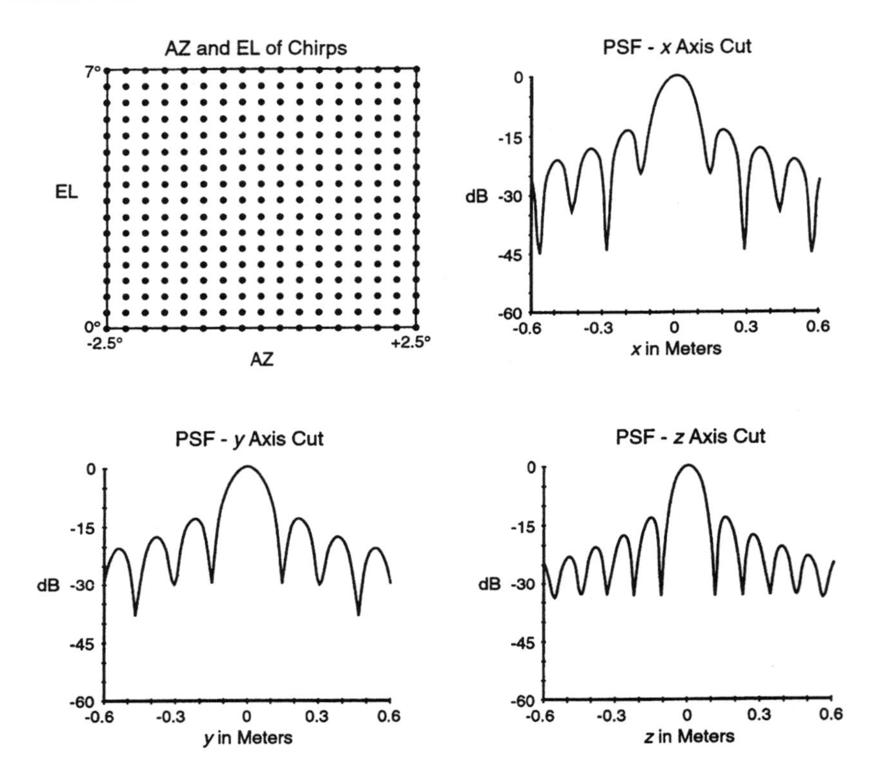

Figure 6.9 Three-dimensional ISAR results for case C. (Courtesy of Veridian ERIM-International.)

The expression for the PSF is again found by assuming that the target is a point at the origin:

$$\begin{aligned} \rho(\mathbf{r}) &= \delta(\mathbf{r}) \\ E(\mathbf{k}) &\sim e^{-2jkR} \\ \hat{\rho}(\mathbf{r}) &\sim \int d\mathbf{k}\, e^{-2jkR} e^{2jk|\mathbf{r}-\mathbf{R}|} \end{aligned} \tag{6.62}$$

The discrete form is

$$\hat{\rho}(r) \sim \sum_{k_x, k_y, k_z} e^{2jk[(R_x-x)^2+(R_y-y)^2+(R_z-z)^2]^{1/2}} e^{-2jkR} \tag{6.63}$$

Graphs of $|\hat{\rho}|^2$ could be prepared for the near-field case, analogous to Figures 6.7 through 6.10.

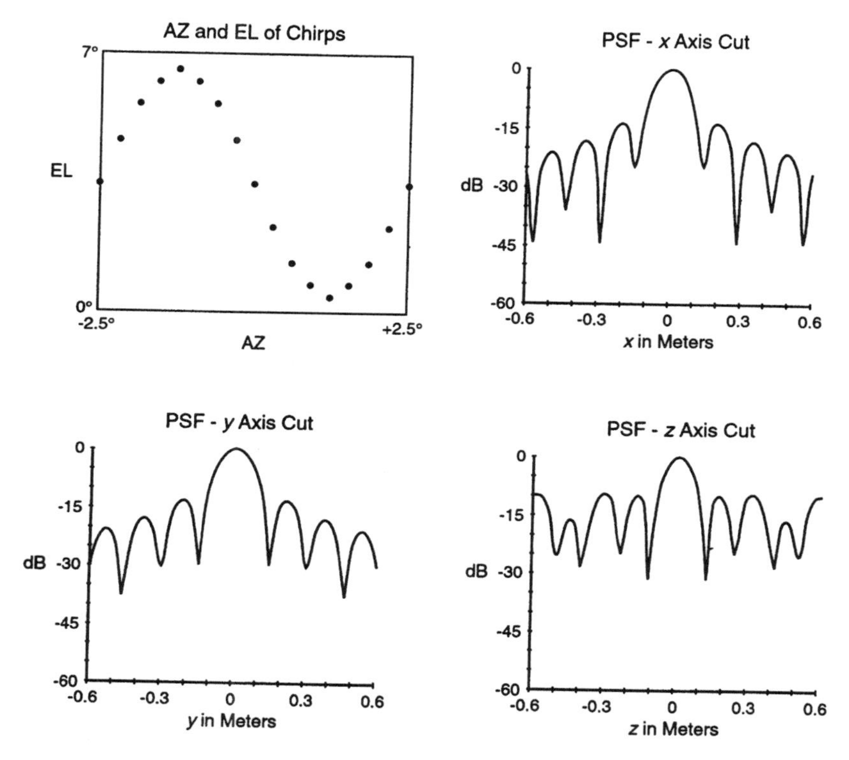

Figure 6.10 Three-dimensional ISAR results for case D. (Courtesy of Veridian ERIM-International.)

Near-field ISAR imaging of large objects may result in images that are unfocused at points distant from the image center, because of variation in the wavefront curvature. Fortuny [11] has developed a technique for three-dimensional near-field ISAR that overcomes that difficulty.

6.12 Target with Unknown Variable Translation and Rotation

Stuff, et al. [12, 13] have presented a method by which a human analyst can obtain a two-dimensional or even three-dimensional ISAR image of a target with unknown motion, as long as the target has several prominent points. The method is applicable to an airborne or surface-based radar observing an airborne or surface-based target.

We make the assumption that the target is the far field and the wavefronts are planar; however, if the target is in the near field such that the wavefronts are spherical (see Section 6.11), the motion still can be determined, although the mathematics become much more cumbersome.

A minimum of four noncoplanar points is required; the points form the vertices of a tetrahedron. The four vertices are denoted by (unknown) vectors $\mathbf{x_i}(t)$, $i = 0, 1, 2, 3$, which move as a rigid body. The first processing step is pulse compression, yielding a plot of down-range distance versus time, $r(t)$, for each prominent point. Stuff assumes that the radar is on the y axis; thus, the observed down-range distance (range) values are $r_i(t) = x_{i2}(t)$, where x_{ij}, $j = 1, 2, 3$, are respectively the x, y, and z components of $\mathbf{x_i}$.

Because the target moves as a rigid body, the temporal behavior (rotation plus translation) of the tetrahedron can be expressed as

$$\begin{aligned} \mathbf{x_i}(t) &= \mathbf{C}(t)x_i(0) + \mathbf{s}(t) \\ r_i(t) = x_{i2}(t) &= \sum_{j=1}^{3} C_{2j}(t)x_{ij}(0) + s_2(t) \end{aligned} \tag{6.64}$$

$\mathbf{C}(t)$ is an orthogonal rotation matrix. $s_1(t)$ and $s_3(t)$, the translation distances along coordinates perpendicular to the LOS, are undeterminable in principle because they have no effect on the radar returns.

We choose a coordinate system that is translating along the y axis (the other translation values are unobservable) but not rotating. It moves in such a way that vertex number 0 is always at its origin. Thus, its motion is $s_2(t) = x_{02}(t)$, which is observable. In the moving coordinate system, $\mathbf{x_0}(t) = 0$. Vertex 0 is fixed at the origin, and the other three vertices rotate about it. The axes of this coordinate system are $\Delta\mathbf{x_i}$, where

$$\Delta\mathbf{x_i}(t) = \mathbf{x_i}(t) - \mathbf{x_0}(t),\ i = 1, 2, 3 \tag{6.65}$$

We then extend the definition as follows:

$$\begin{aligned} \Delta\mathbf{x_4}(t) &= \mathbf{x_2}(t) - \mathbf{x_1}(t) \\ \Delta\mathbf{x_5}(t) &= \mathbf{x_3}(t) - \mathbf{x_1}(t) \\ \Delta\mathbf{x_6}(t) &= \mathbf{x_3}(t) - \mathbf{x_2}(t) \\ \Delta r_k(t) &\equiv \Delta x_{k2}(t),\ k = 1, \ldots, 6 \end{aligned} \tag{6.66}$$

The $\Delta\mathbf{x_k}(t)$ are six vectors representing the temporal behavior of the six edges of the tetrahedron. The relative ranges to vertices 1 through 3 are

$$\Delta r_i(t) = r_i(t) - r_0(t) = \sum_{j=1}^{3} C_{2j}(t)\Delta x_{ij}(0),\ i = 1, 2, 3 \tag{6.67}$$

By using Cramer's rule and the fact that

$$\sum_{j=1}^{3} C_{2j}^2(t) = 1$$

Stuff [13] shows that there exist six invariants of the motion, I_k, such that, for all t,

$$\sum_{k=1}^{6} (I_k)[\Delta r_k(t)]^2 = 1 \tag{6.68}$$

Because the $\Delta r_k(t)$ are observable and therefore considered known, it is necessary in principle to collect range-compressed data from only six pulses (six values of t) to obtain six equations in six unknowns.

The I_k can therefore be determined unless some of the equations are linearly dependent. Stuff [13] states that this case occurs if and only if the radar motion, in target-centered coordinates, is along the surface of an elliptical cone with its vertex at the target. In other words, if the target motion is fairly random, its motion can be determined. However, the simple case of a stationary radar observing a target rotating on a turntable, at some look-down angle, is nondeterminable. That is true even if the magnitude of the angular velocity vector changes from pulse to pulse. The simple case of spotlight SAR (see Section 7.1.1) with the airborne radar flying past a stationary target along a straight line (at constant or nonconstant speed) is also nondeterminable, because the radar moves along the surface of a cone, the base of which is an ellipse with infinite major axis. Thus, to generalize, it appears that this technique is most promising for three-dimensional ISAR on targets (e.g., vehicles) with complicated three-dimensional rotational motion (such as a truck on a bumpy road or an aircraft or ship that is pitching, rolling, and yawing) and not cases where radar and target motion are quite simple.

In practice, one would use many more than six radar pulses and perhaps more than four prominent points. The redundant information provides a test of rigid-body motion; the target passes the test if the results of (6.68) from all the pulses are consistent with six constant values of I_k. It also provides for a least-squares solution that overcomes the fact that noise or clutter inevitably will introduce small errors into each measurement of prominent-point range.

The six $|\Delta x_k|$, the lengths of the edges of the tetrahedron, can be determined from the six I_k [13]. From there, we can determine the motion, that is, the positions of the four corners of the tetrahedron, as a function of time. For a particular time, say $t = 0$, we can visualize the solution (which is performed mathematically) as follows (the position of the end point of a vector is represented by the vector symbol).

- Position of $\mathbf{x_0}$ (translation). We have already determined x_{02} and shown that it is not possible to determine x_{01} or x_{03}.
- Position of $\mathbf{x_1}$. $\mathbf{x_1}$ lies on a circle in a plane normal to the y axis (the LOS) at $y - x_{02} = \Delta x_{12}$. Because the distance from $\mathbf{x_0}$ to $\mathbf{x_1}$ is Δx_1, the radius of this circle is

$$\rho_1 = (|\Delta x_1|^2 - \Delta x_{12}^2)^{1/2} \tag{6.69}$$

  The azimuth of $\mathbf{x_1}$ about the LOS cannot be determined, because it does not affect the radar data. For simplicity, we consider the case such that $\mathbf{x_1}$ is directly "above" the y axis, that is, in the yz plane with positive z value.
- Position of $\mathbf{x_2}$. Because we know the edge distances of $\mathbf{x_2}$ from $\mathbf{x_1}$ and $\mathbf{x_0}$, plus the range of $\mathbf{x_2}$, we can determine the location of $\mathbf{x_2}$, except that it may be either on the right or left side of the vertical yz plane. Thus, we determine its location except for a reflection. We cannot distinguish between a target and its mirror image; if the target were a left-hand metal glove (from a suit of armor), we could not distinguish it from a similar right-hand glove. Those two objects are said to have opposite *parity*, and this method cannot determine the parity of the target. For a given target with given motion, we cannot distinguish between the target and its mirror-image moving with the mirror-image motion, because they produce the same radar echoes.
- Position of $\mathbf{x_3}$. This is completely determined by the range of $\mathbf{x_3}$, the positions of the three other vertices, and the lengths of the six edges of the tetrahedron.

This procedure is performed mathematically for each time of interest (each pulse). Thus, with the exceptions noted, using the Stuff method, we can in principle determine the arbitrary motion of a translating-rotating rigid body. If the radar pulses are LFM (see Section 4.2.6), we can determine the locations in three-dimensional wavenumber (**k**) space corresponding to the frequency samples (see Section 6.8). Therefore, we can perform three-dimensional ISAR and obtain a three-dimensional image.

The image provides information concerning the size and the shape of the target, but, as we have seen, it does not uniquely determine its orientation or parity. However, for the purposes of target recognition, either automated

(Chapter 9) or by a human analyst, orientation generally is not required or it may be determined from context. Furthermore, parity is not usually an issue. Most targets of interest are symmetric and essentially invariant to a parity reversal; even if an image of an asymmetric target, such as an aircraft carrier, were produced with the wrong parity (e.g., showing the superstructure on the port side instead of the starboard side), the error would be immediately obvious to the human interpreter and would not be an impediment to target recognition.

In summary, the Stuff prominent-point method can in principle test a potential target for rigid-body motion, determine the unknown motion of the target, and produce an image of the target, if and only if the following conditions exist:

- At least four prominent points are observable for each sampling time (pulse).
- At least six range-compressible radar pulses are collected, at different times.
- The radar positions, in target-centered coordinates, do not lie on the surface of an elliptical cone with the target at the vertex.

The Stuff prominent-point method has these limitations:

- Only down-range (not cross-range) translation can be determined.
- An ambiguity remains corresponding to a rotation of the target about the LOS.
- Target parity cannot be determined (e.g., left-handed and right-handed targets cannot be distinguished).

Figure 6.11 (from [12]) shows results for a synthetic target (the radar return from which is simulated on a computer) processed both by conventional SAR and by the Stuff method. The method can require substantial human effort to produce a single image, especially in deciding which down-range track corresponds with which prominent point, as points cross each other in the down-range versus time plot. Stuff [13] discusses progress achieved in automating the procedure.

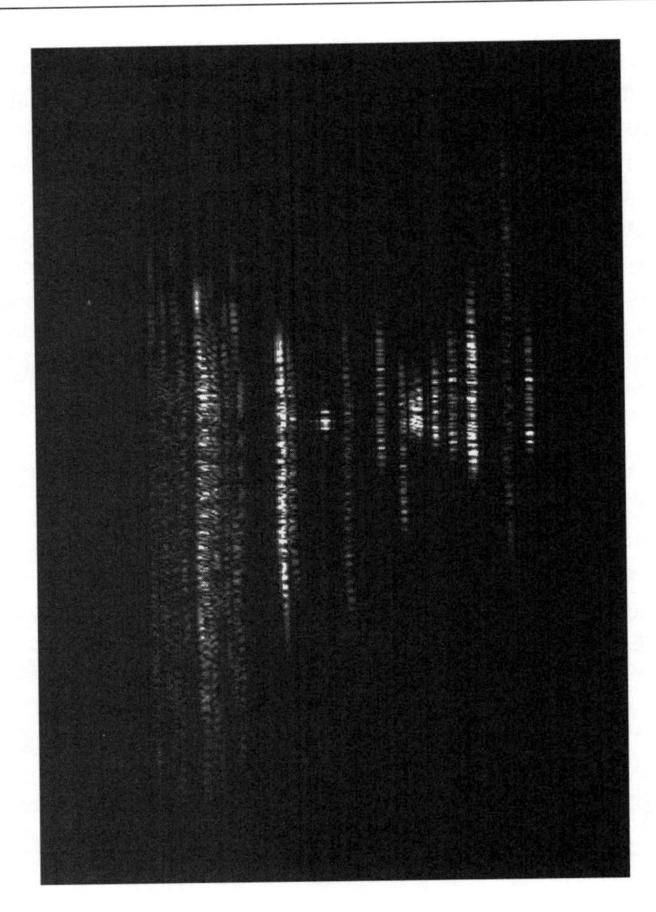

Figure 6.11 ISAR image of synthetic target with unknown translation and rotation: (a) synthetic moving target image after conventional SAR processing and (b) final synthetic target image after complete motion analysis and compensation. (Reproduced from [12], Figures 2 and 6, with permission of SPIE.)

References

[1] Knott, E. F., J. F. Shaeffer, and M. T. Tuley, *Radar Cross Section*, 2nd ed., Norwood, MA: Artech House, 1993.

[2] Papoulis, A., *The Fourier Integral and Its Applications*, New York: McGraw-Hill, 1962.

[3] Mensa, D. L., S. Halevy, and G. Wade, "Coherent Doppler Tomography for Microwave Imaging," *Proc. IEEE*, Vol. 71, No. 2, Feb. 1983, pp. 254–261.

[4] Brigham, E. O., *The Fast Fourier Transform and Its Applications*, Englewood Cliffs, NJ: Prentice Hall, 1988.

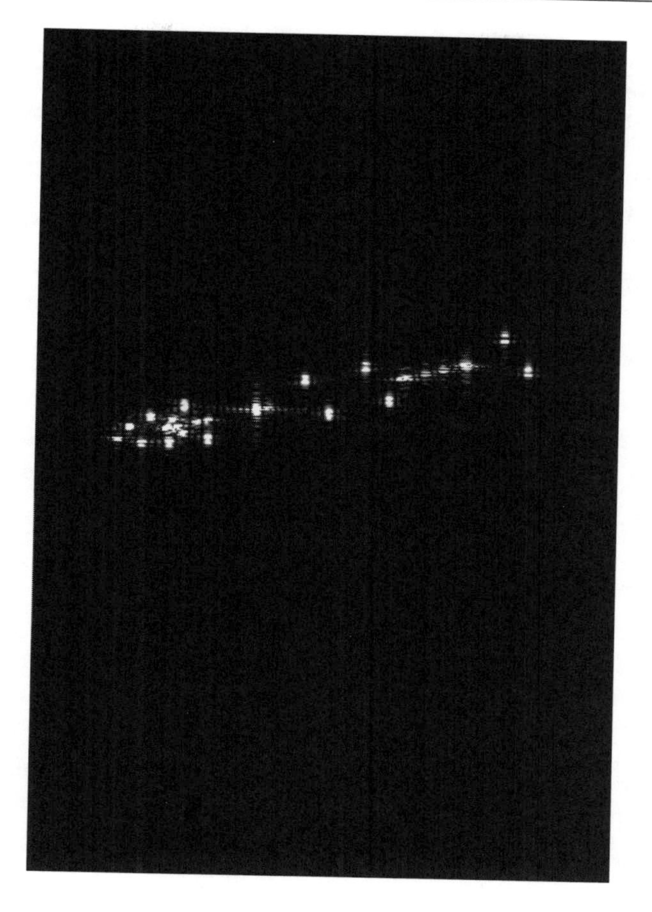

Figure 6.11 (continued).

[5] Carrara, W. G., Goodman, R. S., and Majewski, R. M., *Spotlight Synthetic Aperture Radar*, Norwood, MA: Artech House, 1995.

[6] Munson, D. C., Jr., J. D. O'Brien, and Jenkins, W. K., "A Tomographic Formulation of Spotlight-Mode Synthetic Aperture Radar," *Proc. IEEE*, Vol. 71, No. 8, Aug. 1983, pp. 917–925.

[7] Jakowatz, C. V., Jr., et al., *Spotlight-Mode SAR: A Signal-Processing Approach*, Boston: Kluwer Academic Publishers, 1996.

[8] Mensa, D. L., *High-Resolution Radar Cross Section Imaging*, Norwood, MA: Artech House, 1991.

[9] Ausherman, D. A., et al., "Developments in Radar Imaging," *IEEE Trans.*, AES, Vol. AES-20, No. 4, July 1984.

[10] Norton, S. J., and M. Linzer, "Ultrasonic Reflectivity in Three Dimensions: Exact Inverse Scattering Solutions for Plane, Cylindrical, and Spherical Apertures," *IEEE Trans. Biomedical Engineering*, Vol. BME-28, No. 2, Feb. 1981.

[11] Fortuny, J., "An Efficient 3-D Near-Field ISAR Algorithm," *IEEE Trans. Aerospace and Electronic Systems*, Vol. 34, No. 4, Oct. 1998, pp. 1261–1270.

[12] Stuff, M., et al., "Automated Two and Three Dimensional, Fine Resolution, Radar Imaging of Rigid Targets with Arbitrary Unknown Motion," *SPIE*, Vol. 2230, 1994, pp. 180–189.

[13] Stuff, M., "Three-Dimensional Analysis of Moving Target Radar Signals: Methods and Implications for ATR and Feature-Aided Tracking," *SPIE*, Paper 3721-51, 1999.

Problems

Problem 6.1

Rewrite the first paragraph of Section 6.1 assuming that the waveform consists of a sequence of M LFM pulses transmitted at PRF = f_R/N, each of bandwidth B and pulse width τ, with each pulse sampled N times.

Problem 6.2

In (6.15), show that, if the range estimate is in error by a small distance ΔR, the image will simply be translated by ΔR, otherwise it remains unchanged.

Problem 6.3

For standard two-dimensional ISAR, with a step chirp waveform, show that the minimum required PRF is

$$f_R(\text{min}) = \frac{2NM\delta_{crpn}\Omega}{\lambda}$$

where

N = number of samples per pulse group

M = number of pulse groups

δ_{crpn} = cross-range resolution (peak to first null)

Ω = angular velocity of target

λ = wavelength

Problem 6.4

Using (6.52), calculate δr_{pn} for $B = 0$ (single-frequency imaging), $\lambda = 0.03$m, and $\Delta\phi = 2\pi$. How does that compare with the conventional value of 0.3m for $B = 500$ MHz and $\Delta\phi = 3$ degrees?

Problem 6.5

For the continuous formalism, extend the results to three dimensions:

$$\Delta k << \bar{k},\ \Delta\phi << 1,\ \Delta\theta << 1$$

Show that

$$\rho_0(x, y, z) \sim \mathrm{sinc}(x\Delta k) \cdot \mathrm{sinc}(\bar{k}y\Delta\phi) \cdot \mathrm{sinc}(\bar{k}z\Delta\theta)$$

and

$$\delta_x = \frac{c}{2B},\ \delta_y = \frac{\bar{\lambda}}{2\Delta\phi},\ \delta_z = \frac{\bar{\lambda}}{2\Delta\theta}$$

7

Synthetic Aperture Radar

So far, most of our discussion has concerned RAR, in which the antenna is a physical object that first emits, then collects the radiation. We now turn our attention to the case in which the antenna moves to cover a synthetic aperture (L_{SA}), thus producing SAR. This chapter is a brief overview, based on [1]; more detailed treatments are provided in [2–7]. Good summaries of existing SARs are given in [2, 8].

As discussed in Chapter 6, SAR generally refers to the case of a moving radar and a stationary target—usually an extended scene, such as the surface of the Earth; ISAR refers to the case in which the radar is relatively stationary and a rotating target provides all (or most) of the motion to create the synthetic aperture. Obviously, those distinctions are not fundamental, because they depend on the user's coordinate system. Furthermore, the two concepts are not mathematical inverses, and there are gray areas where they merge.

This chapter assumes an LFM SAR waveform. It also assumes that the Earth's surface is stationary and (except as noted) flat. For a discussion of SAR imaging of the ocean, which is moving, see [9].

7.1 Introduction to SAR

Figure 7.1 compares RAR and SAR. The crossrange resolution (peak to first null) of a SAR is

$$\delta_{cr}(SAR) \cong \frac{\lambda}{2\Delta\theta} \cong \frac{\lambda}{2(L_{SA}/R)} = \frac{R\lambda}{2L_{SA}} \tag{7.1}$$

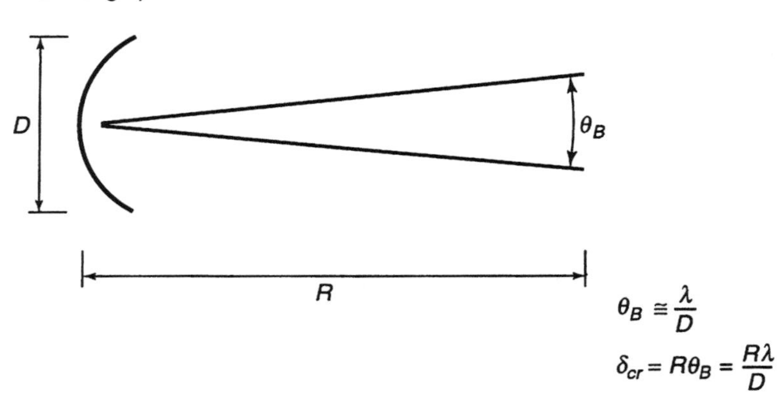

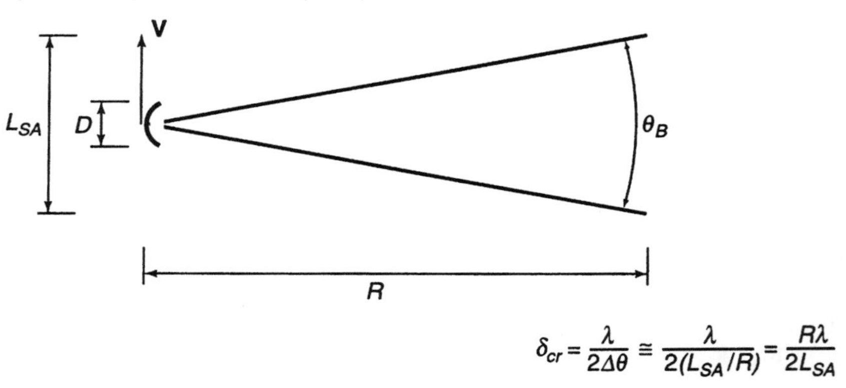

Figure 7.1 Comparison of RAR and SAR: (a) RAR; (b) SAR.

From Section 6.2, that resolution is finer than the resolution for a RAR of equal aperture. An intuitive explanation for that interesting result is that for RAR the echo received at a particular aperture location results from energy transmitted from all locations in the aperture. For SAR, the echo received at a particular aperture location results from energy transmitted from that (known) location in the aperture; that is, more information is received [1, p. 36].

In this book, references to SAR mean focused SAR. Some precursors to focused SAR include the following:

- Side-looking airborne radar (SLAR), an aircraft-mounted RAR pointed perpendicular to the direction of flight (hence *side-looking*), with cross-range resolution $R\lambda/D$;

- Doppler beam sharpening (DBS) ([10], Section 8.1), in which, for an airborne radar, the real-beam return is doppler processed to produce crossrange resolution finer than that provided by the real beam alone; broadside crossrange resolution is $\sim R\lambda/2L_{\mathrm{DBS}}$, where L_{DBS} is the aperture length generated during a target dwell (e.g., for a scanning antenna, the aperture generated during the time that the antenna scans by the target—see Problem 7.1);
- Unfocused SAR (USAR), which involves a short synthetic aperture with a maximum two-way phase shift across the aperture of $\pi/2$; the returned pulses are processed coherently, and crossrange resolution is approximately $1/2(\lambda R)^{1/2}$ (Problem 7.2).

For example, if $\lambda = 0.03$m (X band), $D = 2$m, $R = 100$ km, $L_{\mathrm{DBS}} = 10$m, and $L_{\mathrm{SA}} = 5$ km, then δ_{cr} (SLAR) = 1,500m, δ_{cr} (DBS) = 150m, δ_{cr} (USAR) = 27m, and δ_{cr} (SAR) $\sim \lambda/2\Delta\theta = 0.3$m.

7.1.1 SAR Modes

As shown in Figure 7.2, there are at least three distinct SAR modes. Stripmap SAR (or strip SAR), shown in Figure 7.2(b), is also called search SAR, because it is useful for imaging large areas at relatively coarse resolution. In stripmap SAR, the beam remains at a constant squint angle θ_{sq} to the perpendicular to the flight path (the latter is assumed to be a straight line) and continuously observes a strip of terrain parallel to the flight path. The beam is often broadside ($\theta_{\mathrm{sq}} = 0$); if $\theta_{\mathrm{sq}} \neq 0$, we refer to the radar as squinted SAR.

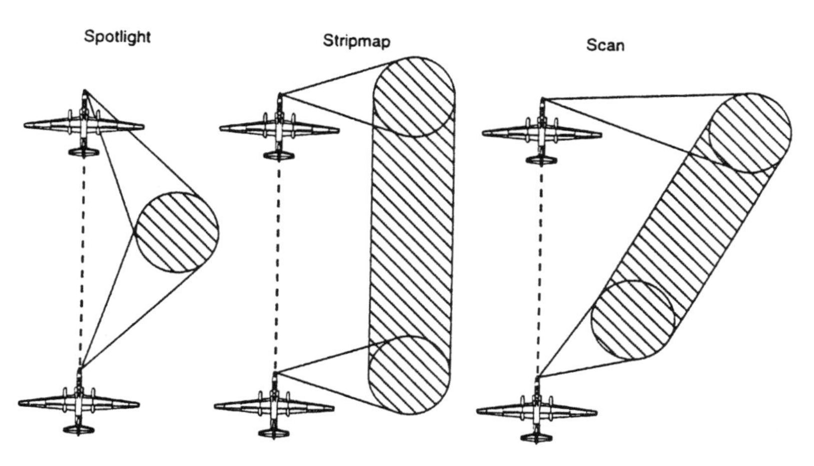

Figure 7.2 SAR modes: (a) spotlight; (b) stripmap; (c) scan. (Reproduced from [1], Figure 1.2, with permission of Artech House.)

For stripmap SAR, the aperture angle is essentially equal to the two-way, real-aperture, 3-dB beamwidth θ_{3dB}:

$$\Delta\theta \cong \theta_{3dB} \cong \frac{\beta\lambda}{D\cos\theta_{sq}} \tag{7.2}$$

$$\delta_{cr} \cong \frac{\lambda}{2\Delta\theta} \cong \frac{D\cos\theta_{sq}}{2\beta} \tag{7.3}$$

For a broadside beam, if $\beta \sim 1$, then $\delta_{cr} \cong D/2$. If $D >> \lambda$ and SNR $>> 1$, then the smaller the physical antenna, the finer the crossrange resolution, independent of range.

Spotlight SAR (or spot SAR), shown in Figure 7.2(a), is used to obtain a relatively fine-resolution image of a known location or target of interest. As the platform passes by the target, the beam direction moves, to keep pointing at the target. In that way, $\Delta\theta$ can be made considerably greater than θ_B, and δ_{cr} (spot) $<$ δ_{cr} (strip). If the radar platform (aircraft) follows the arc of a circle with the target directly below the center of the circle (at a constant LDA), then spotlight SAR is mathematically equivalent to ISAR. Usually the spot-SAR platform follows a straight line past the target. However, the appropriate phase corrections are easily made, and again the processing becomes essentially the same as for ISAR. We may also want to make a correction for the variation in received power ($\sim 1/R^4$) as the range to the target varies slightly over the synthetic aperture.

The aperture time t_A required to collect the data for a stripmap or spotlight SAR image is found as follows:

$$\begin{aligned} \delta_{cr} &\approx \frac{\lambda}{2\Delta\theta} \approx \frac{\lambda R}{2L_{SA}\cos(\theta_{sq})} = \frac{\lambda R}{2Vt_A\cos(\theta_{sq})} \\ t_A &\approx \frac{\lambda R}{2V\delta_{cr}\cos(\theta_{sq})} \end{aligned} \tag{7.4}$$

where V is the platform speed.

One other (and seldom used) mode is scan SAR. The beam observes a straight strip of terrain that is not parallel to the flight path. Clearly such a strip must be of finite length, since eventually the range becomes so great that the SNR is too low to produce clear imagery. In principle, a SAR beam may describe even more complicated patterns on the ground; however, processing difficulty rapidly outweighs the benefits of such schemes.

Figure 7.3 is a SAR image of the Baltimore-Washington International Airport, obtained with the SAR designed for the U.S. Predator unmanned aerial vehicle.

7.1.2 Range and Velocity Contours

Consider an airborne radar with an isotropic antenna, moving parallel to a flat ground, as illustrated in Figure 7.4(a). Using pulse-compression, the radar can distinguish between targets at different ranges. A particular target may be determined to be located on a constant-range contour. In three-dimensional space, the contours are concentric spheres with the radar at the center, as

Figure 7.3 SAR image of Baltimore-Washington International Airport. (Courtesy of Northrop-Grumman Corporation.)

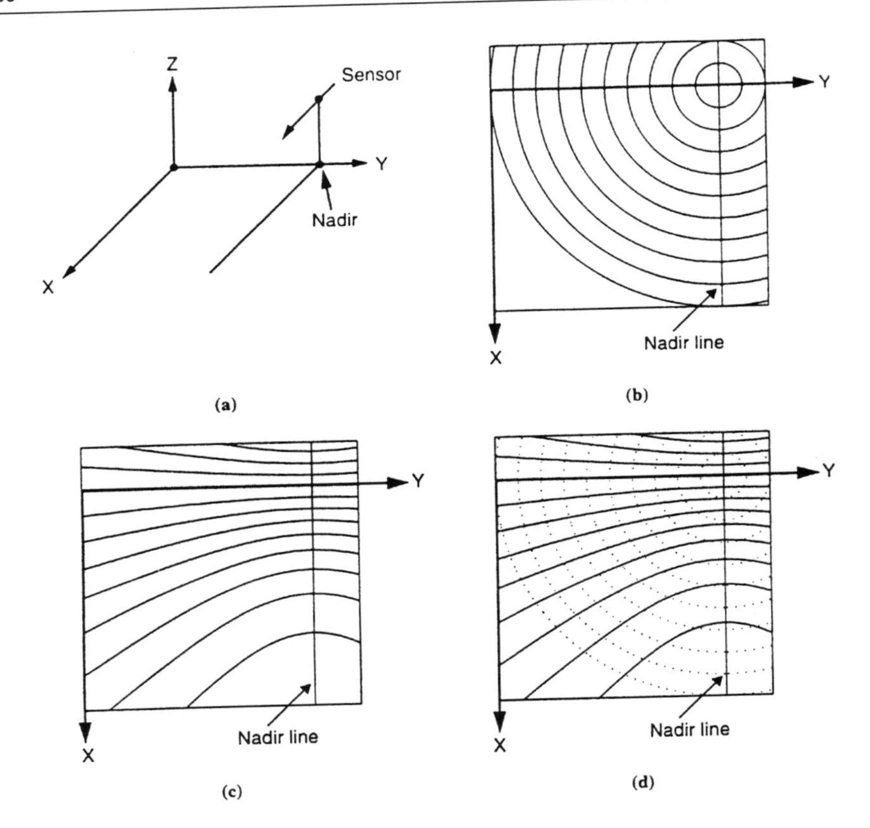

Figure 7.4 Range and velocity contours on the Earth's surface: (a) collection geometry; (b) circles of constant range in the *xy* plane; (c) hyperbolas of constant doppler cone angle in the *xy* plane; (d) combination of constant-range circles and constant doppler hyperbolas. (Reproduced from [1], Figure 2.8, with permission of Artech House.)

shown in Figure 7.5(a). On the ground, the range contours are the intersection of the spheres with the ground, a set of concentric circles with the subradar point at the center, as shown in Figure 7.4(b).

Similarly, using doppler processing, the radar can distinguish between targets of different apparent velocities along the LOS between the radar and the target. If the platform velocity is **V** and the angle between **V** and the LOS to a stationary target is θ, then the apparent LOS velocity is $V_{LOS} = -V\cos\theta$, as shown in Figure 7.5(b). In three-dimensional space, surfaces of constant V_{LOS} are circular cones with axis **V** and generating angle θ, with the radar at the vertex. The intersection between the set of cones and the flat ground is a set of nested hyperbolas, as shown in Figure 7.4(c) (also see Section 11.1).

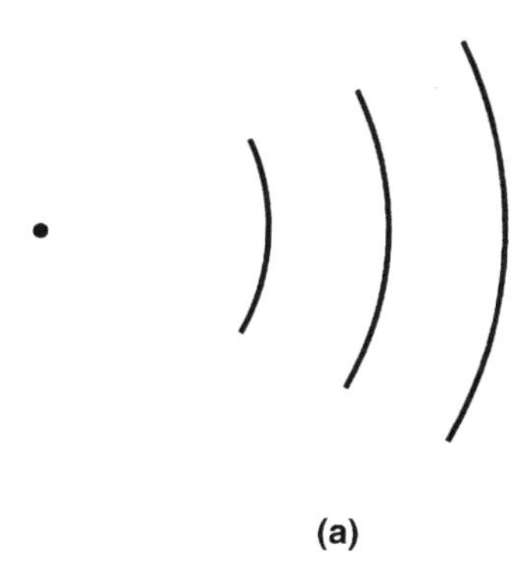

(a)

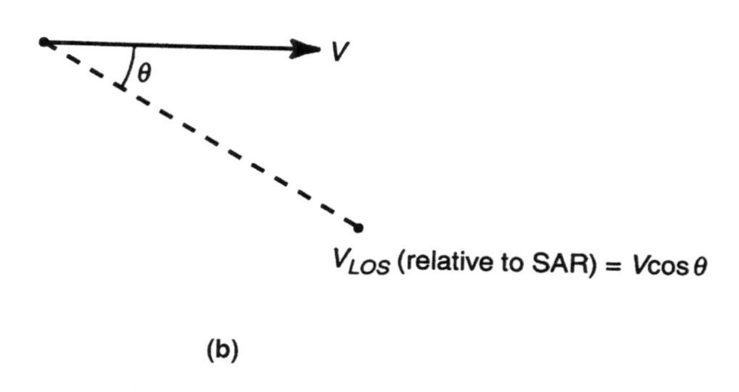

(b)

Figure 7.5 Range and velocity contours in three-dimensional space: (a) constant-range contours (concentric spheres with radar at center); (b) constant-velocity contours (circular cones with radar at vertex).

Figure 7.4(d) shows the combination of constant-range circles and hyperbolas of constant V_{LOS} (called *isodops*). Through appropriate processing, returns from each intersection cell can be distinguished. If the ground range to the target is greater than the SAR altitude, then over a small angle about the broadside direction, the range contours and the isodops are essentially orthogonal to each other. The resulting radar returns can be displayed to yield an image of the ground. At a nonzero squint angle, the isodops are not orthogonal to the range contours; however, additional processing corrections can still result in an essentially undistorted ground image.

7.1.3 Motion Compensation

The basic theory of SAR relies on the assumption that the platform is traveling parallel to the ground in a straight line at constant velocity. That is not exactly true; for successful SAR imaging, it is necessary that the deviations of the platform from the nominal path be measured, recorded, and incorporated into the processing. That procedure is known as motion compensation (mocomp). After appropriate filtering of the data, the true flight path must be estimated to within a fraction of a wavelength. For example, for a particular pulse, if the antenna is estimated to be a distance d away from the straight-line trajectory (along the LOS), the phase

$$\Delta\phi = \frac{4\pi d}{\lambda} = \frac{4\pi df}{c} \tag{7.5}$$

with appropriate sign, is added to the measured phase at the frequency f to produce the best estimate of what the recorded phase would have been if the platform had been on the straight line. Similarly, if the platform speed is not constant, the received data are interpolated to produce the best estimate of what they would have been if the speed had been constant.

When the platform is an aircraft, an onboard inertial navigation system (INS) uses accelerometers and gyroscopes to measure the deviations. Sometimes a smaller inertial measurement unit (IMU) relying on the same general principles is strapped down near the antenna. Without an absolute reference frame, the outputs of any INS or IMU will drift with time as errors accumulate. An absolute frame for position and velocity can be obtained from the global positioning system (GPS), a constellation of at least 24 satellites in polar Earth orbit providing continuous reference signals for determination of precise position and velocity [11].

7.1.4 Slant and Ground Planes

When a SAR image is initially produced, the range pixel size δ_r is usually a constant [it is usually chosen to be somewhat less than $c/2B$, e.g., $0.75(c/2B)$, to ensure adequate sampling]. As illustrated in Figure 7.6, the actual ground locations that correspond to those range samples are not spaced at constant intervals in ground range. Near the scene center they are spaced at

$$\delta_g \cong \delta_r / \cos\psi \tag{7.6}$$

At ranges closer to the radar, they are spaced still farther apart, because of the spherical range contours. The image corresponds to the projection of the ground onto a slant plane; that plane is determined by the LOS and its perpendicular in the ground plane. We often refer to this type of image as a slant-plane image. By appropriate interpolation and resampling, a ground-

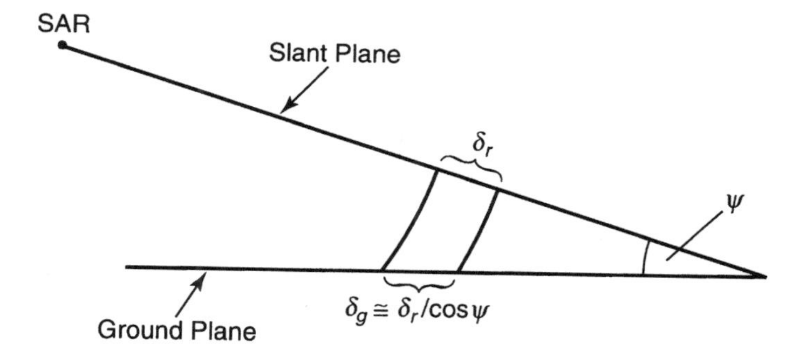

Figure 7.6 Slant and ground planes. $\delta_g > \delta_r$; ground-plane range resolution is coarser than slant-plane range resolution.

plane image, with δ_g = constant = δ_{cr} may be produced. Ground-plane imagery with minimal distortion is necessary if comparison is to be made with maps or with imaging taken from other sensors, such as optical sensors or other SARs.

7.1.5 Pulse Repetition Frequency Requirements for SAR

From Figure 7.7, for broadside squint, the apparent angular velocity of the scene rotation is

$$\Omega = \frac{V}{R} \tag{7.7}$$

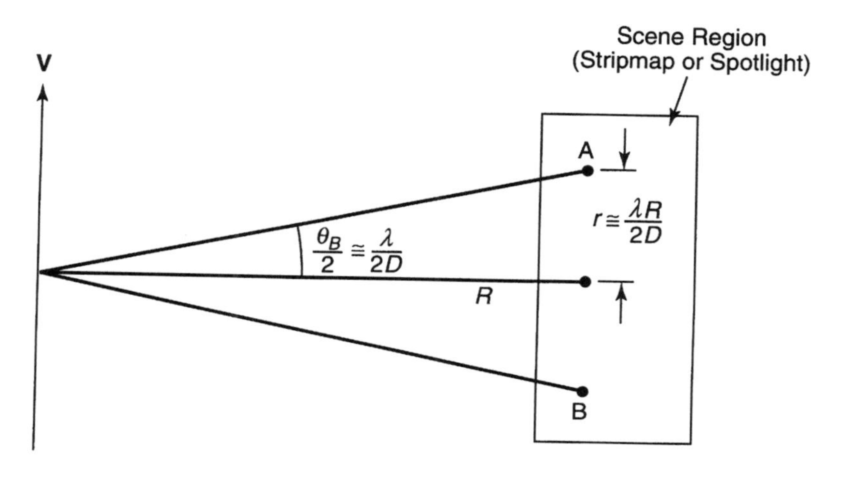

Figure 7.7 Minimum PRF for SAR.

The relative velocity of point A relative to the SAR is

$$v_A = -\Omega r = -\frac{V}{R} \cdot \frac{\lambda R}{2D} = -\frac{\lambda V}{2D} \tag{7.8}$$

Similarly, the relative velocity of point B relative to the SAR is

$$v_B = \Omega r = \frac{\lambda V}{2D} \tag{7.9}$$

Thus, the range of relative velocities in the scene is

$$\Delta v = \Omega r - (-\Omega r) = 2\Omega r = \frac{\lambda V}{D} \tag{7.10}$$

The range of doppler frequencies received from the scene is

$$\Delta f_d = \frac{2}{\lambda} \cdot \frac{\lambda V}{D} = \frac{2V}{D} \tag{7.11}$$

Thus, to avoid velocity ambiguity, the PRF must be at least $2V/D$. Writing

$$f_R(\text{min}) = 2V/D = \frac{1}{t_R(\text{max})} \tag{7.12}$$

we have

$$Vt_R(\text{max}) = \frac{D}{2} \tag{7.13}$$

Thus, the distance traveled by the platform during the time between pulses (t_R) must be no more than $D/2$, and the SAR must transmit at least two pulses as its physical antenna passes a stationary point in space.

We also frequently want range-unambiguous operation, which implies

$$\frac{2V}{D} < f_R < \frac{c}{2R} \tag{7.14}$$

For example, if $V = 180$ m/s (350 kn), $D = 1$ m, and $R = 150$ km, then 360 Hz $< f_R <$ 1,000 Hz. (Also see Problem 7.3.)

7.1.6 Range Migration

As we have seen, a focused SAR can obtain range resolution through pulse compression with $\delta_r = c/2B$ and obtain cross-range resolution through doppler processing, with $\delta_{cr} = \lambda/2\Delta\theta$. If we wanted to prevent range migration (movement of a point target from one range bin to the next during the image formation), we would require that ΔR, the variation of range during the data collection (over the synthetic aperture), be less than δ_r. We have [12]

$$R_{max} \cong R_0 + \frac{V^2(t_A/2)^2}{2R_0}, \quad R_{min} = R_0 \tag{7.15}$$

$$\Delta R = R_{max} - R_{min} = \frac{(Vt_A)^2}{8R_0} = \frac{L_{SA}^2}{8R_0} = \frac{R_0}{8}(\Delta\theta)^2 = \frac{R_0\lambda^2}{32\delta_{cr}^2} < \delta_r \tag{7.16}$$

Parameters for a typical SAR might be $R_0 = 200$ km, $\lambda = 0.03$m, and $\delta_r = 1$m; then $\Delta R = 5.6$m $> \delta_r$, and the condition is not satisfied. Thus, in general, the processor must correct for range migration.

7.2 SAR Waveforms and Processing

The term *fast time* refers to the time increments within one PRI. *Slow time* is used to describe time increments within one aperture time (t_A), that is, over many pulses.

7.2.1 Fast-Time Processing

To obtain fine range resolution, we typically use pulse compression. Most SARs use a waveform such that the full waveform bandwidth, B, is incorporated into a single pulse, as shown in Figures 7.8(a) and 7.8(b). The most common type is LFM: $f = f_c + \gamma t$. Other possible waveforms appropriate for pulse compression are

- Nonlinear FM. $f = f_c + F(t)$.
- PM.
  - Biphase. The phase is switched between 0 and 180 degrees in a coded fashion, for example, Barker code [12–13].

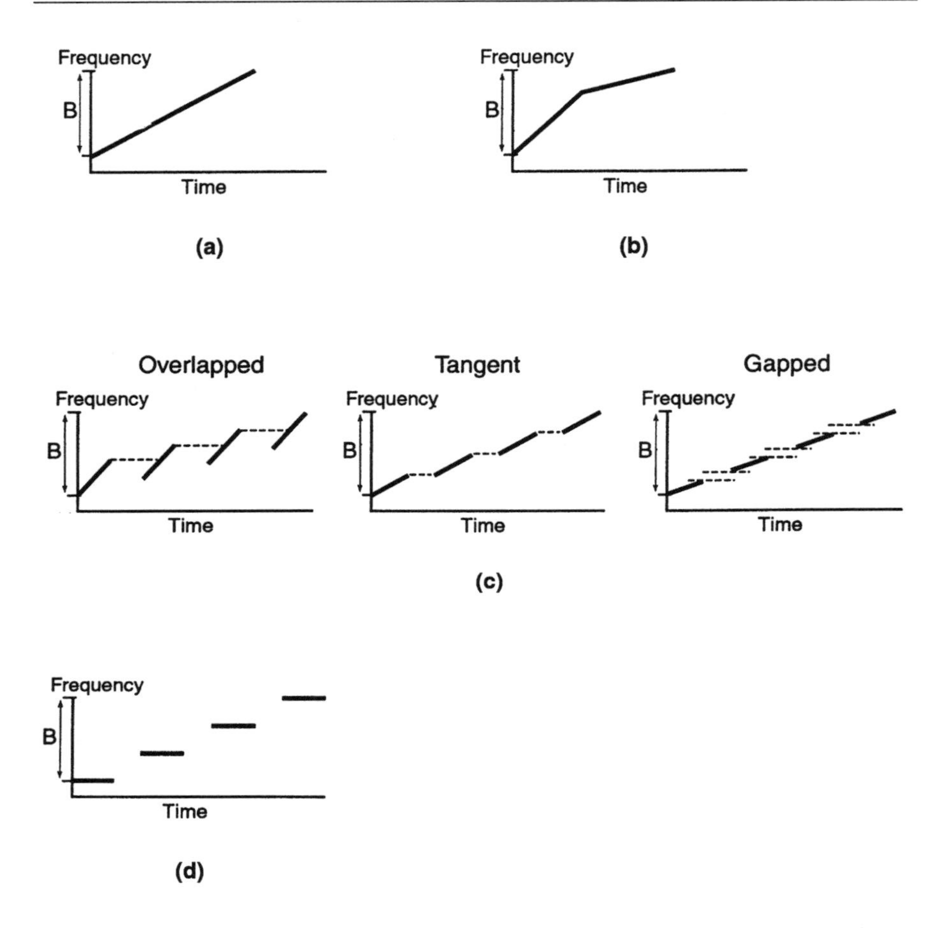

Figure 7.8 Examples of pulse-compression waveforms: (a) LFM; (b) nonlinear FM; (c) FJB; (d) step chirp.

- Polyphase. The phase is switched in a more complicated fashion, for example, quadriphase among 0, 90, 180, and 270 degrees (e.g., Frank code [12]).

- "Chipped" pulses. Pulse subsections (chips) are each a relatively simple FM or PM; chips are ordered in a complicated fashion, perhaps for low probability of intercept (LPI) [14] (also see Section 13.2 of this book) or for a Costas waveform (Section 4.2.7.3).

Another option is a sequence of pulses with overall bandwidth B:

- Frequency-jump burst (FJB) [15]. Each pulse is LFM, and center frequencies vary linearly; individual pulse bandwidths may be overlapped, tangent, or gapped, as shown in Figure 7.8(c).
- Step chirp. FJB with individual pulses monochromatic, as shown in Figure 7.8(d).

7.2.1.1 LFM Processing in SAR

Following Carrara et al. [1, Sec. 2.6], we consider a SAR LFM pulse illuminating a scene containing one point target. The transmitted signal is

$$S_{Tx}(n, t) = A_o \operatorname{rect}\left(\frac{\hat{t}}{t_p}\right) e^{j[2\pi f_c t + \pi\gamma\hat{t}^2]} \tag{7.17}$$

Here, t = time, with the origin at the beginning of the waveform; t_p = pulse width; n = pulse number, starting at zero; T = interpulse period (PRI); and $\hat{t} = t - nT$ (fast time), with the origin at the beginning of the pulse. Note that the carrier frequency is coherent pulse-to-pulse and is referenced to the STALO; hence, the first t in the argument of exponential has no circumflex. The second is $\hat{t}$, with a circumflex; the LFM begins anew at the start of each pulse.

The echo signal received from the point target (at range R_t) is

$$S_{Rx}(n, t) = a_t \operatorname{rect}\left(\frac{\hat{t} - 2R_t/c}{t_p}\right) e^{j2\pi f_c(t-2R_t/c)} e^{j\pi\gamma(\hat{t}-2R_t/c)^2} \tag{7.18}$$

A central point within the scene, at range R_0, is designated as the mocomp point (MCP). The expected signal returned from that point (without amplitude modulation) is proportional to

$$S_{\text{ref}}(n, t) = e^{j2\pi f_c(t-2R_0/c)} e^{j\pi\gamma(\hat{t}-2R_0/c)^2} \tag{7.19}$$

The conjugate of that reference signal is mixed with the echo to yield

$$S_{IF}(n, t) = a_t \operatorname{rect}\left(\frac{\hat{t} - 2R_t/c}{t_p}\right) e^{-j\frac{4\pi\gamma}{c}\left(\frac{f_c}{\gamma}+\hat{t}-\frac{2R_0}{c}\right)(R_t-R_0)} e^{j\frac{4\pi\gamma}{c^2}(R_t-R_0)^2} \tag{7.20}$$

[1, Sec. 2.6] (see Problem 7.4). For N_p pulses,

$$S_{IF}(n, t) = a_t \sum_{n=0}^{N_p-1} \left[\operatorname{rect}\left(\frac{\hat{t} - 2R_t/c}{T_p}\right) e^{j\Phi(n,\hat{t})} \right] \tag{7.21}$$

where

$$\Phi(n, \hat{t}) \equiv -\frac{4\pi\gamma}{c}\left(\frac{f_c}{\gamma} + \hat{t} - \frac{2R_0}{c}\right)(R_t - R_0) + \frac{4\pi\gamma}{c^2}(R_t - R_0)^2 \quad (7.22)$$

The first term is linear in $R_t - R_0$. The second term is called the residual video phase (RVP).

7.2.1.2 The Dechirp Procedure

Figure 7.9 illustrates the relationship between the transmitted LFM pulse and the echoes received at different frequencies (f) and times (t) from a desired scene (swath) of width S in the LOS direction. The time interval during which relevant echoes are returned is $2S/c + t_p$. The relevant information may be described by a data function of (f, t), which is nonzero over a parallelogram-shaped region of (f, t) space, with constant-frequency contours parallel to the

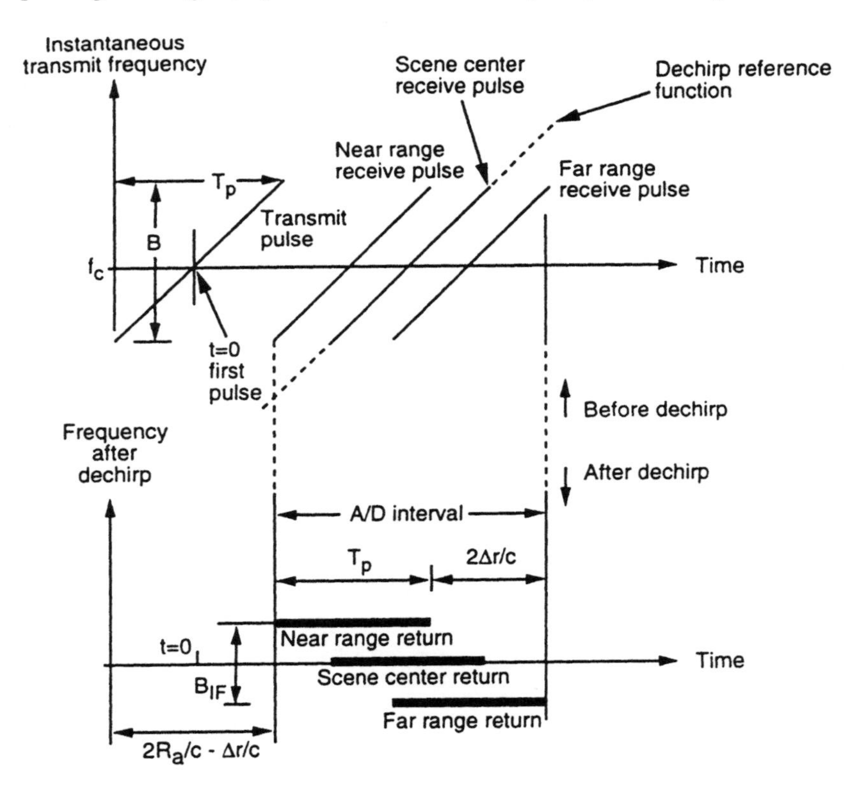

Figure 7.9 LFM pulse echo and dechirp procedure. (Reproduced from [1], Figure 2.10, with permission of Artech House.)

time axis. (We assume that echoes from outside the desired swath are set to zero.)

The echo signal can be processed by various methods [1]. It can be directly digitized and matched filtered to yield a downrange profile. Another commonly used procedure involves dechirping (also called deramping) and deskewing [1].

In the dechirp procedure, the received echo is mixed with a delayed replica of the transmitted pulse. The result is shown in Figure 7.9. The region of nonzero information in (f, t) space is still a parallelogram, but now the constant-range contours are parallel to the time axis.

Although, in principle, dechirping can be performed in either hardware or software, it is often done in hardware, before A/D sampling, to reduce the bandwidth of the echo signal from B to the IF bandwidth, B_{IF}, which is given by (Problem 7.5)

$$B_{IF} = \gamma\left(\frac{2S}{c}\right) \tag{7.23}$$

The bandwidth ratio is

$$\frac{B_{IF}}{B} = \frac{\gamma(2S/c)}{\gamma t_p} = \frac{(2S/c)}{t_p} \tag{7.24}$$

If $2S/c < t_p$, then $B_{IF} < B$. The A/D converter need only operate at frequency $2B_{IF}$ (B_{IF} if a quadrature mixer is used) instead of $2B$. Operating at B_{IF} traditionally has been desirable because as the maximum sampling frequency of an A/D converter (of given number of bits) becomes lower, its cost becomes lower and it becomes more readily available. Evidently, that is not applicable if $2S/c > t_p$. For example, a typical SAR might use the following parameters (t_p = 100 μs):

- Spotlight mode: S = 2 km, B = 600 MHz ($\delta_r \sim$ 0.3m), $2S/c$ = 13.3 μs, B_{IF}/B = 0.13, analog dechirp is used.
- Stripmap mode: S = 10 km, B = 180 MHz ($\delta_r \sim$ 1m), $2S/c$ = 66.7 μs, B_{IF}/B = 0.67, analog dechirp is not used (no significant advantage).

The dechirp procedure for $2S/c < t_p$ is sometimes called *stretch*, because the sampling time is stretched to $2S/c + t_p$; if a very short (compressed) pulse were actually used, the sampling time would be only $\sim 2S/c$.

7.2.1.3 The Deskew Procedure

We want a time sample to include the results of all ranges across the swath, at a single transmitted frequency. Time samples immediately after dechirp do not produce that result; rather, some samples do not contain information about all ranges, and each sample corresponds to multiple transmit frequencies. We want to perform a deskew operation that transforms the nonzero portion of the data function to the desired rectangular form, as shown in Figure 7.10 [1, Sec. C.2].

From (7.22), suppressing n as an argument of Φ and denoting

$$R_\Delta \equiv R_t - R_0 = \text{constant (over fast time)} \tag{7.25}$$

and

$$t_1 \equiv \hat{t} + \frac{f_c}{\gamma} - \frac{2R_0}{c} \tag{7.26}$$

we have

$$\begin{aligned} \Phi(t_1) &= -\frac{4\pi\gamma}{c} t_1 R_\Delta + \frac{4\pi\gamma}{c^2} R_\Delta^2 \\ &= \frac{4\pi\gamma}{c}\left(-R_\Delta t_1 + \frac{R_\Delta^2}{c}\right) \end{aligned} \tag{7.27}$$

The second term is the RVP. From Figure 7.9, we see that, after dechirp, the returned signal from a point target at R_Δ has a constant frequency given by

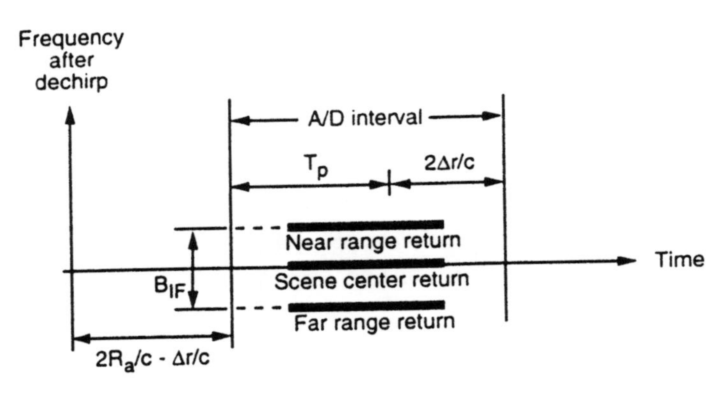

Figure 7.10 Deskew procedure. (Reproduced from [1], Figure C.3, with permission of Artech House.)

$$f_\Delta = \frac{1}{2\pi}\frac{d\Phi}{dt} = \frac{1}{2\pi}\frac{d\Phi}{dt_1} = -\frac{2\gamma}{c}R_\Delta \tag{7.28}$$

corresponding to a frequency-dependent time delay

$$t_d = \frac{2R_\Delta}{c} = -\frac{f_\Delta}{\gamma} \tag{7.29}$$

That is consistent with the slope of the sides of the parallelogram in Figure 7.8 (Problem 7.6). Then

$$\Phi(t_1) = \frac{4\pi\gamma}{c}\left(\frac{cf_\Delta}{2\gamma}t_1 + \frac{cf_\Delta^2}{4\gamma^2}\right) = 2\pi f_\Delta t_1 + \frac{\pi f_\Delta^2}{\gamma} \tag{7.30}$$

In the deskew procedure, we replace $\Phi(t_1)$ with

$$\Phi'(t_1) = \Phi(t_1) - \frac{\pi f_\Delta^2}{\gamma} = 2\pi f_\Delta t_1 \tag{7.31}$$

that is, we multiply $e^{j\Phi}$ by $e^{-j\pi f^2/\gamma}$. That accomplishes the deskew procedure as illustrated in Figure 7.10, making Φ' a separable function of f_Δ and t_1. It corresponds to a frequency-dependent time delay given by (7.29) and transforms the nonzero region of the data function into a rectangle [1, Sec. C.2]. Each time sample now corresponds to one transmitted frequency and all ranges within the swath, just as with a step-chirp waveform; and the PFA or other focusing algorithm can now be performed.

If the scene contains several targets at $R_{\Delta 1}, R_{\Delta 2}, \ldots, R_{\Delta M}$, then

$$\Phi'(t_1) = 2\pi t_1 \sum_{m=1}^{M}\left(-\frac{2\gamma}{c}R_{\Delta m}\right) = 2\pi t_1 \sum_{m=1}^{M} f_{\Delta m} \tag{7.32}$$

An FT of the N time samples of $\Phi'(t_1)$ (with $N > M$) yields the f_Δ's and therefore the ranges of the individual targets, that is, it produces a down-range scene profile. In modern digital processors, the FT is usually performed using the FFT [16].

7.2.1.4 The LFM Ridge

The ambiguity function of a train of LFM pulses was discussed in Table 4.1 and Section 4.2.7.2; here we use the notation of that Section. Figure 4.6 illustrates a schematic example showing, in the delay coordinate τ, resolution (peak to first null) of $1/B$ ($c/2B$ in range) and first ambiguity t_R = PRI; and in Doppler frequency ν, resolution $1/t_{dwell}$ (for a SAR, $\lambda/2\Delta\theta$ in crossrange) and first ambiguity $1/t_R$ = PRF. Furthermore [12, p. 136] the LFM ambiguity function of a single LFM pulse has a "ridge," with slope

$$\frac{d\nu}{d\tau} = \frac{B}{t_p} \tag{7.33}$$

extending to approximately $(-t_P, B)$ in the τ, ν plane. If Figure 4.6 were to be drawn for a single LFM pulse, the contour centered at the origin would intersect the horizontal line $\nu = 1/t_R$ at $\tau = -(t_p/B)(1/t_R) = -(1/B)(t_p/t_R) = -\alpha$, where $\alpha < 1/B$. In fact, in Figure 4.6, the contour at $(0, 1/t_R)$ is displaced slightly to the left by approximately this amount. The displacement is much less pronounced for a real SAR waveform ($t_R >> t_p$) and has no significant effect on the SAR image. (For further details, see Ref. [17] of Chapter 4, pp. 460–469.)

7.2.2 Slow-Time Processing

For cross-range, or slow-time, processing, spotlight SARs typically utilize the PFA (discussed in Section 6.8).

In slow time, the return from a point target at a particular range will exhibit a quadratic phase behavior (i.e., phase varies as the square of the time referenced to the closest approach) that is unique to the target's location on the ground [2, 3, 12]. Some stripmap SARs use a matched-filter approach to take advantage of that phenomenon. In fact, for the echo from a point target in the scene, a close analogy exists between its quadratic phase variation in fast time from a LFM pulse echo and its quadratic phase variation in slow time due to platform motion [4, p. 421]. Other stripmap SARs divide the strip into subpatches and use the PFA for each subpatch [1, Sec. 4.8].

The newer range migration algorithm (RMA) [1, Chap. 10], originally developed for seismic applications, provides the most theoretically correct solution to the stripmap image problem. It does not make a far-field approximation but treats the wavefronts as spherical. It is particularly applicable to very wideband SARs. RMA involves substantial computational complexity; however, as processors become more sophisticated, that limitation is disappearing. A simpler, faster version of RMA is the chirp-scaling algorithm (CSA) [1, Chap. 11].

Once an image is obtained, it usually contains distortions due to mocomp errors or other effects. Autofocus is the use of information in the (complex) image itself to estimate and correct phase errors, then reprocess and sharpen the image. Table 7.1 summarizes some autofocus algorithms; details are presented in [1, Chap. 6].

Table 7.1
Summary of Autofocus Algorithms

Algorithm	Phase Error Constraints	Advantages	Disadvantages
Mapdrift	Quadratic	Simple implementation Robust Good performance on clutter Large pull-in range Highly accurate Low computation	Quadratic only
Multiple aperture mapdrift	Up to fifth order	Good performance on clutter Large pull-in range	High computation Performance degrades with increased model order
Phase difference	Quadratic	Simple implementation Robust Good performance on clutter Large pull-in range Highly accurate Noniterative	Quadratic only Moderate computation
Phase gradient autofocus	Arbitrarily high	Large pull-in range Robust Good performance on clutter Not model based	Moderate to high computation
Prominent point processing	None	Highly accurate Estimates errors caused by any type of unknown motion or other error source Estimates azimuth scale factor (resolution)	Interactive (not automatic) Performance is signature dependent (requires up to three prominent points) High computation for multiple-point algorithm

Source: Reproduced from [1], Table 6.1, with permission of Artech House.

7.3 SAR Image Quality

It is clearly important that a SAR produce high-quality imagery. Image quality typically is measured using several image-quality metrics (IQMs), described in the following sections. More detailed discussion of SAR imagery is given by [17, 18].

7.3.1 Impulse Response

A point target can be considered an impulse input to a SAR processor, and the PSF in the image can be regarded as an IPR. The primary IQM for most SARs is the 3-dB IPR width.

In Section 2.3.2, in connection with antenna beam forming, we saw that the FT of a rectangular function is a sinc function, with 3-dB width of $(0.886)\theta_{pn}$ (where θ_{pn} = peak-to-first-null angle) and first sidelobe -13.3 dB below the peak. From Section 2.3.4, when a tapering, or weighting function multiples the rectangular input, we learned that the result is a function with broader mainlobe and lower sidelobes than the sinc.

From Section 6.8, FTs, especially the highly efficient FFT, are also used extensively in SAR and ISAR processing. If uniform weighting (i.e., no weighting) is used in SAR processing, then the IPR function (image of a point target) will be a sinc with 3-dB width (0.886) $(c/2B)$ in range and $\sim(0.886)(\lambda/2\Delta\theta)$ in cross-range, with first sidelobes -13.3 dB below the peak. A typical weighting function used in SAR processing is Taylor weighting, with first sidelobe constrained to be -35 dB below the peak and "nbar = 5" [1, Sec. D.2], which produces a widened mainlobe of 3-dB IPR value $(1.19)\theta_{pn}$. Another choice is Hann [12] (or "Hanning" [1]) weighting, which results in an even wider mainlobe of $(1.43)\theta_{pn}$; the first sidelobe is -31.7 dB below the peak, and the far sidelobes are very low compared with uniform or Taylor weighting. An excellent discussion of more than 20 weighting functions (not including Taylor) is given by Harris [19].

7.3.2 Signal-to-Noise Ratio

Because SAR processing is essentially matched-filter processing, we have from Section 4.2.2

$$\text{SNR} = \frac{E}{kT_{\text{sys}}} \tag{7.34}$$

$$E = P_{Rx\text{-avg}} t_A \tag{7.35}$$

where $P_{Rx\text{-avg}}$ = average received power and t_A = aperture time. Furthermore, from Section 1.11,

$$E = \frac{P_{Tx\text{-avg}} G^2 \lambda^2 \sigma}{(4\pi)^3 R^4 L} \cdot t_A = \frac{P_{Tx\text{-avg}} A^2 \eta^2 \sigma}{4\pi R^4 \lambda^2 L} \cdot t_A \tag{7.36}$$

For SAR, at $\theta_{sq} = 0$, from (7.4)

$$t_A = \frac{\lambda R}{2V\delta_{cr}} \tag{7.37}$$

Thus [2, 3],

$$\text{SNR} = \frac{P_{Tx\text{-avg}} G^2 \lambda^3 \sigma}{2(4\pi)^3 R^3 k T_{\text{sys}} L V \delta_{cr}} = \frac{P_{Tx\text{-avg}} A^2 \eta^2 \sigma}{8\pi R^3 \lambda k T_{\text{sys}} L V \delta_{cr}} \tag{7.38}$$

If clutter is being observed, then

$$\sigma = \sigma^0 \delta_{cr} \delta_r / \cos\psi \tag{7.39}$$

where δ_r = pixel width in slant range. Then the clutter-to-noise ratio (CNR) is

$$CNR = \frac{P_{Tx\text{-avg}} G^2 \lambda^3 \sigma^0 \delta_r}{2(4\pi)^3 R^3 k T_{\text{sys}} L V \cos\psi} = \frac{P_{Tx\text{-avg}} A^2 \eta^2 \sigma^0 \delta_r}{8\pi R^3 \lambda k T_{\text{sys}} L V \cos\psi} \tag{7.40}$$

It is useful to consider the noise-equivalent clutter ($NE\sigma^0$), defined as the clutter level (σ^0) that produces a received power equal to the thermal noise power, that is, a CNR of unity. We set CNR = 1 and have

$$NE\sigma^0 = \frac{2(4\pi)^3 R^3 k T_{\text{sys}} L V \cos\psi}{P_{Tx\text{-avg}} G^2 \lambda^3 \delta_r} = \frac{8\pi R^3 \lambda k T_{\text{sys}} L V \cos\psi}{P_{Tx\text{-avg}} A^2 \eta^2 \delta_r} \tag{7.41}$$

For example, if R = 200 km, T_{sys} = 580K, L = 5, V = 180 m/s, ψ = 10 degrees, P_{avg} = 700W, G = 34 dB, λ = 0.03m (X band), and δ_r = 0.3m, then $NE\sigma^0 = -22.0$ dB (see also Problem 7.7).

A clear SAR image must have a CNR greater than 5 to 15 dB. From Figure 3.15, we see that this example SAR could image wooded hills, that is, $\sigma^0 \sim -17$ dB with CNR ~ 5 dB, but could not image "flatland" (perhaps desert) at $\sigma^0 \sim -27$ dB.

7.3.3 Integrated Sidelobe Ratio

An actual PSF typically resembles the theoretical sinc(x) or similar function but is somewhat different, especially in the sidelobes, due to phase noise, mocomp imperfections, and other real-world effects. A useful figure of merit is the integrated sidelobe ratio (ISLR), defined as [1]

$$ISLR = \frac{Integral\ over\ PSF\ Sidelobes}{Integral\ over\ PSF\ Mainlobe} \tag{7.42}$$

ISLR usually is measured in decibels; a typical value might be –20 dB.

7.3.4 Multiplicative Noise Ratio

Thermal noise is often referred to as additive noise, because it adds to the scene independent of the scene content. The other type of noise is multiplicative noise, which is roughly proportional to the average scene intensity. The multiplicative noise ratio (MNR) [1] of a SAR image is defined as the ratio of the image intensity in no-return area (NRA) (not including thermal noise) divided by the average image intensity in a relatively bright surrounding area (in principle not including thermal noise). An NRA is an area with essentially zero return—usually a very smooth area, such as a calm lake or a specially constructed large sheet of aluminum.

If MNR were zero, the image of an NRA would be dark, with sharp edges. MNR includes the effects of

- ISLR of points in the surrounding area;
- Ambiguities, that is, range and velocity;
- Quantization noise; and
- Miscellaneous image artifacts.

An actual NRA image has nonsharp edges (Figure 7.11). Most practical NRAs are not large enough for the center to be free of ISLR effects. Because it depends on the size of the NRA, MNR is not well defined. Despite that fact, MNR is relatively easy to measure and is a widely used IQM for SAR. If thermal noise is included in the numerator and denominator of the ratio, then the ratio is called the contrast ratio and can be measured directly from the image.

7.3.5 Comparison of SAR and Optical Imagery

The human eye is a system for producing images using visible light. The light hits the lens and is focused on the retina, and the resulting image is transmitted to the brain. Over many millennia, humans have become fully accustomed to

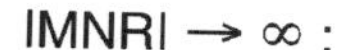

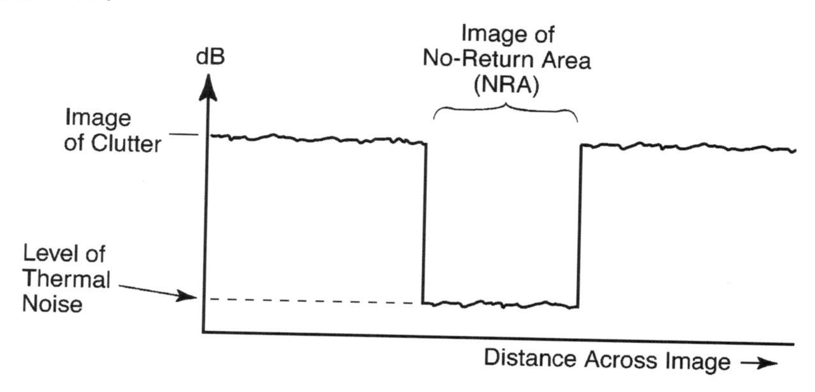

|MNR| < |CNR|:

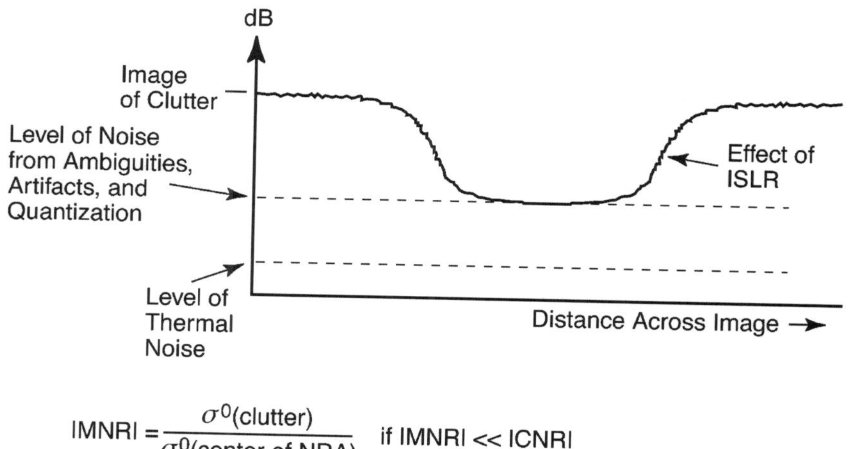

$$|MNR| = \frac{\sigma^0(\text{clutter})}{\sigma^0(\text{center of NRA})} \quad \text{if } |MNR| \ll |CNR|$$

Figure 7.11 Multiplicative noise ratio: (a) |MNR| >> |CNR|; (b) |MNR| < CNR.

seeing and processing that visible imagery. Therefore, on seeing a SAR image, we may instinctively assume that it has certain properties of a visible image, which, in fact, it does not possess. Optical imagery is based on an angle-angle principle, whereas SAR imagery is based on a downrange-crossrange procedure. SAR imagery is not visible imagery, and we should not expect it necessarily to look like visible imagery.

When a SAR image is displayed, the direction to the radar should be at the top, especially for a broad swath with significant variation in grazing angle.

Figure 7.12(a) illustrates the appearance of a flat landscape to the human eye. The terrain is illuminated by sunlight (perhaps diffused thorough clouds). At the eye (or camera), each pixel subtends the same azimuth and elevation angles. Thus, pixels farther from the eye are larger (coarser resolution), in both down-range and cross-range, than pixels closer to the eye.

Figure 7.12(b) shows that for a SAR image the situation is quite different (assuming adequate SNR). The ground range pixel size δ_g is

Optical Image of Terrain

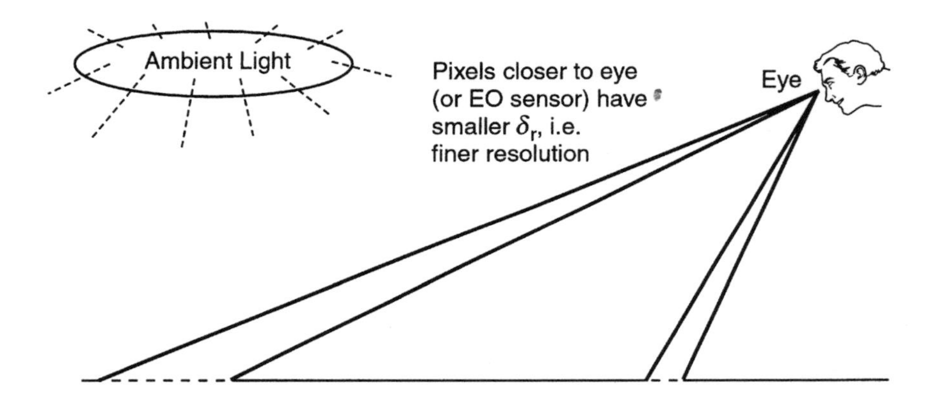

SAR Image of Terrain

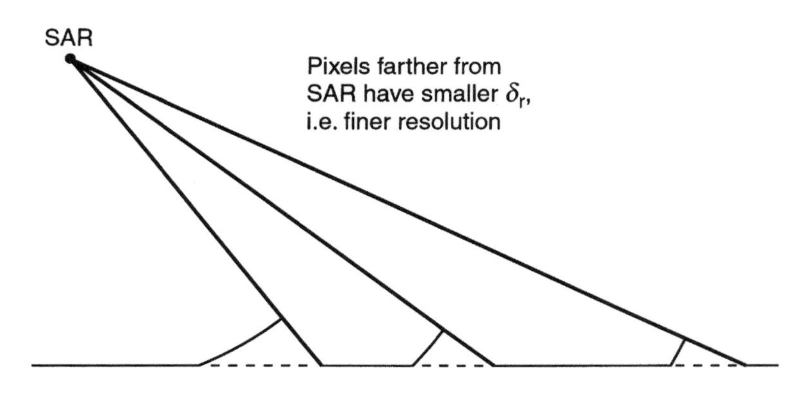

Figure 7.12 Comparison of SAR and optical imagery: (a) optical image of terrain; (b) SAR image of terrain ("natural" direction for viewing image is with SAR at top).

$$\delta_g \cong \frac{c}{2B} \cdot \frac{1}{\cos\psi} \tag{7.43}$$

Pixels farther from the SAR are smaller in range (finer down-range resolution) than pixels closer to the SAR; and cross-range resolution is approximately independent of range.

When we display a SAR image, especially of a large landscape, it is usually most satisfying to display it with the SAR direction at the top. The finer-resolution pixels are at the bottom, just as they are with a naturally oriented optical image. Such an orientation tends to look most natural to a human observer.

Because SAR imagery and optical imagery are collected using entirely different physical principles, we should not be surprised if they look different. A good example is provided in a SAR image of the Washington Monument, courtesy of Veridian ERIM-International. Figure 7.13(a) illustrates the collection geometry (and the result) of optical imagery of the Washington Monument, with the monument's shadow pointed toward the observer. For clarity, we assume that the sun is to the south of the monument and the observer is to the north. The image shows a shadow on the north side cast by the sun. The portion of the monument visible in the image is the north side, illuminated by diffusely scattered sunlight. In comparison, Figure 7.13(b) shows the geometry and the result of SAR imagery, again with the shadow on the north side. This time, the shadow is cast by the SAR itself. The portion of the monument visible in the image is the south side. Figure 7.14 shows the SAR image. It does not look entirely like an optical image, nor should it.

Another difference between SAR and optical images is the presence of speckle [17, Sec. 2-5.1] in SAR images. When terrain, especially vegetation, is imaged, the amplitude (voltage) of a particular pixel is the magnitude of the phasor sum of the coherent returns from many scatterers within the area represented by the pixel. In another nearby pixel, even if the terrain is nominally the same as in the first pixel, the coherent returns will add differently and the pixel magnitude will be somewhat different. That phenomenon, characteristic of coherent imagery, causes SAR imagery of terrain to exhibit more pixel-to-pixel fluctuation (speckle) than corresponding optical imagery.

7.4 Summary of Key SAR Parameters

Tables 7.2 to 7.4 summarize the possible parameters for a potential SAR.

A review of the basic equations of SAR follows.

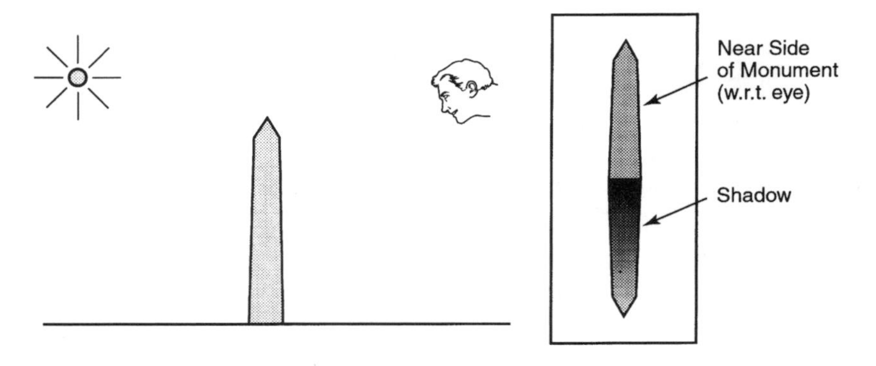

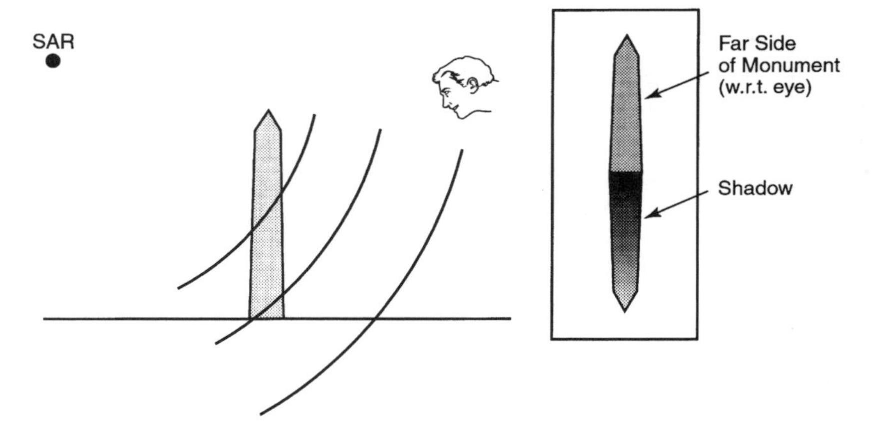

Figure 7.13 Principles of imaging: (a) optical image of the Washington Monument; (b) SAR image of the monument.

- Range resolution = $\delta_r \cong c/2B$ (c = speed of light, B = pulse bandwidth).
- Cross-range resolution = $\delta_{cr} \cong \lambda/2\Delta\theta$ (λ = wavelength, $\Delta\theta$ = angle subtended by synthetic aperture).
- Physical beamwidth $\cong \lambda/D$ (D = antenna size).
- For stripmap SAR:

$$\delta_{cr} \cong \frac{\lambda}{2\Delta\theta} \cong \frac{\lambda}{2(\lambda/D)} = \frac{D}{2} \tag{7.44}$$

Figure 7.14 SAR image of the Washington Monument. (Copyright 1996 by Veridian ERIM-International.)

- $t_A = \lambda R / (2V \delta_{cr} \cos \theta_{sq})$
- $f_R = PRF \geq 2V/D$ (V = platform velocity).
- For unambiguous range, $f_R < c/2R$; thus,

$$\frac{2V}{D} \leq f_R < \frac{c}{2R} \tag{7.45}$$

$$CNR = \frac{P_{avg} G^2 \lambda^3 \sigma^0 \delta_r}{(2)(4\pi)^3 R^3 k T_{sys} L V \cos \psi} = \frac{P_{avg} A^2 \eta^2 \sigma^0 \delta_r}{8 \pi R^3 \lambda k T_{sys} L V \cos \psi} \tag{7.46}$$

Table 7.2
Example SAR Hardware Parameters

Parameter	Description	Value
f_c	Center frequency	10 GHz
λ_c	Center wavelength	0.03m
B	RF bandwidth	185 MHz
PRF	Pulse repetition frequency	1,000 Hz
γ	FM chirp rate	4.8 MHz/μs
T_p	Transmitted pulse length	38.5 μs
β_a	Azimuth beamwidth	5.25 degrees
β_e	Elevation beamwidth	6.25 degrees
G	Antenna gain	29 dB
F_s	A/D complex sample rate	45 MHz
K	Complex samples per pulse	2,048
N_p	Pulses per synthetic aperture	1,950
PCR	Pulse compression ratio	7,120

Source: Reproduced from [1], Table 2.4, with permission of Artech House.

Table 7.3
Example SAR Performance Parameters

Parameter	Description	Value
ρ_r	Range resolution	1m
ρ_a	Azimuth resolution	1m
ISLR	Integrated sidelobe ratio	–21 dB
K_r	Mainlobe-broadening factor in range	1.23
K_a	Mainlobe-broadening factor in azimuth	1.30
	Aperture weighting function (Taylor)	–35 dB; $\bar{n}$ = 5
β_s	Azimuth synthetic beamwidth	0.0057 degrees
BCR	Beam compression ratio	916
MNR	Multiplicative noise ratio	–18 dB
σ_n	Additive noise coefficient	–35 dB

Source: Reproduced from [1], Table 2.5, with permission of Artech House.

7.5 Special SAR Applications

In this section we briefly discuss several specific aspects of SAR, specifically moving targets, vibrating targets, measurement of object height, forward-look SAR, foliage-penetration SAR, polarimetric SAR, and interleaved SAR modes.

Table 7.4
Example SAR Collection Geometry and Platform Parameters

Parameter	Description	Value
H	Altitude	5,000m
V_a	Forward velocity	100 m/s
R_a	Slant range	10,000 m
θ_{ac}	Squint angle	90 degrees
Δ_r	Range scene size (slant plane)	1,050 m
W_a	Azimuth scene size	916 m
T_a	Synthetic aperture time	1.95 sec
L	Synthetic aperture length	195m
γ_a	Azimuth chirp rate	66.66 Hz/s
$\Delta\theta$	Coherent integration angle	1.12 degrees
B_a	Azimuth bandwidth after dechirp	610 Hz
TB_a	Azimuth time bandwidth product	1190
B_d	Doppler bandwidth of a single scatterer before azimuth dechirp	130 Hz
B_{IF}	Range bandwidth after dechirp	33.62 MHz

Source: Reproduced from [1], Table 2.6, with permission of Artech House.

7.5.1 Moving Targets

The basic theory of SAR assumes that the ground (scene) is stationary. A moving target in the scene will have a "wrong" relationship between its location and its LOS velocity. If the target motion is smooth, the target image will be displaced in cross-range by

$$r_{\text{displ}} = \frac{V_{\text{LOS}}}{\Omega} = \frac{V_{\text{LOS}} R}{V} \tag{7.47}$$

Such information can be utilized to detect, locate, reposition, and even image (Section 11.4.5) moving targets in a SAR image. This topic is an active area of current research.

7.5.2 Vibrating Targets

Consider a SAR observing a scene that contains a point target [1] vibrating along the LOS such that the differential radar-target distance is

$$R_{\text{tgt}} = d \sin(2\pi f_{\text{vib}} t) \tag{7.48}$$

In addition to the phase appropriate to a stationary scene, the pixel containing the target will produce an additional echo with a periodic phase error

$$\phi_e = \frac{4\pi d}{\lambda}\sin(2\pi f_{\text{vib}}t) = \phi_0 \sin(2\pi f_{\text{vib}}t) \tag{7.49}$$

We assume $d << \lambda$; thus, $\phi_0 << 1$. Then the corresponding phasor is

$$e^{j\phi_e} = e^{j\phi_0 \sin(2\pi f_{\text{vib}}t)} \tag{7.50}$$

$$\cong 1 + j\phi_0 \sin(2\pi f_{\text{vib}}t) \tag{7.51}$$

$$= 1 + \frac{\phi_0}{2}\left(e^{j2\pi f_{\text{vib}}t} - e^{-j2\pi f_{\text{vib}}t}\right) \tag{7.52}$$

Let the doppler phasor corresponding to a stationary pixel be $e^{j2\pi f_d t}$. Then the phasor corresponding to the vibrating target is

$$e^{j\phi_{\text{dop}}} = e^{j2\pi f_d t} e^{j\phi_e} \tag{7.53}$$

and

$$e^{j\phi_{\text{dop}}} = e^{j2\pi f_d t} + \frac{\phi_0}{2}\left(e^{j2\pi t(f_d + f_{\text{vib}})} - e^{j2\pi t(f_d - f_{\text{vib}})}\right) \tag{7.54}$$

In the SAR processing, the vibrating point target will appear in three locations: the main target image still appears at the correct location, while a small fraction ($\phi_0/2$) will appear in each of two pixels separated by f_{vib} in doppler frequency. The two additional returns are known as paired echoes.

The corresponding velocity separation is $\Delta v = \pm f_{\text{vib}}\lambda/2$, and the cross-range displacement is then

$$\Delta r = \frac{\Delta v}{\Omega} = \frac{\Delta v}{V}R = \pm\frac{f_{\text{vib}}\lambda_{\text{avg}}R}{2V} \tag{7.55}$$

where λ_{avg} = average wavelength (assuming low fractional bandwidth). The relative amplitude of each of the paired echoes is

$$\text{Voltage: } \frac{\phi_0}{2} = \frac{2\pi d}{\lambda_{\text{avg}}} \tag{7.56}$$

$$\text{Power: } \left(\frac{\phi_0}{2}\right)^2 = \left(\frac{2\pi d}{\lambda_{\text{avg}}}\right)^2 \tag{7.57}$$

Thus, the amplitude of the paired echoes is proportional to the square of the vibration amplitude, and cross-range displacement is proportional to the vibration frequency.

For bright, pointlike targets, additional terms should be retained in (7.51), describing the fact that a series of paired echoes, of decreasing amplitude, may appear in crossrange (Problem 7.8).

Figure 7.15 shows a SAR image of a scene that includes a vibrating target—a truck with the engine running; the image contains paired echoes.

7.5.3 Measurement of Object Height

The basic theory of SAR assumes a flat scene. To the extent that the scene is not flat, distortions in the SAR image will result. In some cases, they can be used to measure the height of elevated objects above a flat terrain.

7.5.3.1 Shadows

The simplest method of measuring object height is to observe the length, L, of the shadow of the object cast by the SAR and calculate the object height from the known SAR altitude, H and ground range, R_g:

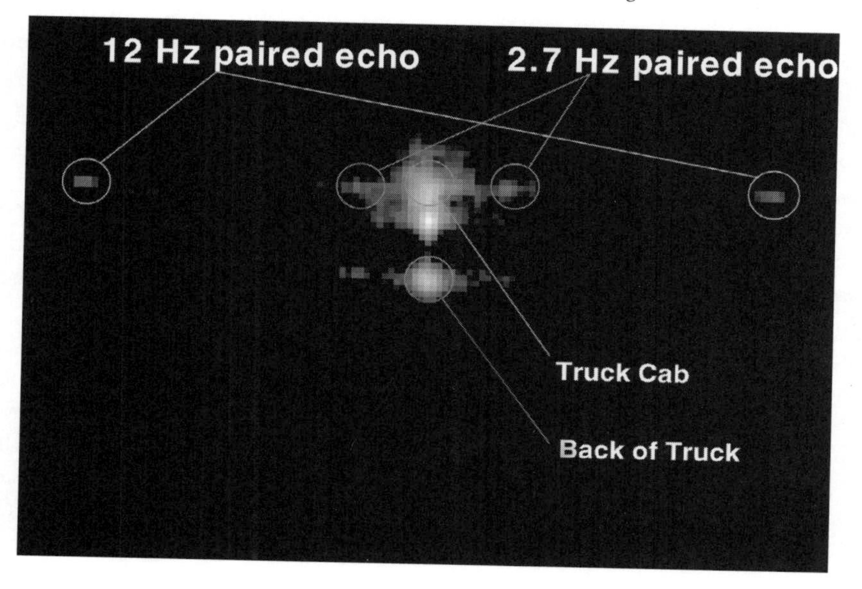

Figure 7.15 SAR scene containing vibrating target. (Courtesy of Northrop-Grumman Norden Systems.)

$$h_{\text{tgt}} = L_{\text{shadow}} \cdot \frac{H}{R_g} \tag{7.58}$$

Equation (7.58) assumes flat Earth and may be generalized to curved-Earth if R_g is relatively large; see Problem 7.9. The shadow method works only for isolated, relatively high objects on essentially flat terrain (e.g., Figure 7.14).

7.5.3.2 Layover

SAR processing sorts target returns into bins (pixels) depending on the range, R, and apparent velocity, v, of the target relative to the platform. If two or more targets have the same R and v, then they will be placed at the same location in the SAR image.

We shall define a layover contour as the locus of points in three-dimensional space such that an object at any of the points will be assigned to the same location in a SAR image. From Figure 7.16, a layover contour is the intersection of a constant-range sphere of range R and a constant velocity cone of generating angle $\beta = \cos^{-1}(-v/V)$, that is, a circle of radius $R\sin\beta$ ahead of the platform. ($\beta > 90$ degrees corresponds to targets behind the platform.) We shall therefore call the contour the layover circle. If the top of an elevated object, such as a tower, is on the layover circle and if the ground is flat, then the top of the tower will appear in the SAR image at the same position as a point on the ground where the layover circle intersects the ground. The tower will be "laid over," hence the nomenclature.

As shown in Figure 7.16, let us consider a SAR at altitude H forming a spot image, the center of which is at slant range $R_s >> H$ and squint angle θ_{sq}. The size of the spot image is small compared with R_s, and flat Earth is assumed. Suppose a tower of height h ($h << H$) is in the area that is imaged. We describe locations within the image by a coordinate system (x_1, y_1). If the base of the tower is at (x_{10}, y_{10}), we want to ascertain the image coordinates of the top of the tower.

Figure 7.16(a) illustrates a perspective view. Because $R_g >> H$, the isodop (y_1 – axis) makes an angle θ_{sq} with the y axis. The image center is a distance $S \cong R_g \cos\theta_{sq}$ from the x axis, where R_g = ground range. Figure 7.16(b) shows a view from the $+x$ axis, indicating the layover circle and showing that the image location of the tower top is located a distance (the layover distance) $d \cong hH/(R_g \cos\theta_{sq})$ from the image location of the tower base. Figure 7.16(c) then depicts the view in image coordinates (x_1, y_1). The image coordinates of the tower top are $x_{11} = x_{10} + d\sin\theta_{sq}$, $y_{11} = y_{10} - d\cos\theta_{sq}$.

For example, if R_g = 100 km, H = 5 km, h = 100m, and $\theta_{sq} = 0$, then d = 5m, $x_{11} = x_{10}$, and $y_{11} = y_{10}$ – 5m. The tower top appears in the image

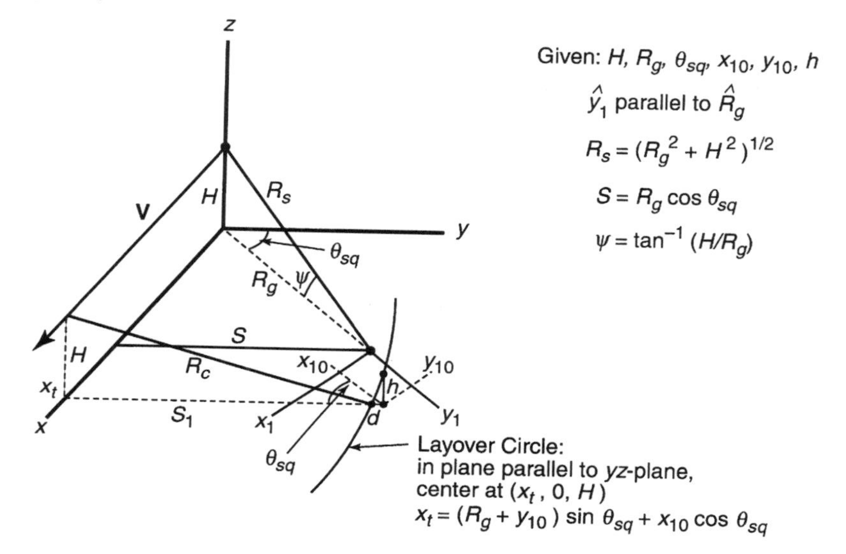

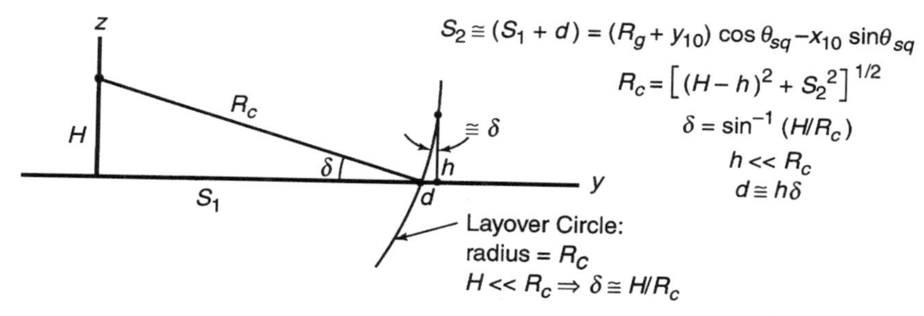

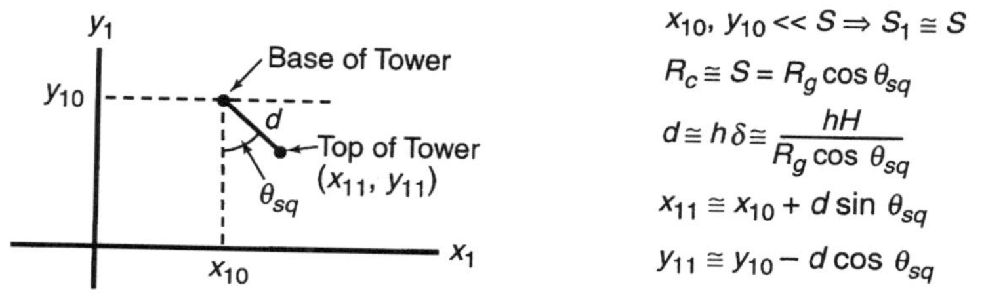

Figure 7.16 Layover: (a) perspective view; (b) view from $+x$ axis; (c) view in $x_1 y_1$ (image) system.

5m closer to the radar than the tower base. This principle can sometimes be used to estimate the height of isolated, towerlike structures on relatively level ground:

$$h \cong \frac{dR_g \cos\theta_{sq}}{H} \tag{7.59}$$

The intersection of the doppler cone and the ground is a hyperbola. If H is not << R, the isodop direction will not be parallel to the down-range direction. The geometry is more complicated but still can be computed to determine layover distance and estimated object height (Problem 7.10).

7.5.3.3 Stereo

Two SAR images of the same scene can be obtained from somewhat different locations. Noncoherent comparison of the two—the stereo technique—can enable estimation of object height. The technique is analogous to the method by which we humans use two eyes to help estimate the distance of the objects that we see. In fact, the two SAR images can be printed on the same page using two different colors, with the viewer using special glasses so the left eye sees only one image and the right eye only the other image, with the brain processing the two together so the scene is perceived in three dimensions.

7.5.3.4 Interferometric SAR

Interferometric SAR (IFSAR) involves two SAR images taken from antennas at slightly different locations and compared coherently to obtain fine-resolution information regarding the height of terrain or targets in the image [1, Sec. 9.3]. IFSAR can be performed using a single platform with two antennas (single-pass IFSAR) or by the same platform making two passes over the same terrain (two-pass IFSAR). It is essential that the relative locations of the two antennas be rather precisely known. The advantages and disadvantages of the two types of IFSAR are as follows.

Two-Pass IFSAR

- No special hardware is required; a conventional SAR may be flown twice over the designated terrain.
- A long baseline (distance between antenna locations) provides fine vertical resolution (but challenging ambiguities).
- The scene may change between passes due to wind, etc.
- Mocomp is challenging; the relative position of the antennas must be known precisely; this position changes somewhat over the synthetic aperture.

Single-Pass IFSAR

- More sophisticated (expensive) hardware is required: two antennas, two receiver channels, two sets of A/D converters.
- Baseline is relatively well known, providing consistency throughout synthetic aperture.
- Scene is same for both images.
- On-board, real-time processing is a possibility.

To understand the theory of IFSAR, we first consider two antennas, A and B, separated by baseline L, observing a point target at range vector $\mathbf{R}$ and angle ψ_1 from the baseline (Figure 7.17). We consider two possibilities: (1) one antenna transmits and each receives ($n = 1$), and (2) antenna A transmits, then receives; then antenna B transmits, then receives ($n = 2$). For a single pulse of wavelength λ, the difference in the phases observed by the two antennas is

$$\phi_1 = \frac{2\pi ns}{\lambda} = \frac{2\pi nL\sin\psi_1}{\lambda} \tag{7.60}$$

If the point target moves a small distance z perpendicular to R so as now to be described by angle ψ_2, then the phase difference is

$$\phi_2 = \frac{2\pi nL\sin\psi_2}{\lambda} \tag{7.61}$$

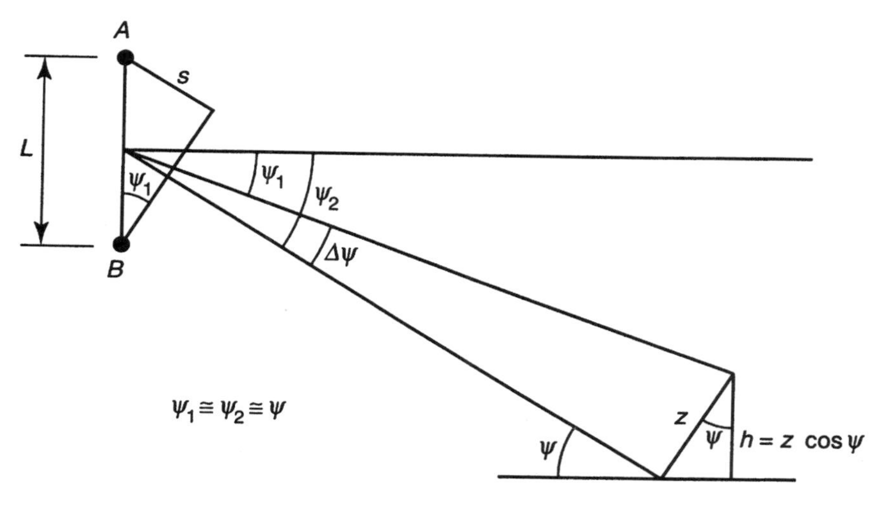

Figure 7.17 IFSAR: vertical antenna separation.

We consider the quantity

$$\Delta\phi \equiv \phi_2 - \phi_1 = \frac{2\pi nL}{\lambda}(\sin\psi_2 - \sin\psi_1) \tag{7.62}$$

$$\cong \frac{2\pi nL}{\lambda}\cos\psi_1 \Delta\psi = \frac{2\pi nLz\cos\psi}{\lambda R} \tag{7.63}$$

We now consider $\delta\phi$, the change in $\Delta\phi$ due to a change in z given by δz:

$$\delta\phi = \frac{2\pi nL}{\lambda R}\delta z \cos\psi \tag{7.64}$$

Antennas A and B can be considered as separated in vertical distance on an aircraft. From Figure 7.17, the relationship between $\delta\phi$ and a variation in terrain altitude $\delta h = \delta z \cos\psi$ is

$$\delta h = \frac{\lambda R \delta\phi}{2\pi nL} \tag{7.65}$$

We now postulate an aircraft with two antennas separated horizontally by L. The aircraft is banking at angle $\gamma > 0$ and collecting data from the ground at grazing angle ψ (Figure 7.18; we make the flat-Earth approximation). Then the effective aperture (perpendicular to the LOS) is $L\sin(\psi + \gamma)$ instead of $L\cos\psi$. We consider terrain elevated at height $h(x)$ above the nominal flat-Earth surface. After some mathematics, we find

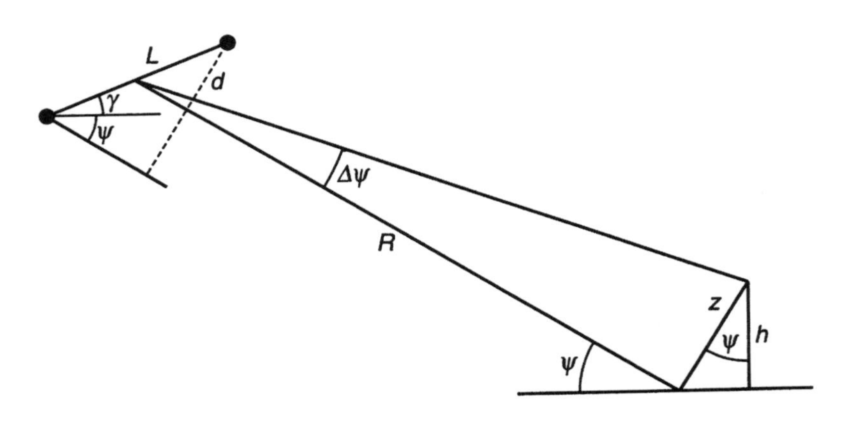

Figure 7.18 IFSAR: horizontal antenna separation.

$$\delta h = \delta z \cos\psi = \frac{\lambda R |\delta\phi| \cos\psi}{2\pi n L \sin(\psi + \gamma)} \qquad (7.66)$$

In either geometry, because both channels are noisy, the minimum resolvable phase difference (1-sigma) is given by [12]

$$\delta\phi = \frac{1}{\sqrt{\text{SNR}}}\sqrt{2} \qquad (7.67)$$

Thus, the theoretical resolution for terrain altitude measurement is

- Vertical antenna separation (no banking):

$$\delta h = \frac{\lambda R}{\pi n L \sqrt{2\text{SNR}}} \qquad (7.68)$$

- Horizontal antenna separation:

$$\delta h = \frac{\lambda R \cos\psi}{\pi n L \sin(\psi + \gamma)\sqrt{2\text{SNR}}} \qquad (7.69)$$

Furthermore, when the phase moves through an interval of 2π, an ambiguity occurs in terrain altitude measurement; the corresponding altitude difference is

- Vertical antenna separation (no banking):

$$\Delta h = \frac{\lambda R}{nL} \qquad (7.70)$$

- Horizontal antenna separation:

$$\Delta h = \frac{\lambda R \cos\psi}{nL \sin(\psi + \gamma)} \qquad (7.71)$$

Although (7.68) through (7.71) were derived for a single monochromatic pulse, they can be shown [1, Sec. 3.0 and Sec. 9.3] to be true for SAR pixels also, with λ replaced by c/f_{avg}.

An IFSAR image of the stadium at the University of Michigan in Ann Arbor, viewable with two-color glasses, was used for the cover of the *Proceedings of the 1996 National Radar Conference* [20]. The National Aeronautics and Space Administration (NASA) in February 2000 performed X/C-band polarimetric IFSAR (single-pass) for the shuttle radar topography mission (SRTM), to produce a complete three-dimensional map between 60 degrees north latitude and 56 degrees south latitude (nearly 80% of the Earth's surface), with best vertical accuracy of 6m on a 30-m horizontal grid [21].

7.5.4 Forward-Look SAR

The SAR process can distinguish the returns from any of a set of concentric spheres and from any of a set of common-axis cones, assigning each (point) target return to the appropriate bin (pixel), with generalized coordinates corresponding to sphere number and cone number. If the image is of a flat ground, that reduces to a distinction among returns from any of a set of concentric circles (range contours) and from any of a set of common-axis hyperbolas (isodops). For a broadside image, the range contours and isodops are orthogonal, and an undistorted raster-scan display of the bins produces an image with relatively little processing. (As discussed in Section 7.1.4, resampling in the range direction is necessary to transform from slant plane to ground plane.) That is consistent with the discussion in Section 5.1, illustrating that when a radar image is obtained of a rotating object with axis of rotation perpendicular to the LOS, the image appears as if viewed from along the axis of rotation.

For a nonzero squint angle θ_{sq}, the range contours and isodops are not orthogonal, and additional processing is necessary to produce an undistorted (ground-plane) image.

If we form a SAR image of an object (such as a vehicle) in the forward-look direction at θ_{sq} = 90 degrees, the image can still be formed; the apparent rotation axis is now parallel to the ground. The apparent view is thus from the direction parallel to the ground and perpendicular to the platform velocity—from along the y axis in Figure 7.4(a). If the vehicle is oriented along the x axis, the view will be a side view rather than the top view that would be obtained in a broadside image. Interpretation of the return from the ground requires consideration of the intersection of the set of common-axis cones and the ground; that is, range-resolution is good but cross-range resolution is minimal. Examples and additional discussion are given in [1, Sect. 9.2].

7.5.5 Foliage-Penetration SAR

Although higher frequency ($\gtrsim$ 2 GHz) microwaves do not penetrate foliage well, lower frequency microwaves do, as illustrated in Figure 7.19 [22] (see

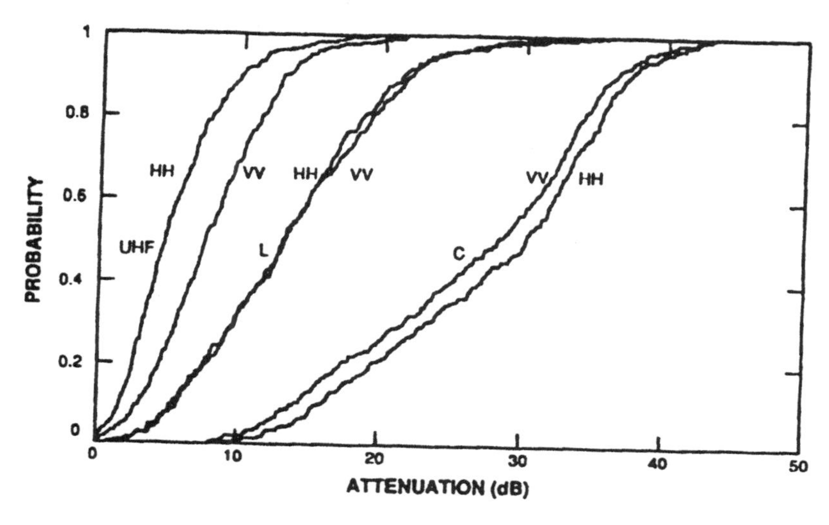

Figure 7.19 Penetration of foliage by microwaves. (Reproduced from [22], Part I, Figure 3, copyright 1996 IEEE.)

also [23, Sec. 21-6]). For each frequency shown, the cumulative probability of attenuation is given versus the two-way attenuation value (decibels). For example, for C band, the attenuation varies from ~10 dB to ~40 dB; the probability that the attenuation is less than 20 dB is about 0.2. On the other hand, for UHF radiation, attenuation varies from 0 to ~20 dB; half the time, it is less than ~7 dB. Thus, for foliage penetration (FOPEN), UHF radiation is necessary; shorter waves will not penetrate the foliage, and for airborne applications, longer waves would require prohibitively large antennas. (The specific values of attenuation, in decibels per meter, vary with tree type, leaf density, and moisture content; however, the preceding statement is a valid summary of those results.)

The aperture time t_A required to collect sufficient data for a SAR image is found from (7.4). For example, for R = 100 km, V = 180 m/s (350 kn), θ_{sq} = 0, and δ_{cr} = 1m: for f = 10 GHz (X band, λ = 0.03m), t_A = 8.3 sec, and the fractional bandwidth B/f_c = 0.015. On the other hand, for f_c = 0.5 GHz (UHF, λ = 0.6m), t_A = 167 sec = 2.8 min, and B/f_o = 0.3. Such a high fractional bandwidth (an ultrawideband SAR) presents challenges in designing hardware components, such as antennas, that are reasonably linear over the full frequency range [24]. Furthermore, the long aperture time presents mocomp challenges, and the wide real-beam angle adds to processing difficulties, very likely requiring RMA processing.

7.5.6 Polarimetric SAR

Usually, when a radar transmits a pulse at a particular polarization (e.g., horizontal—H), it receives the echoes at the same polarization. Some radars are capable of transmitting at one polarization and receiving at two orthogonal polarizations (e.g., H and V). Furthermore, some radars can transmit at either of two orthogonal polarizations (e.g., H and V, or R and L; see Section 3.1.4.3) and receive at either of the transmitted polarizations; the choice of transmitted and received polarizations can be varied from pulse to pulse. Such a radar is fully polarimetric. We can designate the choice of polarizations as follows: HV is "transmit H, receive V," and so forth. Several fully polarimetric SARs have been demonstrated [25, 26]. For example, [25] includes HH and HV X-band SAR images taken of the same scene at the same time using HH and HV modes interleaved on a pulse-to-pulse basis.

7.5.7 Interleaved SAR/ISAR Modes

SAR has proved to be of great benefit in producing extensive imagery of the ground in day/night all-weather conditions. Using an agile-beam ESA (see Section 2.4), it is possible to collect more imagery, in some cases using less power, than with conventional fixed-beam SAR.

Referring to an airborne monostatic SAR, we shall use the following terminology, expanding on the definitions given in Section 7.1.1.

- Spotlight SAR is the mode that produces a SAR image (usually relatively fine-resolution) of a specific predesignated ground location at a squint angle θ_{sq}, measured in the ground plane, to broadside (i.e., the broadside direction corresponds to $\theta_{sq} = 0$).
- Search SAR is the mode that produces imagery of a ground swath of indefinite length parallel to the platform flight path (assumed to be straight), regardless of the hardware or software procedures.
- Stripmap SAR is a specific case of search SAR such that the physical beam remains at a fixed orientation assumed to be broadside.

As was shown in (7.40), the CNR for a broadside-looking SAR (any mode) can be expressed as

$$CNR = \frac{P_{Tx\text{-avg}} A^2 \eta^2 \sigma^0 \delta_r}{8\pi R^3 \lambda k T_{sys} L V \cos\psi} = \frac{P_{Tx\text{-avg}} L_h^2 L_v^2 \eta^2 \sigma^0 \delta_r}{8\pi R^3 \lambda k T_{sys} L V \cos\psi} \tag{7.72}$$

where the (rectangular) antenna area is expressed as L_h (horizontal dimension) $\cdot$ L_v (vertical dimension). To maintain a given CNR, we can write [27]

$$P_{Tx\text{-avg}} = \frac{KR^3}{L_h^2 \delta} \tag{7.73}$$

We now have assumed that the range and cross-range resolutions are equal (the usual case) and denoted by δ and that all parameters of the radar, platform, and scene are constant except perhaps R, L_h, and δ. K is a constant representing the other parameters.

7.5.7.1 Agile-Beam Search SAR

If the SAR is operating in conventional stripmap mode at broadside ($\theta_{sq} = 0$), then, from (7.3),

$$\delta \approx \frac{L_h}{2} \tag{7.74}$$

and (replacing the $\approx$ with an equality)

$$P_{\text{fixed}} = \frac{KR^3}{4\delta^3} \tag{7.75}$$

Thus, we might conclude that the power required for broadside stripmap SAR operation is directly proportional to R^3 and inversely proportional to δ^3. We use the notation P_{fixed} to indicate that the beam is stationary in the broadside direction, to distinguish this case from the one to be considered next.

Mrstik [27] shows that, if the antenna is an ESA with agile-beam capability, it is possible to observe the same swath in search mode using less power by employing a longer antenna to obtain a larger aperture area and collecting a series of spotlight images. Assume that, for the agile-beam case, the antenna length is increased by a factor of N, from 2δ to $2N\delta$, and the radar performs spotlight SAR. Because (7.72) applies to both stripmap and spotlight SAR (Section 7.3.2), (7.73) becomes

$$P_{\text{agile}} = \frac{KR^3}{4N^2\delta^3} \tag{7.76}$$

The beam is only $1/N$ as wide as before. However, we can use the agile-beam feature to interleave N beams and simultaneously collect N spotlight images over the same swath segment previously covered by one stripmap beam. The resolution will be the same, because, from (7.3),

$$\delta = \frac{\lambda}{2\Delta\theta} \tag{7.77}$$

where $\Delta\theta$ is the angle subtended during the period that the clutter patch is being illuminated. Furthermore, from (7.76), when N beams are interleaved to achieve search SAR, the required power is decreased by N compared with conventional stripmap SAR (7.75).

From (7.11), for each spotlight image the required PRF is reduced by a factor N from the stripmap case:

$$f_R(\text{one spot}) = \frac{2V}{L_h} = \frac{2V}{2N\delta} = \frac{V}{N\delta} \tag{7.78}$$

Therefore, the overall PRF required to support the N simultaneous spots is simply the original PRF:

$$f_R(N\text{ spots}) = \frac{V}{\delta} \tag{7.79}$$

In summary, compared with a conventional fixed-beam stripmap SAR, an agile-beam SAR with an antenna N times longer can collect imagery in search mode on a swath of the same width, with the same resolution and PRF, using only $1/N$ the average power. (The processing could be more complicated if each spotlight image required significant range-walk correction, etc.)

Mrstik assumes the following approximate radar parameters of the Global Hawk unmanned aerial vehicle (UAV):

- UAV altitude ~ 19 km (~62,000 ft)
- UAV speed ~ 175 m/s
- Wavelength ~ 3 cm (frequency ~10 GHz)
- Losses and noise figure ~ 17 dB
- Antenna height ~ 0.5m
- Slant range ~ 100 km
- $\sigma^0 \sim -15$ dB

- CNR ~ 10 dB (NEσ^0 ~ −25 dB)

Using those parameters, Mrstik calculates the required average power as a function of resolution; it is reproduced here as Figure 7.20. Mrstik also points out that, by virtue of its narrower instantaneous beam, the agile-beam search-mode SAR may be more effective against jammers by keeping them in the sidelobes and, by using multiple subapertures, potentially employing space-time adaptive processing (STAP—Chapter 12) jammer-cancelation techniques more effectively.

7.5.7.2 Interleaved Search and Spotlight Modes

Mrstik further shows that, using the agile-beam approach, search and spotlight SAR modes can be interleaved on a pulse-to-pulse basis. We assume that R is the slant range to the near edge of the search swath, which has a slant width S. We define τ as the transmit pulse width and t_{switch} as the beam switching time. Furthermore, we make the conservative assumption of range nonambiguity. The search resolution is δ; the spotlight resolution depends on the spotlight synthetic aperture length. Then, during the PRI of δ / V, the SAR must transmit the pulse, wait for its return (simultaneously switching to receive mode), receive the echo, and switch back to transmit mode. Within a PRI, there still remains a dead time of

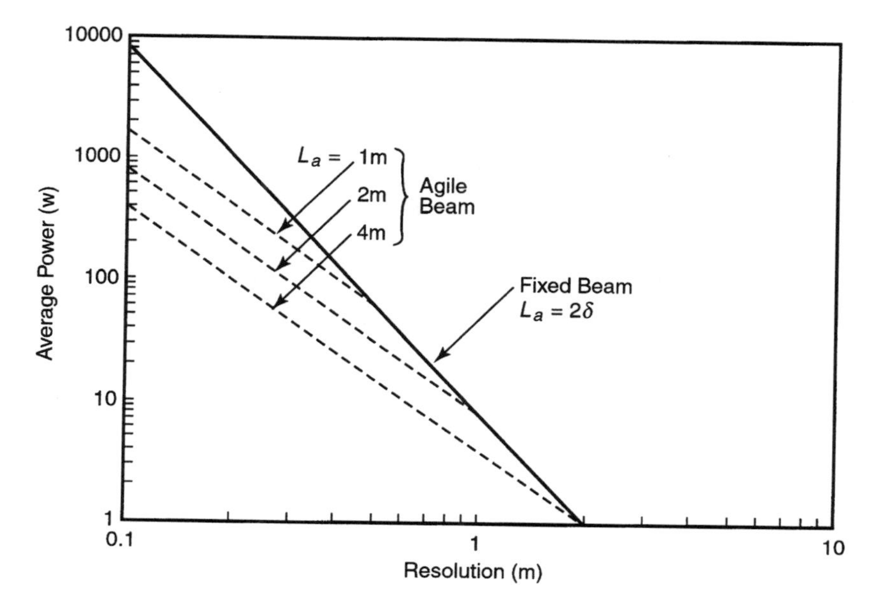

Figure 7.20 Agile-beam search SAR. (After [27], Figure 4, copyright 1998 IEEE.)

$$t_{\text{dead}}(\text{search}) = \frac{\delta}{V} - \left(\frac{2(R+S)}{c} + \tau + t_{\text{switch}}\right) \tag{7.80}$$

The time to collect one pulse in a spotlight collection is

$$t_{\text{coll}}(\text{spot}) = \frac{2R}{c} + \tau + t_{\text{switch}} \tag{7.81}$$

where, for simplicity, we have assumed that the spots are at the same range as the swath, though not necessarily at broadside (see Problem 7.11), and that they have a negligible width.

We can now divide (7.80) by (7.81) to obtain the number of spots that can be interleaved during a single PRI of search mode. Furthermore, each spot is sampled at a rate lower than that of the search mode; the spotlight PRI can be as long as $L_h/2V\cos\theta_{sq}$, where θ_{sq} is the squint angle off broadside. Within each spot PRI, there will be several search PRIs. Therefore, the dead time in each search PRI can be used for different spots. The total number of spots that can be simultaneously interleaved with a search mode is

$$n_{\text{spots}} = \text{Int}\left(\frac{L_h}{2\delta\cos\theta_{sq}}\right) \cdot \text{Int}\left[\frac{\frac{\delta}{V} - \left(\frac{2(R+S)}{c} + \tau + t_{\text{switch}}\right)}{\frac{2R}{c} + \tau + t_{\text{switch}}}\right] \tag{7.82}$$

where *Int* indicates the integer part of its argument.

Figure 7.21 ([27, Fig. 6]) illustrates, for the assumed Global Hawk parameters, the potential number of spots that could be simultaneously interleaved with a 1m-resolution broadside search having a swath width of 10 km. For an antenna length of 7m (not inconceivable) at a range of 100 km, more than 20 spots could be simultaneously interleaved with the search.

7.5.7.3 Agile-Beam Spotlight SAR

We now consider the case of spotlight imaging only, without search. The PRI for a single spot is

$$t_{PRI\text{-single}} = \frac{L_h}{2V\cos\theta_{sq}} \tag{7.83}$$

Division of (7.83) by (7.81) yields the number of spots that can be simultaneously maintained at a given range:

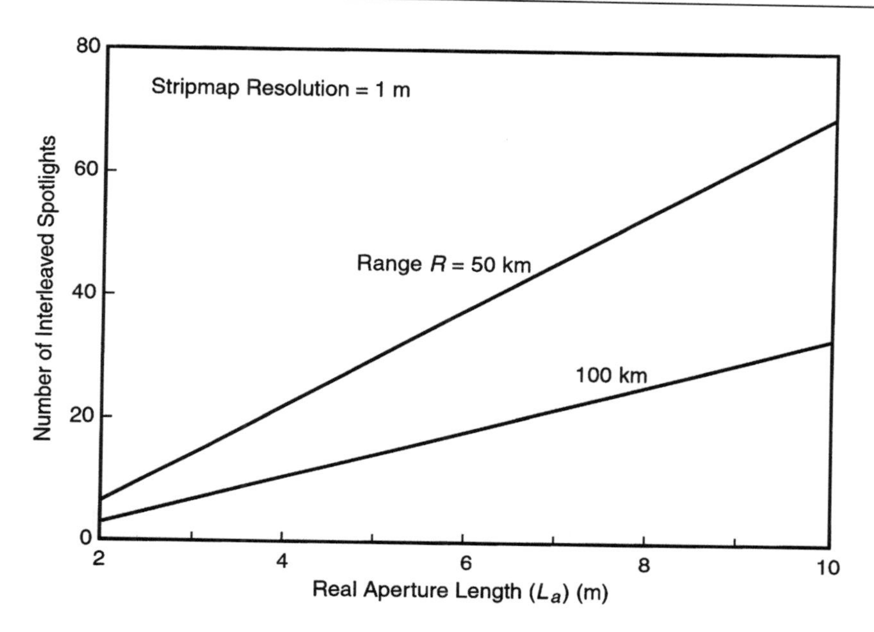

Figure 7.21 Interleaved search and spotlight SAR. (After [27], Figure 6, copyright 1998 IEEE.)

$$n_{\text{spots}}(\text{no search}) = \frac{L_h}{2V(\cos\theta_{\text{sq}})\left(\dfrac{2R}{c} + \tau + t_{\text{switch}}\right)} \tag{7.84}$$

The dwell time required to collect a spot is (Section 7.1.1)

$$t_{\text{spot}} = \frac{\lambda R}{2V\delta\cos\theta_{\text{sq}}} \tag{7.85}$$

Dividing (7.84) by (7.85) yields the spots that can be collected per unit time:

$$\dot{n}_{\text{spots}}(\text{no search}) = \frac{L_h\delta}{\lambda R\left(\dfrac{2R}{c} + \tau + t_{\text{switch}}\right)} \tag{7.86}$$

(In (7.84), we should have taken just the integer part of the right side; however, following [27], we have preserved the fractional part to show that the right side of (7.86) is essentially independent of V and θ.)

Figure 7.22 [27, Fig. 7] illustrates examples of results for 0.3m-resolution spots collected using the assumed Global Hawk parameters. At a range of 40 km, a nonagile beam SAR can collect only about five spots/minute, whereas an agile-beam SAR can, in principle, collect over a hundred. Of course, the power requirement is high (Problem 7.12).

7.5.7.4 Other Agile-Beam Cases

Mrstik [27] also considers the case of an agile-beam radar obtaining simultaneous interleaved ISAR images on a set of moving ground vehicles that are traveling at the same speed and turning with the same radius. Because of the highly restrictive nature of that assumption, we leave the case as a problem (Problem 7.13).

In principle, successive pulses can be made different, for example, by using phase coding, a procedure known as pulse tagging (Section 10.4). In this case, pulses corresponding to different spots could be distinguished, and it would not be necessary to wait for the return from pulse n before transmitting pulse $(n + 1)$. The near-range return from the latter would not be confused with the far-range return from the former, because of the different pulse tags. Thus, it might not be necessary to require range nonambiguity. Eclipsing and

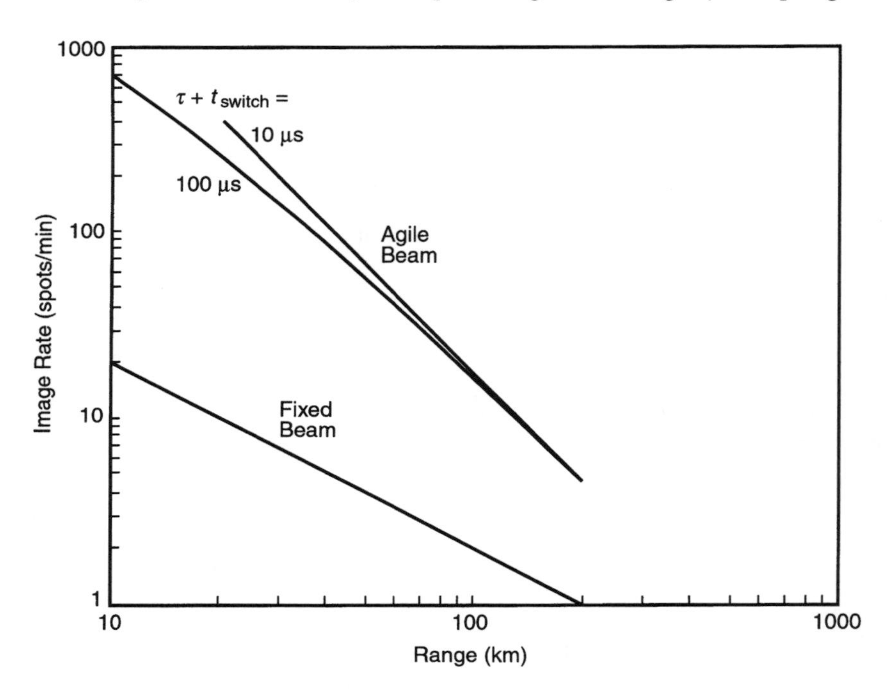

Figure 7.22 Agile-beam spotlight SAR. (After [27], Figure 7, copyright 1998 IEEE.)

beam switching, however, still would be limiting factors. That possibility is explored in Problem 7.14.

References

[1] Carrara, W. G., R. S. Goodman, and R. M. Majewski, *Spotlight Synthetic Aperture Radar*, Norwood, MA: Artech House, 1995.

[2] Curlander, J., and R. McDonough, *Synthetic Aperture Radar*, New York: Wiley, 1991.

[3] Cutrona, L. J., "Synthetic Aperture Radar," in M. Skolnik, *Radar Handbook*, 2nd ed., New York: McGraw-Hill, 1990.

[4] Stimson, G. W., *Introduction to Airborne Radar*, 2nd ed., Mendham, NJ: SciTech, 1998.

[5] Jakowatz, C. J., Jr., et al., *Spotlight-Mode SAR: A Signal-Processing Approach*, Boston: Kluwer Academic Publishers, 1996.

[6] Hovanessian, S. A., *Introduction to Synthetic Array and Imaging Radars*, Norwood, MA: Artech House, 1980.

[7] Harger, R. O., *Synthetic Aperture Radar Systems: Theory and Design*, New York: Academic, 1970.

[8] Birk, R., et al., "Synthetic Aperture Radar Imaging Systems," *IEEE AES Magazine*, Nov. 1995, pp. 15–23.

[9] Zurk, L. M., and W. J. Plant, "Comparison of Actual and Simulated SAR Image Spectra of Ocean Waves," *J. Geophysical Research*, Vol. 101, No. C4, April 15, 1996, pp. 8913–8931.

[10] Schleher, D. C., *MTI and Pulsed Doppler Radar*, Norwood, MA: Artech House, 1991.

[11] Kaplan, E. D., *Understanding GPS: Principles and Applications*, Norwood, MA: Artech House, 1996.

[12] Levanon, N., *Radar Principles*, New York: Wiley-Interscience, 1988.

[13] Skolnik, M., *Introduction to Radar Systems*, 2nd ed., New York: McGraw-Hill, 1980.

[14] Wehner, D. R., *High-Resolution Radar*, 2nd ed., Norwood, MA: Artech House, 1995.

[15] Maron, D. E., "Frequency-Jumped Burst Waveforms with Stretch Processing," *Rec. 1990 IEEE International Radar Conf.*, 1990, pp. 274–279.

[16] Brigham, E. O., *The Fast Fourier Transform and Its Applications*, Englewood Cliffs, NJ: Prentice Hall, 1988.

[17] Henderson, F. M., and A. J. Lewis (eds.), *Principles and Applications of Imaging Radar*, New York: Wiley, 1998.

[18] Oliver, C., and S. Quegan, *Understanding Synthetic Aperture Radar Images*, Norwood, MA: Artech House, 1998.

[19] Harris, Frederic J., "On the Use of Windows for Harmonic Analysis With the Discrete Fourier Transform," *Proc. IEEE*, Vol. 66, No. 1, Jan. 1978, pp. 51–83.

[20] Adams, G. F., et al., "The ERIM Interferometric SAR: IFSAR," *Proc. 1996 IEEE National Radar Conf.*, 1996, pp. 249–254.

[21] Scott, W. B., "Flight to Radar-Map Earth from Space," *Aviation Week and Space Technology*, Sept. 20, 1999, pp. 50–53.

[22] Fleischman, J. G., et al., "Foliage Penetration Experiment" (series of three papers), *IEEE Trans. Aerospace and Electronic Systems*, Vol. 32, No. 1, Jan. 1996, pp. 134–166. (This series of papers was awarded the 1996 M. Barry Carlton Award; see *IEEE Trans. Aerospace and Electronic Systems*, Vol. 35, No. 4, Oct. 1999, p. 1472.)

[23] Ulaby, F. T., R. K. Moore, and A. K. Fung, *Microwave Remote Sensing*, 3 vols., Norwood, MA: Artech House, 1986.

[24] Ayers, E. L., et al., "Antenna Measures of Merit for Ultra-Wide Synthetic Aperture Radar," *Proc. 1998 IEEE Radar Conf.*, 1998, pp. 331–336.

[25] Sullivan, R. J., et al., "Polarimetric X/L/C-Band SAR," *Proc. 1988 IEEE National Radar Conf.*, 1988, pp. 9–14.

[26] Held, D. N., W. E. Brown, and T. W. Miller, "Preliminary Results From the NASA/JPL Multifrequency, Multipolarization SAR," *Proc. 1988 IEEE National Radar Conf.*, 1988, pp. 7–8.

[27] Mrstik, V., "Agile-Beam Synthetic Aperture Radar Opportunities," *IEEE Trans. Aerospace and Electronic Systems*, Vol. 34, No. 2, April 1998, pp. 500–507.

Problems

Problem 7.1

For doppler beam sharpening (DBS) of an airborne real-beam radar, show that

Doppler resolution = $\Delta f = 2V(\cos\theta)\Delta\theta/\lambda$

DBS angular resolution = $\Delta\theta_{\text{DBS}} = \lambda/(2Vt_d\cos\theta)$

where V = aircraft speed, θ_{sq} = squint angle off broadside, λ = wavelength, and t_d = dwell time. If the radar points at a constant squint angle as it passes the target, show that

$$\Delta\theta_{\text{DBS}} = \frac{D}{2\beta R}$$

where D = real aperture length, real beam width $\theta_B = \beta\lambda/D$, and R = range to target. If the radar is performing a fast scan at angular velocity Ω, show that

$$\Delta\theta_{\text{DBS}} = \frac{\Omega D}{2V\cos\theta_{sq}}$$

Problem 7.2

An unfocused SAR has a relatively short synthetic aperture, in which the maximum two-way phase-shift is $\leq \pi/2$. Show that the along-track resolution is $\Delta y = (1/2)(\lambda R_0)^{1/2}$. (Hint: See [12].)

Problem 7.3

What is the minimum PRF for using a LFM waveform and two FFTs with uniform weighting, to obtain an ISAR image of a target of length L rotating at angular velocity Ω perpendicular to the radar LOS?

Problem 7.4

Show that when the received signal (7.18) is mixed with the reference signal (7.19), the result is the IF signal (7.20).

Problem 7.5

Verify (7.23): For a deramped SAR, show that the IF bandwidth, B_{IF}, is given by $B_{IF} = \gamma(2S/c)$, where γ = chirp slope and S = scene (swath) width.

Problem 7.6

Show that, to produce Figure 7.10 from Figure 7.9, we apply a frequency-dependent time delay/advance of $t_d = -f/\gamma$.

Problem 7.7

Assume the following SAR parameters:

- P_{avg} = 500W
- Gain = 35 dB
- f = 10 GHz
- δ_r = 1m (stripmap mode)
- V = 200 m/s
- FL = 10 dB
- Altitude = 15 km

Find the noise-equivalent terrain reflectivity (NEσ_n^0) at R = 200 km, that is, the value of σ^0 for which SNR = 1. (Neglect atmospheric absorption.)

Problem 7.8

Express (7.50) as a series of Bessel functions. Using either that series or the conventional series

$$e^x = 1 + x + \frac{x^2}{2!} + \frac{x^3}{3!} + \ldots$$

compute the position and relative strength of the second- and third-order paired echoes resulting from a vibrating target. (Hint: See [1, p. 213].)

Problem 7.9

Generalize (7.58) and compute h_{tgt} as a function of L_{shadow}, H, and R_g, using curved-Earth geometry (assume 4/3 Earth).

Problem 7.10

From Figure 7.16, compute the equation for the hyperbolic isodop and calculate the angle between the isodop and the down-range direction at the scene center as a function of H, R_g, and θ_{sq}. Show that this angle goes to zero as H/R_g goes to zero. (Also see Section 11.1.)

Problem 7.11

Recompute Figures 7.21 and 7.22 under the assumption that spot slant ranges are randomly distributed between 20 km and 200 km and their squint angles are randomly distributed between ±45 degrees. Assume $\cos(\psi) \sim 1$.

Problem 7.12

Produce a companion plot to Figure 7.21 with average power as the ordinate instead of spot image rate.

Problem 7.13

Consider an agile-beam radar obtaining simultaneous interleaved ISAR images on a set of moving ground vehicles that are traveling at the same speed V_{tgt} and turning with the same radius ρ at slant ranges of 10, 50, and 100 km. Produce a plot, analogous to Figure 7.22, that depicts ISAR images per minute versus target angular velocity ω ($\hat{\omega}$ is assumed perpendicular to the radar LOS). Compare with the fixed-beam case. (Hint: See [27, Sec. VI].)

Problem 7.14

Recompute Figures 7.21 and 7.22 and the plot of Problem 7.12, assuming pulse tagging to obviate the requirement for unambiguous range. What are the theoretical limitations of beam-switching time, and eclipsing?

8

SAR/ISAR Digital Imagery

Chapters 6 and 7 showed that, using either a step-chirp waveform or a deramped/deskewed LFM waveform, we can obtain in wavenumber space (**k**-space) a series of complex samples of radar echoes corresponding to variations in $|\mathbf{k}|$ and $\hat{\mathbf{k}}$ (magnitude and direction of **k**) and that, through the PFA, the RMA, or other focusing procedure, we can transform those data into a form such that performing a weighted FT produces an image of the target area. This chapter investigates more carefully this step of image formation from digitized focused **k**-space data and discusses techniques for obtaining imagery with improved resolution and reduced sidelobes compared with imagery obtained simply using the weighted FT.

We also introduce two other definitions [1]:

- Signal processing, which is the processing of the **k**-space signal to form an image;
- Image processing, which is additional processing performed on the image to improve the contrast, resolution, and so on.

8.1 Digital Image Formation (Signal Processing)

We now investigate the production of a discrete image by means of discrete signal processing.

8.1.1 Real and Complex Imagery

As the received signal enters the antenna, it is characterized by its electric field vector $E(t)$, a sinusoid possessing both amplitude and phase. Its power is proportional to E^2 (Section 1.3). As the signal passes through the LNA and associated electronics (Section 2.2), and is digitized by the A/D converter, the signal is characterized by a complex voltage with both amplitude and phase, and the power is proportional to the square of the magnitude of the voltage. When the image is formed by FT processing, it is characterized by a (usually two-dimensional) set of pixels, each possessing an amplitude V and phase ϕ. We refer to that as a complex image. When we display a SAR/ISAR image, we usually are interested only in the pixel magnitude. We can display $V(x, y)$, which we shall refer to as a voltage image.

It is often more useful to store or display a SAR/ISAR image using $V^2(x, y)$, which we call the power image (this was assumed in Section 7.3.1). In the signal domain, the power of the recorded signal is proportional to the square of its voltage. In the image domain, we again say that the power of the image pixel is proportional to the square of the magnitude of its voltage, even though the term *energy* might be more appropriate for describing the recorded value of the V^2 of a pixel. Whenever a SAR/ISAR image is displayed, it should be made clear whether the display shows the voltage image or the power image. An advantage of the power image is that Parseval's theorem [2, p. 23] applies: With the proper multiplicative normalization constant (independent of image content), the sum of V^2 (~ energy) in the signal domain equals the sum of V^2 in the image domain. Often the phase is discarded, because only V or V^2 is necessary for ordinary image display; such an image is often referred to as a real image. However, throughout this chapter, the term *image* is used to mean the complex image.

From Sections 6.4 and 7.3.1, we recall that the SAR/ISAR image of a point target is referred to as a PSF or IPR function, characterized by a mainlobe and sidelobes. For uniform weighting (i.e., no weighting), the voltage-image PSF is a sinc function, and the power-image PSF is a sinc-squared function with peak-to-first-null values of

$$\text{Range: } \delta_{rpn} = \frac{c}{2B} \tag{8.1}$$

$$\text{Cross-range: } \delta_{crpn} = \frac{\lambda}{2\Delta\theta} \; (\Delta\theta << 2\pi) \tag{8.2}$$

We refer to those collectively as δ_{pn}. Furthermore, the 3-dB mainlobe width of the sinc-squared function is $(0.886)\delta_{pn}$, and the first sidelobe is 13.3 dB below the mainlobe peak. When a weighting function is applied (Section 6.3.1), the mainlobe is broadened and the sidelobes are lowered relative to the sinc-squared function.

8.1.2 Discrete Fourier Transform

Section 2.1 introduced the continuous FT: The functions in both signal and image domain are defined over a continuous variable. Chapters 2 to 7 referred to the FT without specifying whether it was continuous or discrete. Modern radars generally collect and process a finite set of data samples (digital technology). It is therefore most appropriate to use the discrete FT (DFT), which can be calculated using the highly efficient FFT algorithm. The DFT and its inverse (IDFT) are given by [2, p. 97, using notation therein]:

$$\text{DFT:} \quad G\left(\frac{n}{NT}\right) = \sum_{k=0}^{N-1} g(kT)e^{-j2\pi nk/N} \tag{8.3a}$$

$$\text{IDFT:} \quad g(kT) = \frac{1}{N}\sum_{n=0}^{N-1} G\left(\frac{n}{NT}\right)e^{j2\pi nk/N} \tag{8.3b}$$

Equations (8.3a) and (8.3b) are the discrete form of (2.1). The notation reflects the assumption of time and frequency domains:

- N = number of samples processed (assumed to be sampled uniformly in each domain).
- $k = 0, 1, \ldots, N - 1$ = index identifying time-domain sample.
- T = time interval between samples (overall time interval $\approx NT$).
- $g(kT)$ = discrete function in time domain.
- $n = 0, 1, \ldots, N - 1$ = index identifying frequency-domain sample.
- $G(n/NT)$ = discrete function in frequency domain (overall frequency interval $\cong 1/T$ for $N >> 1$).

The FFT requires much less computation time than a direct calculation of the DFT. If N is a power of 2, then direct calculation requires N^2 complex multiplications, whereas an FFT of the same N requires only $N(\log_2 N)/2$ complex multiplications [2, pp. 134–135]. For example, if $N = 2^{10} = 1{,}024$, then the FFT requires less than 1/200 the computation time of the direct calculation.

8.1.3 Zero Padding

We often want to sample the frequency-domain function more finely, to see its structure more clearly. A simple way to accomplish that is to use zero-padding. When M zeroes are added to the N time-domain samples, the frequency-domain function will be sampled at $M + N$ locations over the same interval $1/T$. No information is added. The frequency-domain function will have the same general shape, but it will be sampled more finely and will appear smoother. Figures 8.1, 8.2, and 8.3 illustrate examples of zero-padding using Mathcad®, a software package manufactured by MathSoft, Inc., Cambridge, Massachusetts. The reader can easily verify the procedure with a computer (Problem 8.1). Zero padding is often performed in SAR/ISAR signal processing to produce more finely sampled images.

8.1.4 Formation of a Digital SAR/ISAR Image

From Section 6.3, we have

$$E(\mathbf{k}) \sim \int d\mathbf{r}\rho(\mathbf{r})e^{-2j\mathbf{k}\cdot\mathbf{r}} \tag{8.4a}$$

$$\rho(\mathbf{r}) \sim \int d\mathbf{k}E(\mathbf{k})e^{2j\mathbf{k}\cdot\mathbf{r}} \tag{8.4b}$$

where $E(\mathbf{k})$ is the complex returned electric field associated with wavenumber $\mathbf{k}$ and $\rho(\mathbf{r})$ is the three-dimensional image. It is convenient to write (8.4) in terms of Cartesian coordinates, because the kernels (the exponential factors) are separable:

$$E(k_x, k_y, k_z) \sim \int e^{-2jk_xx}\int e^{-2jk_yy}\int e^{-2jk_zz}\rho(x, y, z)dxdydz \tag{8.5a}$$

$$\rho(x, y, z) \sim \int e^{2jk_xx}\int e^{2jk_yy}\int e^{2jk_zz}E(k_x, k_y, k_z)dk_xdk_ydk_z \tag{8.5b}$$

Here (x, y, z) are target-centered coordinates, which are fixed relative to the target structure and rotate with the target. Thus, even for constant k, for a rotating target k_x, k_y, k_z will vary.

Equations (8.5a) and (8.5b) were written under the assumption that a continuum (infinite number) of frequencies is transmitted over a continuum of target rotation angles. In reality, the frequencies and angles make up a discrete set rather than a continuum. Thus, after polar-formatting and interpolation, the continuously variable vectors $\mathbf{k}$ and $\mathbf{r}$ are replaced by the discrete sets $(u, v, w)\Delta k$ and $(U, V, W)\Delta r$, where u, v, w, U, V, W are integers. We also assume that the target consists of a discrete set of scatterers, each with voltage reflectivity, q_i, proportional to the square root of its radar cross-section. For this discrete case, (8.5a) and (8.5b) become

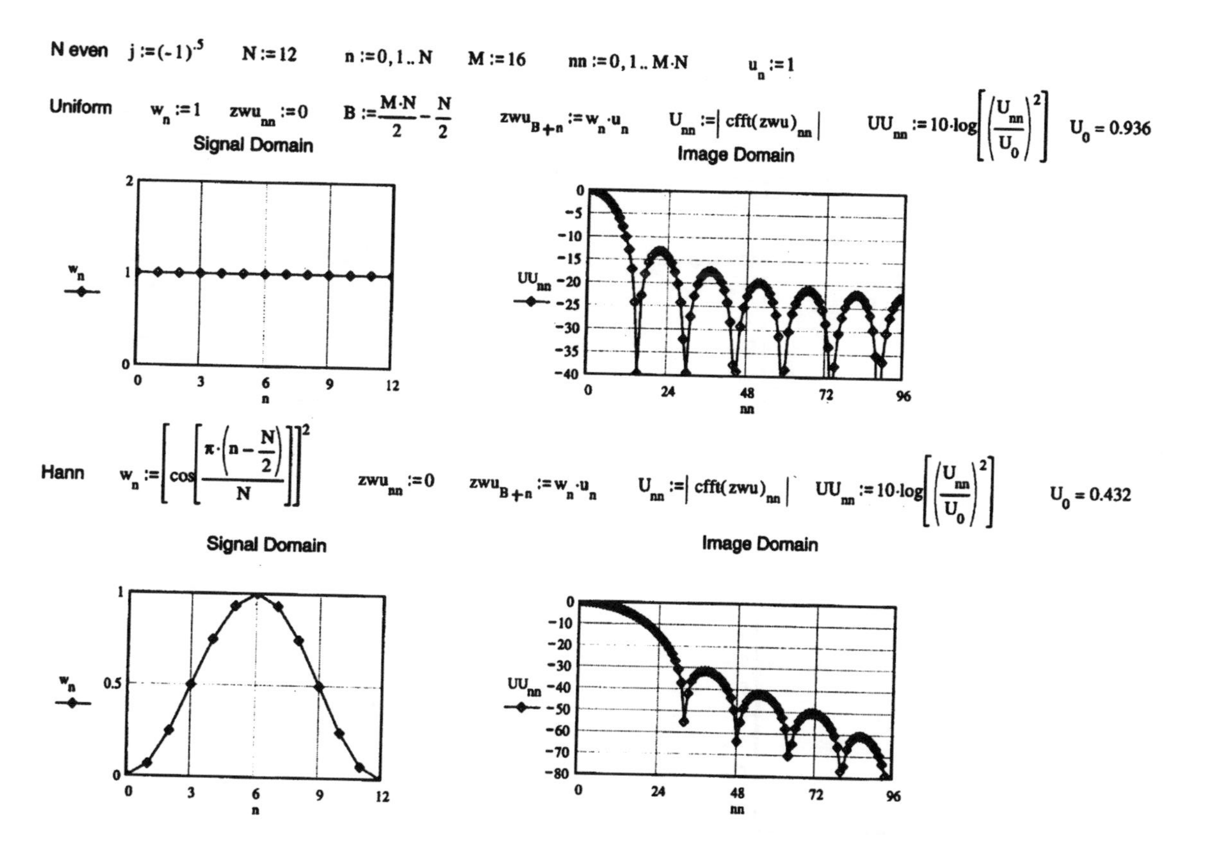

Figure 8.1 Zero-padding procedure.

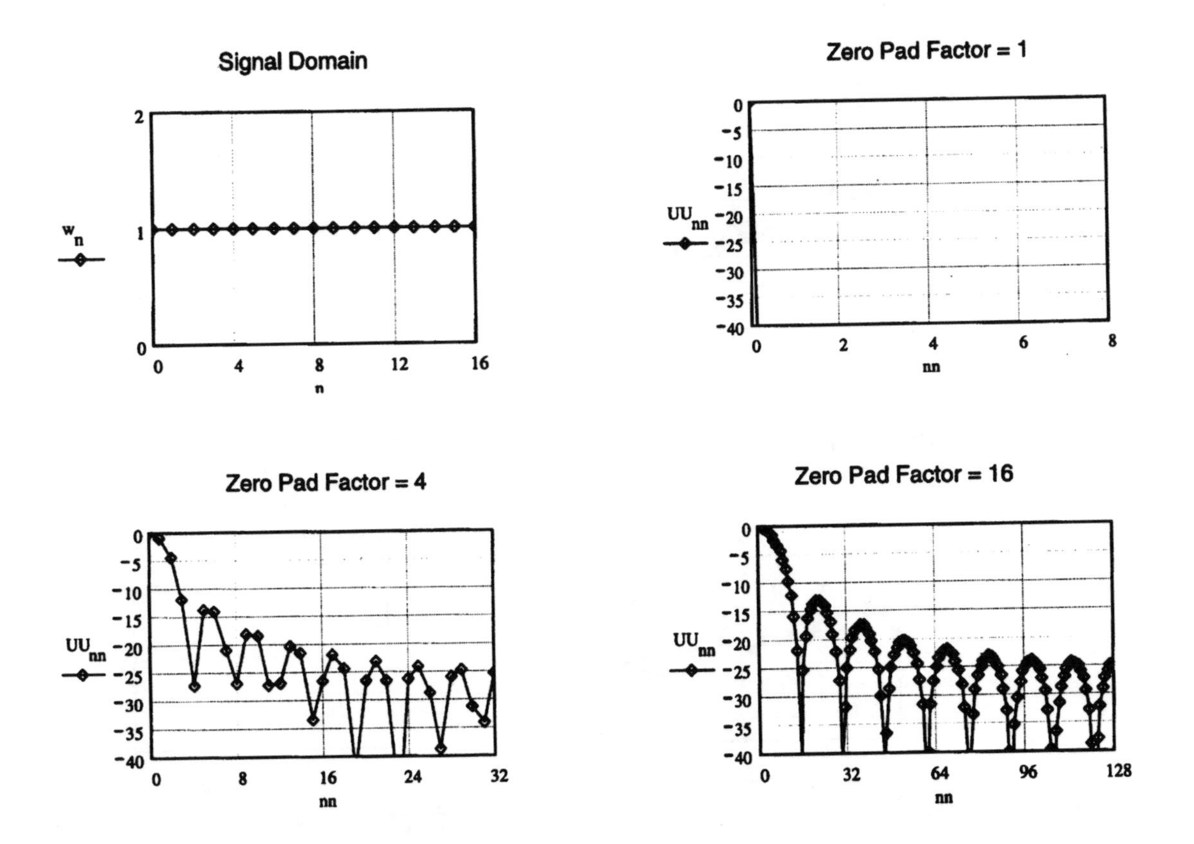

Figure 8.2 Zero-padding, uniform weighting.

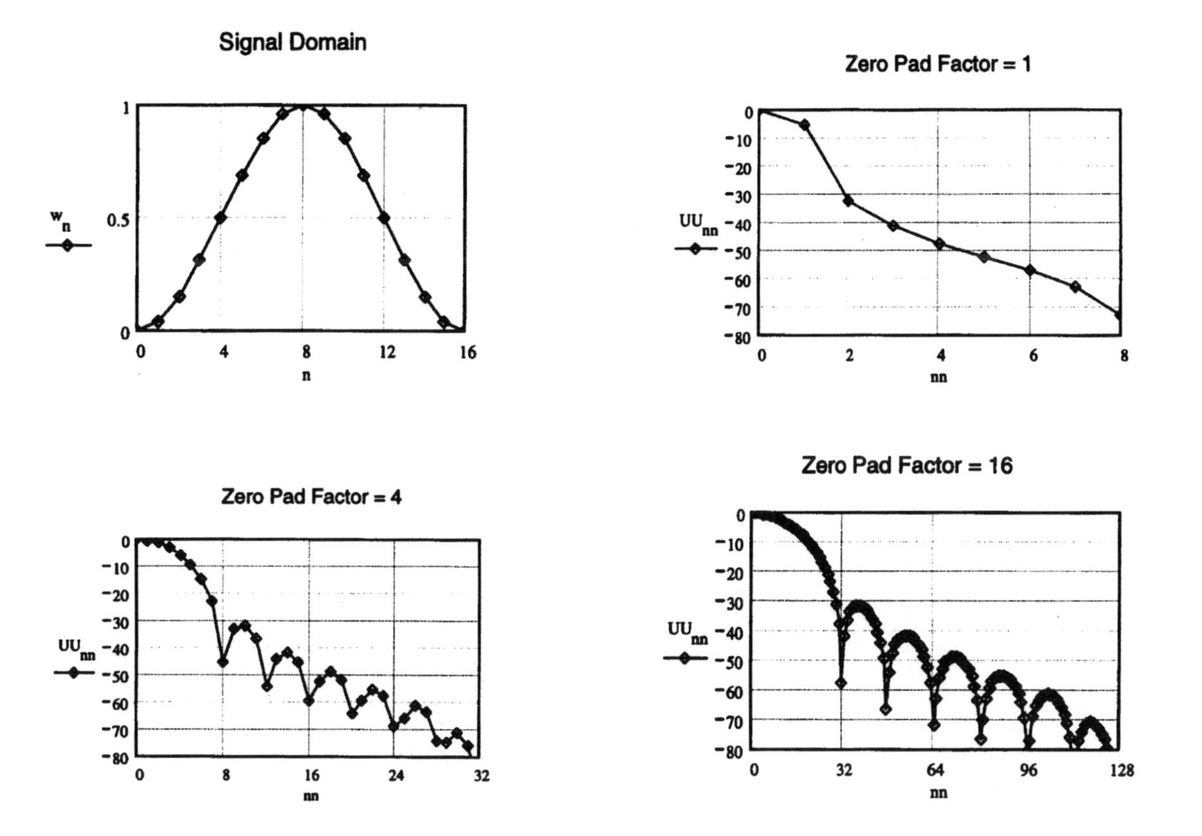

Figure 8.3 Zero-padding, Hann weighting.

$$\text{Signal: } E_{uvw} \sim \sum_i q_i e^{-2j\Delta k(ux_i + vy_i + wz_i)} \tag{8.6}$$

$$\text{Image: } h(U, V, W) = \sum_{u,v,w} E_{uvw} e^{2j\Delta k \Delta r(uU + vV + wW)} \tag{8.7}$$

$$= \sum_i q_i \sum_u e^{-2ju\Delta k(x_i - U\Delta r)} \sum_v e^{-2jv\Delta k(y_i - V\Delta r)} \sum_w e^{-2jw\Delta k(z_i - W\Delta r)} \tag{8.8}$$

(we suppress the normalization constants). (8.8) of a three-dimensional far-field image is a complex function of three space dimensions.

It can be readily shown that the signal history (8.6) can be reconstructed from the image (8.8) (Problem 8.2).

8.1.5 Point-Spread Function

Of particular interest is the PSF, the image resulting from a target consisting of a single point scatter. We let q for the scatterer = 1 and consider only the downrange cut of the PSF. Thus,

$$h_1(U, x) = \sum_{u=u_1}^{u_2} e^{-2ju\Delta k(x - U\Delta r)} \tag{8.9}$$

Here x is the scatterer location (continuous), U is the pixel location (discrete), and u_1 and u_2 are integers. We let $u_2 - u_1 = N - 1$, so there are N values of u. Thus,

$$h_1(U, x) = e^{-2j\Delta k u_1(x - U\Delta r)} \sum_{u=0}^{N-1} e^{-2j\Delta k u(x - U\Delta r)} \tag{8.10}$$

From Section 2.4.1, we recall

$$\sum_{n=0}^{N-1} e^{jn\psi} = e^{j(N-1)\psi/2} \frac{\sin(N\psi/2)}{\sin(\psi/2)} \tag{8.11}$$

and thus, with $\psi = -2\Delta k(x - U\Delta r)$,

$$h_1(U, x) = e^{-2ju_1\Delta k(x - U\Delta r)} e^{-j(N-1)\Delta k(x - U\Delta r)} \frac{\sin[N\Delta k(x - U\Delta r)]}{\sin[\Delta k(x - U\Delta r)]} \tag{8.12}$$

As $N \rightarrow \infty$, $|h_1(U, x)|^2$ becomes (for a given scatterer location x) a sinc-squared function of U with the usual −13.3 dB sidelobes. Note that, if the scatterer lies exactly on the pixel location U, then at the peak of the PSF ($x = U\Delta r$), the phase is zero, independent of U. The processing is such that, for each pixel, the peak of the PSF for a scatterer at that exact pixel location will have zero phase.

The pixel phase is given by

$$\Phi(U, x) = -2\left(u_1 + \frac{N-1}{2}\right)\Delta k(x - U\Delta r) = -2\bar{k}(x - U\Delta r) \quad (8.13)$$

Thus, for fixed U, $\delta\Phi = -2\bar{k}\delta x$, where $\bar{k}$ is the average wavenumber. That justifies (at least for one-dimensional images) the statement made in Section 7.5.3.4 concerning IFSAR that, for a given pixel of an image of a single point scatterer, if the scatterer is shifted by a small distance from one image to another, the pixel phase change is the same as for a monochromatic wave at the average transmitted frequency.

As discussed in Section 7.3.1, the signal data can be multiplied by a weighting function prior to the DFT/FFT, resulting in increased mainlobe width and reduced sidelobe levels.

8.1.6 Range Window

We want to form our downrange image over a window of length L; thus, $\Delta r = L/N$. It is useful to choose $\Delta k = \pi/L$, so that $\Delta k \Delta r = \pi/N$, $\Delta f = c\Delta k/2\pi = c/2L$, $L = c/2\Delta f$, and $\Delta r = c/2N\Delta f \cong c/2B$. Then, from (8.10),

$$h_1(U, x) = e^{-2j\Delta k u_1(x - U\Delta r)} \sum_{u=0}^{N-1} e^{-2\pi jux/L} e^{2\pi juU/N} \quad (8.14)$$

From (8.3b), substituting $k \rightarrow U$, $n \rightarrow u$, $g(kT) \rightarrow h_1(U)$ and $G(n/NT) \rightarrow e^{-2\pi jux/L}$ we see that (8.14) is equivalent to performing a DFT (FFT) on the discrete set of complex signal samples. The peak-to-first-null downrange resolution is $\delta_{rpn} = \Delta r = c/2B$.

We set $x \equiv R + r$, where R is the distance to the near edge of the window of length L, and $0 < r < L$. In general, R is many times the size of L. Then, from (8.14),

$$h_1(U, R, r) = e^{-2j\Delta k u_1(R + r - U\Delta r)} \sum_{u=0}^{N-1} e^{-2\pi juR/L} e^{-2\pi jur/L} e^{2\pi juU/N} \quad (8.15)$$

We set $R = (M + \alpha)L$, with M an integer and $0 < \alpha < 1$. Then

$$h_1(M, \alpha, U, r) = e^{-2\pi j u_1 M} e^{-2\pi j u_1[\alpha + (r - U\Delta r)/L]} \sum_{u=0}^{N-1} e^{-2\pi j u M} e^{-2\pi j u\left(\alpha + \frac{r}{L} - \frac{U}{N}\right)} \tag{8.16}$$

Because u, u_1, and M are integers, the first factor and the first factor after the sum are each equal to unity. Thus, the image of each scatterer (and, more generally, the image of the scene) is independent of M and depends only on the position of the scatterer in the window, not on the absolute distance from the window to the radar (assuming no significant variation in R^4, or other complication, such as multipath).

This concept is illustrated in Figure 8.4. We may consider the radar located at the first of a series of points in space separated by a distance L (the author refers to those points as "fence posts"). For a radar on the ground with pulse width τ illuminating the ground, the swath width of the observed scene resulting from each signal sample will be $c\tau/2$ (Section 1.17). Figure 8.5 illustrates various relationships between the image window and the observed swath.

We now assume downconversion to baseband, specifically $u_1 = 0$, and $\alpha = 0$. Then

$$h_1(U, r) = \sum_{u=0}^{N-1} e^{-2\pi j u\left(\frac{r}{L} - \frac{U}{N}\right)} \tag{8.17}$$

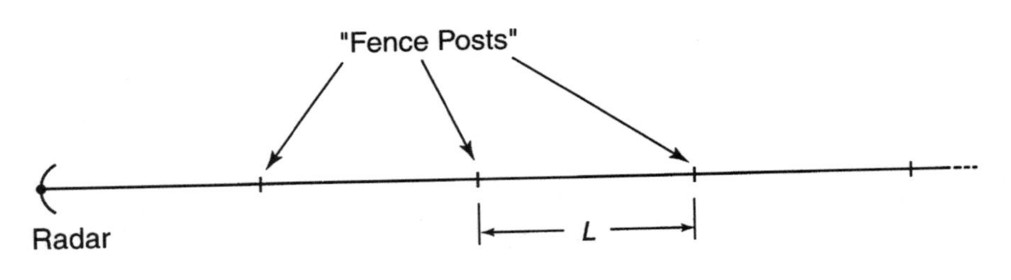

Δf = frequency increment

L = range window $= \dfrac{c}{2\Delta f}$

N = number of frequencies

$\Delta r = \dfrac{L}{N-1} = \dfrac{c}{2(N-1)\Delta f} = \dfrac{c}{2B}$

Figure 8.4 Range windows.

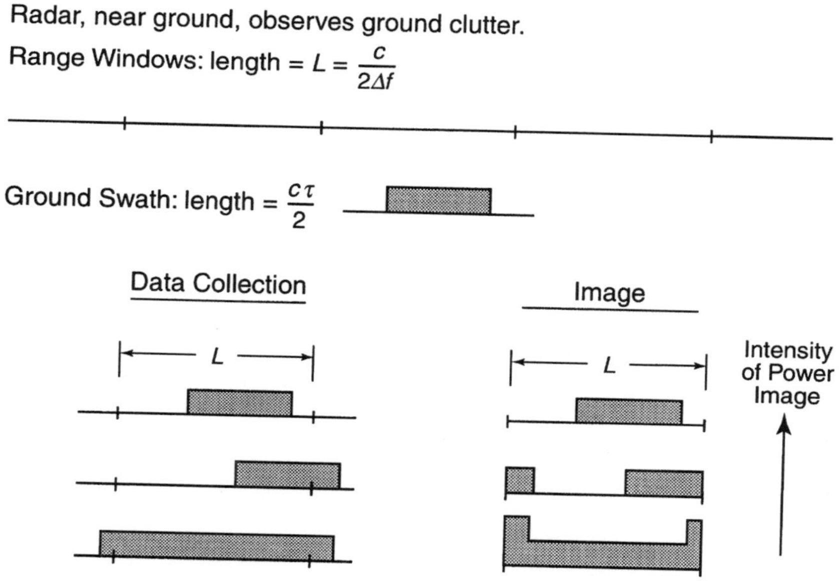

Radar, near ground, observes ground clutter.
Range Windows: length = $L = \frac{c}{2\Delta f}$

Ground Swath: length = $\frac{c\tau}{2}$

Data Collection

Image

L

L

Intensity of Power Image

Figure 8.5 Relationship between range windows and pulse width.

From (8.11), the phase of that quantity is

$$\Phi(U, r) = -(N-1)\pi\left(\frac{r}{L} - \frac{U}{N}\right) \tag{8.18}$$

For fixed r,

$$\left(\frac{\delta\Phi}{\delta U}\right)_r = \left(\frac{N-1}{N}\right)\pi \tag{8.19}$$

Thus, for large N, the phase change between adjacent pixels is approximately π. For $r = 0$,

$$\Phi(U, 0) = \pi U\left(\frac{N-1}{N}\right) \tag{8.20}$$

Table 8.1 illustrates examples of pixel phase for some simple cases.

Table 8.1
Pixel Phase

$$|\Phi(U, 0)| = \pi U\left(\frac{N-1}{N}\right) \tag{8.20}$$

$N = 8$:									
U:	0	1	2	3	4	5	6	7	"8"
$\|\Phi/\pi\|$	0	7/8	14/8	21/8	28/8	35/8	42/8	49/8	7
$\|\Phi/\pi\|$ mod 2	0	7/8	14/8	5/8	12/8	3/8	10/8	1/8	1
$N = 7$:									
U:	0	1	2	3	4	5	6	"7"	
$\|\Phi/\pi\|$	0	6/7	12/7	18/7	24/7	30/7	36/7	6	
$\|\Phi/\pi\|$ mod 2	0	6/7	12/7	4/7	10/7	2/7	8/7	0	

If the sequence of pixels is extended to $U = N$, then $|\Phi(N)|$ mod $2\pi = \pi$ for N even, 0 for N odd.

8.2 Digital Image Enhancement (Image Processing)

We now assume that we have an unweighted SAR/ISAR complex image processed with uniform (i.e., no) weighting, with each point scatterer represented by a sinc PSF. We investigate ways of reducing the mainlobe width and reducing the sidelobes of those PSFs.

We introduce the following definitions.

- *Mainlobe narrowing* is a processing procedure that narrows mainlobes.
- *Sidelobe reduction* is a processing procedure that reduces sidelobes.
- *Resolution* is the minimum distance between two point targets such that they can be individually resolved (observed) in the image. A resolution criterion must be stated. A simple criterion is the Rayleigh criterion [3, p. 8–28], which states simply that the resolution is equal to Δr. A more sophisticated criterion could invoke the discussion of Section 4.3 and would depend on the SNR and the PSF shapes, including sidelobe levels.
- *Superresolution* is a processing procedure that improves resolution beyond the limits of the DFT.

Mainlobe narrowing and superresolution are not necessarily the same thing.

8.2.1 Introduction to Superresolution and Sidelobe Reduction Techniques

DeGraaf [4] has produced a useful review of the aforementioned image enhancement techniques. Except as noted, the remainder of this chapter follows [4]. The techniques are collectively known as spectral estimation techniques, reflecting the fact that the processing implicitly makes an estimate of portions of the input signal not explicitly present, to obtain an improved output, that is, a narrowed mainlobe and/or reduced sidelobes for point scatterers.

Following DeGraaf, we define **X** as a column vector containing the radar signal history samples. For a two-dimensional image consisting of p_x by p_y pixels, we use a $p_x p_y$ by 1 column vector. We assume that resampling from any focusing algorithm (PFA, RMA, etc.) has already occurred, so that, for conventional signal processing, only a two-dimensional weighted DFT remains to be performed.

We also introduce the expected value of a signal or image parameter a, denoted by $\mathrm{E}(a)$. The expected value refers to the average value of the parameter that would be obtained if the data collection were performed many times under nominally (but not exactly) identical conditions (an ensemble of data collections). If that were actually done, the specific measured values of noise and clutter voltages would vary from collection to collection; if desired, the variations could be modeled according to a probability density function (see Section 4.1.2). $\mathrm{E}(a)$ can be regarded as a theoretical ideal. More practically, it can be estimated, for example, by averaging a over an entire signal history, then (1) processing assuming that the average is the expected value, or (2) processing only a portion of the signal history but using $\mathrm{E}(a)$ as estimated from the entire history.

Using $\mathrm{E}(a)$, we define the signal history covariance matrix:

$$\begin{aligned} \mathbf{R} &= E(\mathbf{X}\mathbf{X}^{\mathbf{H}}) \\ R_{ij} &= E(x_i x_j^*) \end{aligned} \tag{8.21}$$

Here the superscripted variable $\mathbf{X}^{\mathbf{H}}$ refers to the Hermitian adjoint, or conjugate transpose [5], of the matrix (vector) **X**; x_i refers to the ith component of **X**; and * represents complex conjugation. For the diagonal elements of **R**, $i = j$ and

$$R_{ij} = E(x_i x_i^*) = E(|x_i|^2) \tag{8.22}$$

Because measured voltages may be either positive or negative, we assume x_i is a zero-mean variable; $\mathrm{E}(|x_i|^2)$ is its variance. More generally, $E(x_i x_j^*)$ is the covariance between x_i and x_j; hence, the nomenclature.

8.2.2 DFT (FFT) Processing

Let us consider data resulting from illuminating a point target at range r with a waveform utilizing N frequencies separated by Δf, with echo signals downconverted to baseband. The transmitted wavenumbers (minus the carrier) are $k_n = n\Delta f$, $n = 0, 1, \ldots, N - 1$. From (8.6), the echo complex samples are

$$W(n, r) = e^{-2jnr\Delta k} \tag{8.23}$$

which may be represented by a column vector $\mathbf{W}(r)$. We define $\mathbf{x}$ as the image (again a column vector containing all pixels, perhaps usually displayed in two dimensions). The image resulting from processing with an unweighted DFT then can simply be expressed as (Problem 8.2)

$$\mathbf{x(r)} = \mathbf{W^H(r)X} \tag{8.24}$$

A weighted DFT image may be written

$$\mathbf{x(r)} = \mathbf{W^H(r)AX} \tag{8.25}$$

where $\mathbf{A}$ is a real-valued diagonal matrix corresponding to a weighting function. To quote from [4, p.733]:

> Unweighted FFT image formation corresponds to evaluating a bank of matched filter outputs, each filter being matched to a point target at a particular spatial location. In the simple case of a single point target in white Gaussian noise or clutter, this matched filter maximizes SIR [signal-to-interference ratio].

For the simple case of a scene containing a single point target exactly on a pixel location r_0,

$$\begin{aligned} \mathbf{X} &= \mathbf{W(r_0)} \\ \mathbf{x(r)} &= \mathbf{W^H(r)W(r_0)} = \delta_{rr_0} \end{aligned} \tag{8.26}$$

(Problem 8.3). Here δ_{ab} is the Kronecker delta [6]; $\delta_{ab} = 1$ if $a = b$, $\delta_{ab} = 0$ if $a \neq b$, where a and b are integers.

8.2.3 Periodogram

A simple extension of the weighted DFT is the periodogram, which is the expected value of the power image over an ensemble of data collections. Thus,

$$\begin{aligned} \mathbf{x}(\mathbf{r}) &= \mathbf{E}(|\mathbf{W}^{\mathbf{H}}\mathbf{A}\mathbf{X}|^2) = \mathbf{E}(\mathbf{W}^{\mathbf{H}}\mathbf{A}\mathbf{X}\mathbf{X}^{\mathbf{H}}\mathbf{A}\mathbf{W}) \\ &= \mathbf{W}^{\mathbf{H}}\mathbf{A}\mathbf{E}(\mathbf{X}\mathbf{X}^{\mathbf{H}})\mathbf{A}\mathbf{W} = \mathbf{W}^{\mathbf{H}}\mathbf{A}\mathbf{R}\mathbf{A}\mathbf{W} \end{aligned} \tag{8.27}$$

However, to improve on a weighted DFT, a periodogram requires an independent estimate of **R**, which generally is not available.

To appreciate one application of a periodogram, recall the concept of *speckle* in a SAR image (Section 7.3.5). Many SAR images of clutter exhibit pixel-to-pixel intensity fluctuations greater than are seen in comparable optical images. That phenomenon, known as speckle, occurs because, in ground clutter (notably vegetation), although neighboring pixels may have similar numbers of scatterers, each with similar values of radar cross-section, the overall magnitudes of the pixels may vary substantially because of differences in the detailed manner by which the scatterer phases add together. Speckle is usually considered undesirable, and one way to reduce it is for the SAR, as it passes each region of the scene, to "look" several times at each region, forming a SAR image for each look. The images are then averaged noncoherently, a procedure known as multilook.

A generalization of the multilook procedure is to collect data over a particular aperture (the "full" aperture) and compute **R** for the full aperture, then form a periodogram image over a subaperture using **R** as determined for the full aperture. However, that results in reduced crossrange resolution. According to [4], "for this reason, the periodogram is of little practical interest."

8.2.4 Minimum-Variance Method

The minimum-variance method (MVM) is also known as the maximum-likelihood method and Capon's method. For a point target at a particular pixel location, it is a theoretically ideal procedure for maximizing the energy in the correct pixel (mainlobe) and minimizing the energy outside it (sidelobes).

The most general linear procedure for determining the pixel value $\mathbf{x}(\mathbf{r})$ at the target location $\mathbf{r}$ is

$$\mathbf{x}(\mathbf{r}) = \mathbf{A}^{\mathbf{H}}(\mathbf{r})\mathbf{X} \tag{8.28}$$

We want to find the optimum $\mathbf{A}(\mathbf{r})$. We impose the constraint that

$$\mathbf{A}^{\mathrm{H}}(\mathbf{r})\mathbf{W}(\mathbf{r}) = 1 \tag{8.29}$$

to ensure that the point target signal passes the filter with unit (i.e., maximum) gain. We then minimize

$$E(|\mathbf{x}(\mathbf{r})|^2) = E(|\mathbf{A}^{\mathrm{H}}\mathbf{X}|^2) = \mathbf{A}^{\mathrm{H}}\mathbf{R}\mathbf{A} \tag{8.30}$$

to minimize the sidelobes. It can be shown [4 and references therein] that

$$\mathbf{A} = \frac{\mathbf{R}^{-1}\mathbf{W}}{\mathbf{W}^{\mathrm{H}}\mathbf{R}^{-1}\mathbf{W}} \tag{8.31}$$

(see also Section 9.2). The numerator is a vector and the denominator is a scalar. Therefore,

$$\mathbf{x}(\mathbf{r}) = \mathbf{A}^{\mathrm{H}}\mathbf{X} = \frac{\mathbf{W}^{\mathrm{H}}\mathbf{R}^{-1}\mathbf{X}}{\mathbf{W}^{\mathrm{H}}\mathbf{R}^{-1}\mathbf{W}} \tag{8.32}$$

Determining **R** requires an estimate of the ensemble of data collections. Computing $\mathbf{R}^{-1}$ dominates the computational complexity. A typical two-dimensional SAR spot image may have several thousand pixels on a side; thus, the column vector **A** may have several million rows, and the square matrix **R** may have over a trillion elements. Current research focuses on methods for reducing the level of computational complexity while preserving the basic advantages of the procedure.

8.2.5 High-Definition Vector Imaging

Benitz [7] has developed a procedure, referred to as high-definition vector imaging (HDVI), for working with a covariance matrix that has reduced rank relative to the **R** matrix referred to above (the rank of a matrix is the number of independent rows [5]). (DeGraaf calls the technique the "reduced-rank minimum-variance method.") The weighting coefficients **W** are constrained in a manner described in [6], resulting in reduced computational requirements relative to the MVM.

Of particular interest is the vector aspect of HDVI. A series of matched filters is applied to the phase history data, each tuned for a different elementary target type, including the point scatterer, flat plate, dihedral, trihedral, and cylinder on a ground plane. HDVI then processes the image according to each matched filter. The output is an image with each pixel represented by a multicomponent vector, which then can be further compared with the expected target type. Figure 8.6 illustrates the principle of HDVI, and Figure 8.7 shows some illustrative results.

8.2.6 Adaptive Sidelobe Reduction

DeGraaf [4] has developed a still simpler approximation to the MVM, which is referred to as adaptive sidelobe reduction (ASR). Although the method can easily be used for two-dimensional data and imagery, to illustrate it we discuss the simpler one-dimensional case. Thus, **r** becomes r. With complete generality, we can write

$$x(r) = \mathbf{A}^{\mathrm{H}}\mathbf{X} = \sum_{k=0}^{K-1} A_k(r) X_k e^{2\pi jrk/N} \tag{8.33}$$

Simplicity is achieved by requiring that the A_k be real. If they were constant and independent of the pixel, that is, if $A_k(r) = A_k$, the procedure would be a weighted DFT interpolated by a zero-padding factor $R = N/K$. ASR is more general than the weighted DFT because the weighting coefficients are dependent on the pixel value; hence, the term *adaptive*.

We now set

$$A_k(r) = 1 + \sum_{m=1}^{M} a(r, m)\cos(2\pi mk/K), \; M << K \tag{8.34}$$

That is actually similar to Taylor weighting [1, Sec. D.2], except that for Taylor weighting the $a(r, m)$ becomes $a(m)$, independent of the pixel. We require that R be an integer. Then, substituting (8.34) into (8.33), we find (Problem 8.4)

$$x(r) = y(r) + \sum_{m=1}^{M} \frac{a(r, m)}{2}[y(r - Rm) + y(r + Rm)] \tag{8.35}$$

where $y(r)$ is the unweighted DFT image for which a point target is represented by a sinc function.

Equation (8.35) is in the form of a convolution, reflecting the fact that multiplication in the signal domain (8.33) yields convolution in the image domain. If $R = 1$, the convolution utilizes the M nearest pixels on each side of the pixel being computed. If $R > 1$, then every Rth pixel is used. In either

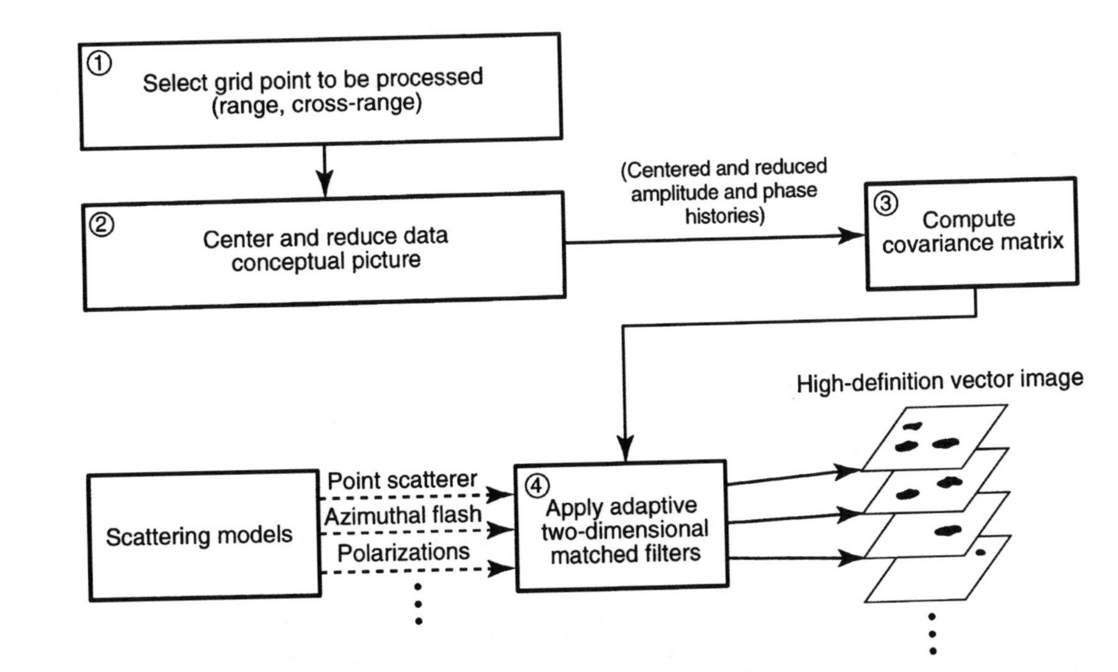

Figure 8.6 Principles of HDVI. (After [6], with permission of MIT Lincoln Laboratory, Lexington, MA.)

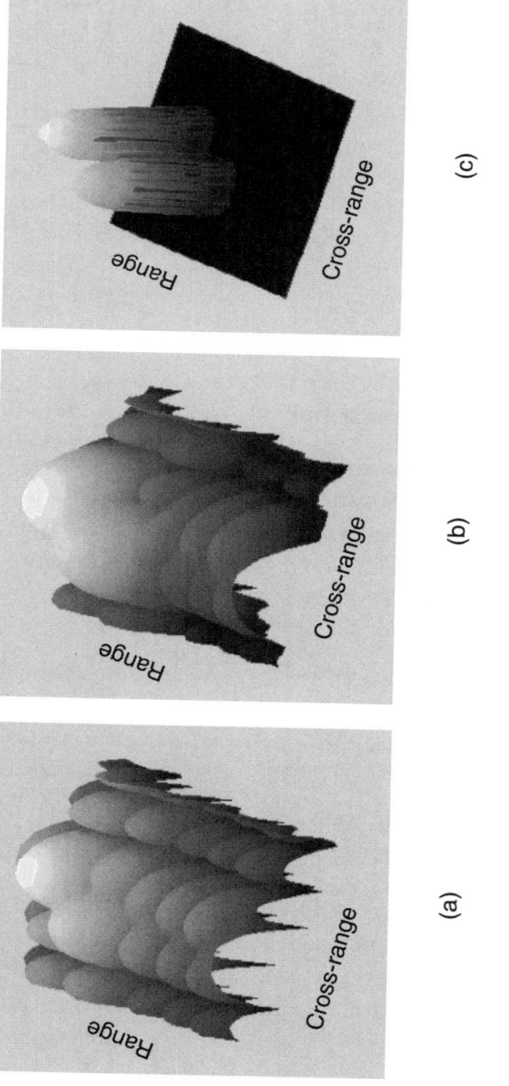

Figure 8.7 HDVI results. (Reproduced from [6], with permission of MIT Lincoln Laboratory, Lexington, MA.)

case, $2M$ pixels are used to compute the convolution. ASR thus can be performed by first generating the unweighted DFT image, then performing a convolution on the complex pixel values.

DeGraaf [4] presents an example of an ASR-processed image with an "order" (M) of 2. Computation of the coefficients $a(r, m)$ is explained in [4].

For a two-dimensional image, we can apply ASR successively to rows and columns (the separable method) or utilize a two-dimensional generalization of (8.34) as follows:

$$A(n_x, n_y, k_x, k_y) = 1 + \sum_{m_x=0}^{M} \sum_{m_y=0}^{M} a(n_x, n_y, m_x, m_y) \cos\left(\frac{2\pi m_x k_x}{K_x}\right) \cos\left(\frac{2\pi m_y k_y}{K_y}\right) \tag{8.36}$$

The sum does not include the term $m_x = m_y = 0$.

8.2.7 Spatially Variant Apodization

Stankwitz et al. [8] (also discussed in [1, Sec. D.3]) have developed a simple form of ASR called spatially variant apodization (SVA). *Apodization*, a term originally coined to refer to sidelobe reduction in optical systems, comes from the Greek α, *take away*, and $\pi o \delta o \varsigma$, *foot*, meaning literally *removing the feet from* [9, 10]. SVA is simply ASR with $M = 1$, $R = 1$, and $a(r) > 0$. Specifically

$$\begin{aligned} A_k(r) &= 1 + a(r)\cos(2\pi k/K) \\ x(r) &= \sum_{k=0}^{K-1} A_k(r) X_k e^{2\pi j r k/K} \\ x(r) &= y(r) + \frac{a(r)}{2}(y(r-1) + y(r+1)) \end{aligned} \tag{8.37}$$

SVA involves a convolution of the unweighted FFT complex image using only the nearest neighbors of each pixel. That simplicity means very little computational burden. Although SVA is efficient at lowering sidelobes, it cannot reduce mainlobe width. Again, the procedure for calculating the $a(r)$ is given in [8].

8.2.8 Super-SVA

Super-SVA, also developed by Stankwitz et al. [11], is a bandwidth-extrapolation procedure that has been found experimentally to reduce the mainlobe width, lower the sidelobes, and produce "pleasing" SAR imagery. As shown in Figure 8.8, the procedure is as follows:

1. Begin with unweighted data in the signal domain.
2. Produce an unweighted FFT image.
3. Perform SVA (or complex dual apodization [1]).
4. Perform an inverse FFT; multiply by an inverse weighting function, which produces extrapolated signal data of essentially constant amplitude over a wider bandwidth than the original data and truncates the signal data beyond the point where it has significant amplitude.
5. Replace the central portion of the modified signal data with the original signal data.
6. Again produce an unweighted FFT image.
7. Again perform SVA, and so on.

The procedure can be iterated.

Stankwitz and Kosek [12] have also shown that super-SVA can be used to interpolate over gaps in the bandwidth of a received signal. Such gaps could result from, for example, legal restrictions on the transmitted signal or the necessity to discard some portion of the received signal due to interference in a particular frequency window. They present examples of simulated imagery obtained both with the original, nongapped received signal and with the gapped received signal using super-SVA.

8.2.9 Other Spectral Estimation Techniques and Applications

Many other spectral estimation techniques for improving imagery are discussed in the literature and summarized in [4]. Some of these include the following:

- Autoregressive linear prediction (ARLP);
- Pisarenko's method;
- Eigenvector (EV) and multiple signal classification (MUSIC);
- Tufts-Kumaresan ARLP; and
- Parametric maximum likelihood (PML).

Furthermore, by increasing matrix dimensions, some spectral estimation techniques can be extended to include

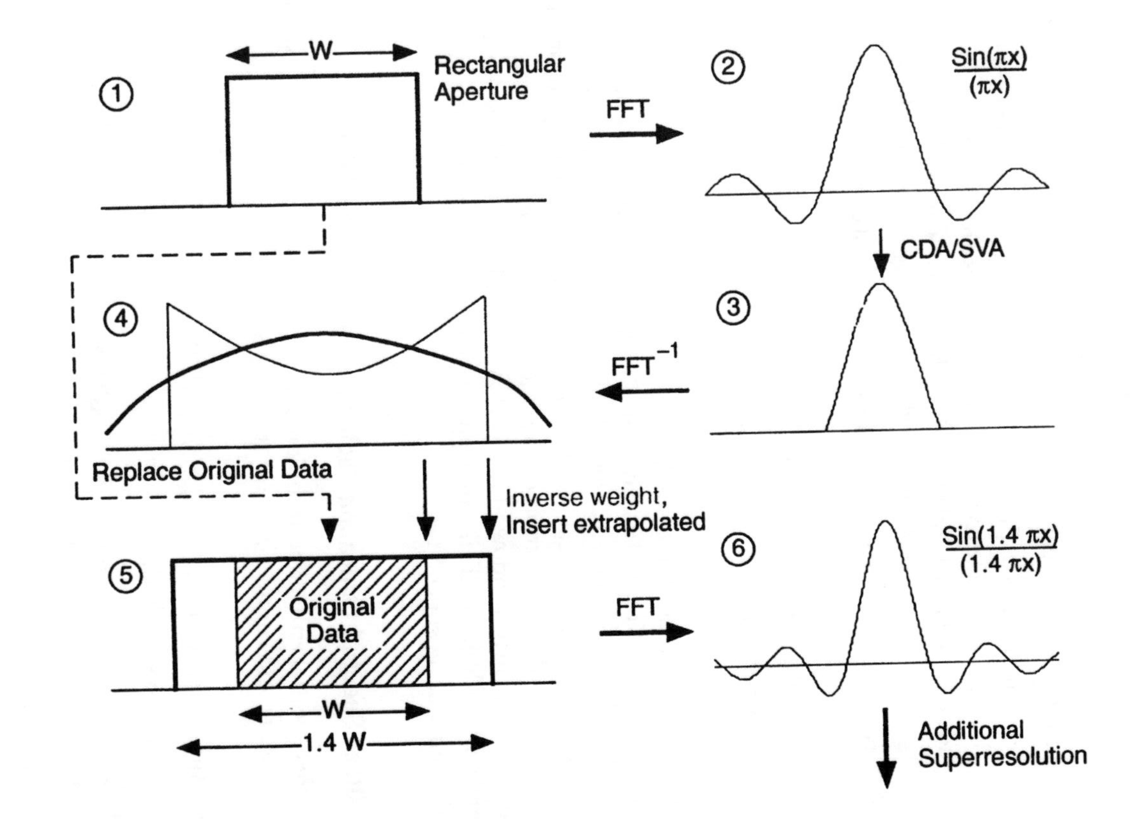

Figure 8.8 Super-SVA. (Courtesy of Veridian ERIM-International.)

- IFSAR, thereby improving vertical, as well as horizontal, resolution;
- Polarimetric SAR imagery with two to four channels representing different combinations of polarizations, for example, HH (transmit H, receive H), HV, VH, and VV, or their circular-polarization analogs.

8.2.10 Example Results

Figure 8.9 compares PSFs for a number of spectral estimation algorithms. The data are from an actual SAR image of a trihedral corner reflector on an asphalt causeway. The top portion of the figure summarizes results of the "easy" methods, that is, those methods that do not involve extensive computation: sinc, Taylor, SVA, ASR, and super-SVA. Regarding the mainlobe, Taylor is widest, sinc and SVA are the same, and second widest, ASR is third widest, and super-SVA is narrowest. Sinc, of course, has the highest sidelobes; the Taylor sidelobes are quite regular and generally second highest; the sidelobes

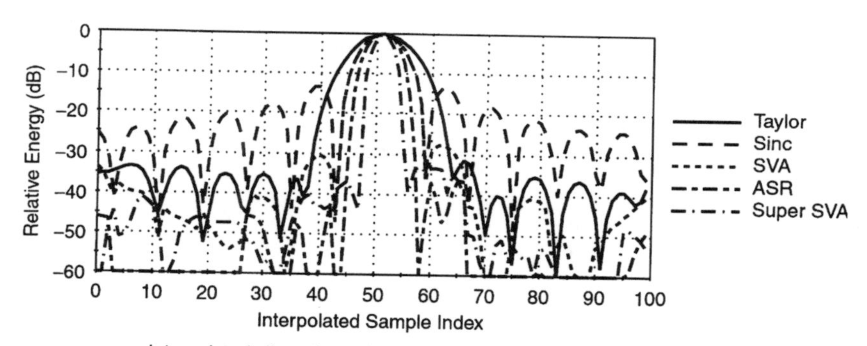

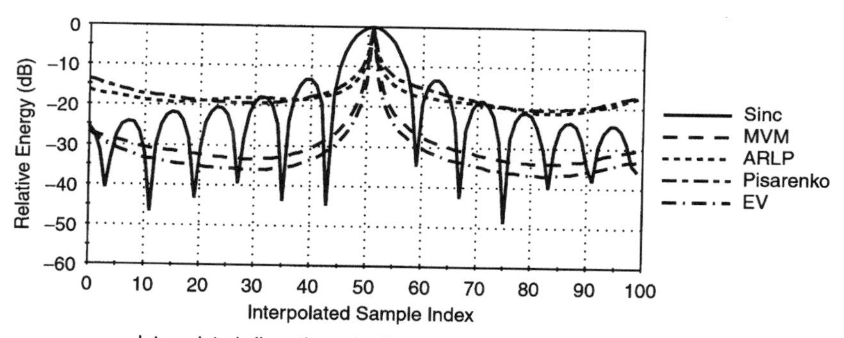

Figure 8.9 PSFs for spectral estimation techniques. (Reproduced from [4], Figures 17 and 18, copyright 1998, IEEE.)

of the other methods are lower and quite irregular. The bottom portion of the figure shows the results of several "hard" methods, compared with sinc. All have mainlobes much narrower than those previously discussed. ARLP and Pisarenko have relatively high sidelobes; MVM and EV have lower and very smooth sidelobes.

DeGraaf [4] presents images of the same simulated scene, with target-to-clutter ratio of 13 dB, processed via many different algorithms. He also includes six actual SAR images of two ships at a dock near Toledo, Ohio, processed according to different algorithms, and provides a detailed discussion of those results.

8.2.11 Comment on Spectral Estimation Techniques

As the name implies, spectral estimation involves estimating signal data. In most cases of interest, such estimation may be justified, but there may be pathological cases where it is not. The user should be appropriately cautious, especially where fine details of images are concerned.

References

[1] Carrara, W. G., R. S. Goodman, and R. M. Majewski, *Spotlight Synthetic Aperture Radar*, Norwood, MA: Artech House, 1995.

[2] Brigham, E. O., *The Fast Fourier Transform and Its Applications*, Englewood Cliffs, NJ: Prentice Hall, 1988.

[3] Wolfe, W. L., and G. Zissis (eds.), *The Infrared Handbook*, rev. ed, Ann Arbor, MI: Environmental Research Institute of Michigan, 1989.

[4] DeGraaf, S. R., "SAR Imaging via Modern 2-D Spectral Estimation Methods," *IEEE Trans. Image Processing*, Vol. 7, No. 5, May 1998, pp. 729–761.

[5] Strang, G., *Linear Algebra and Its Applications*, 3rd ed., San Diego: Harcourt Brace Jovanovich, 1988.

[6] Korn, G. A., and T. M. Korn, *Mathematical Handbook for Scientists and Engineers*, New York: McGraw-Hill, 1961.

[7] Benitz, G. R., "High-Definition Vector Imaging," *Lincoln Laboratory J.*, Vol. 10, No. 7 (Special issue on Superresolution), 1997, pp. 147–170.

[8] Stankwitz, H. C., R. J. Dallaire, and J. R. Fienup, "Non-Linear Apodization for Sidelobe Control in SAR Imagery," *IEEE Trans. Aerospace and Electronic Systems*, Vol. 31, No. 1, Jan. 1995, pp. 267–279.

[9] Hecht, E., and A. Zajac, *Optics*, Reading, MA: Addison-Wesley, 1979.

[10] Jacquinot, P., and B. Roizen-Dossier, "Apodization," in E. Wolf (ed.), *Progress in Optics*, Vol. 3, Amsterdam: North-Holland, 1964, p. 29.

[11] Stankwitz, H. C., and M. R. Kosek, "Super-Resolution for SAR/ISAR RCS Measurement Using Spatially Variant Apodization (Super-SVA)," *Proc. Antenna Measurement and Techniques Association (AMTA) Symp.*, Williamsburg, VA, Nov. 1995.

[12] Stankwitz, H. C., and M. R. Kosek, "Sparse-Aperture Fill for SAR Using Super-SVA," *Proc. 1996 IEEE National Radar Conference*, Ann Arbor, MI, 1996, pp. 70–75.

Problems

Problem 8.1

a. Using a computer, verify the principle shown in Figure 8.1, that is, choose a function in the signal domain, compute the $|\text{FFT}|^2$ in the image domain, and verify that adding $M/2$ zeroes before and after the signal results in a more finely sampled version of the image domain function.

b. Revise the equations in Figure 8.1 to make them appropriate for the case where N is odd.

Problem 8.2

Verify (8.24), that is, for a one-dimensional image, $\mathbf{x}(r) = \mathbf{W}^{\mathbf{H}}(r)\mathbf{X}$ corresponds to performing an FFT on the signal data $\mathbf{X}$. (Note: This extends to two- and three-dimensional images.)

Problem 8.3

Verify (8.26), that is, for an unweighted, unpadded one-dimensional image of a point target that lies exactly on a pixel location, the image value is zero for all other pixel locations.

Problem 8.4

Verify (8.35) by substituting (8.34) into (8.33), that is, one-dimensional ASR can be performed by first forming an unweighted image, then modifying each pixel by computing a complex correction factor that is a function of M pixels on each side.

9

Target Recognition in SAR/ISAR Imagery

Chapter 4 summarized the theory of detection of a target in noise and/or clutter, assuming that the result of a single measurement is a single voltage or test statistic to be compared with a threshold to determine whether the target should be declared present or absent. This chapter discusses an extension of that theory to the case in which several complex voltages are considered simultaneously, in the presence of noise or clutter, and a decision is made regarding whether that set of complex voltages represents a target. The set of complex voltages (often called cells) can represent pixels in a complex SAR/ISAR image. More generally, it can represent any set of complex voltages collected by a radar, including range versus azimuth for a real-beam radar, doppler bin versus angle, polarimetric or interferometric dimensions (Chapter 7), and so forth.

We first consider the theoretical problem of deciding whether a set of received complex voltages does or does not represent a target of a given type. In reality, even if we know the target type (e.g., a certain type of vehicle), its representation would depend on its orientation (azimuth and elevation of the radar in target-centered coordinates); whether it is articulated (i.e., whether a portion of it is moved relative to its "normal state," such as an open door or hatch); whether it is partially obscured by foliage, nets, weather, or other media; whether it is dirty; and so forth. Although actual automatic target recognition (ATR) must contend with those real-world factors, Section 9.2 makes the (unrealistic) assumption that the signature of the postulated target, versus cell, is known exactly, except for an overall multiplicative constant (positive scalar) corresponding to the overall target signature strength.

Because applying the exact theory to actual targets in a SAR/ISAR image is extremely difficult, Sections 9.3 and 9.4 consider some less theoretically exact procedures currently in use.

9.1 Constant False-Alarm Ratio

Chapter 4 assumed that the probability density distributions for interference (noise and clutter) and signal (target return) were constant. In many cases of interest, the interference background varies with space or time. In such cases, we usually utilize constant false-alarm ratio (CFAR) processing [1]. The threshold varies with the interference level in such a way as to keep the probability of false alarm constant. In cell-averaging CFAR, the threshold for testing for a target at a particular cell is determined by evaluating the mean value of nearby cells. Details are provided in [1, Sec. 5.1].

9.2 The Adaptive Matched Filter

The ideal target detection procedure is based on the adaptive matched filter (AMF), which is discussed extensively in the literature ([2–5] and many references therein). We begin by quoting (with slight changes in notation) the seminal theorem proved by Brennan and Reed in 1973 [2, p. 241]:

> Assume a radar (or sonar) transmits a waveform and receives n space-time samples. For noise alone, say hypothesis H_0, the receiver observes the vector $\mathbf{Z} = \mathbf{N}$, where $\mathbf{N}$ is [an] n-component column vector. . . . For signal plus noise, say hypothesis H_1, the receiver observes the vector $\mathbf{Z} = \mathbf{S} + \mathbf{N}$, where $\mathbf{S}$ is [another] n-component column vector. . . . Let $\mathbf{W}$ be the filter vector, [an] n-component column vector of weights. . . . Then the response of the filter to the received observables is
>
> $$x = \sum_{k=1}^{n} w_k^* z_k = \mathbf{W}^{\mathbf{H}}\mathbf{Z}$$
>
> where w_k and z_k are the kth components of $\mathbf{W}$ and $\mathbf{Z}$ respectively, and $\mathbf{W}^{\mathbf{H}}$ denotes the [hermitian adjoint, or conjugate transpose] of vector $\mathbf{W}$. Finally, suppose that the components of $\mathbf{N}$ are jointly distributed Gaussian variates, and that the covariance matrix is $\mathbf{M} = E(\mathbf{N}\mathbf{N}^{\mathbf{H}})$. . . . With the above definitions and assumptions,

the filter which gives the maximum probability of detection P_D for a fixed probability of false alarm P_{FA} is $\mathbf{W} = \beta \mathbf{M}^{-1}\mathbf{S}$, where β is a nonzero complex number.

Brennan and Reed go on to derive P_D as a function of **S**, **M**, and P_{FA} and show that the filter also maximizes the SNR.

Following [2], we present the proof of that theorem, as follows. (Table 9.1 is a very brief summary of relevant matrix types [6]; also see Problem 9.1.) For a general filter **W**, the test statistic is defined as the filter output $|x| = |\mathbf{W}^{\mathbf{H}}\mathbf{Z}|$, which is to be compared with a threshold (determined by P_{FA}) to arrive at the declaration of H_1 or H_0. The expected value of the observed return is the expected value of the target signal:

$$E(x) = \sum_{k=1}^{n} w_k^* E(z_k) = \sum_{k=1}^{n} w_k^* E(s_k) = \mathbf{W}^{\mathbf{H}}\mathbf{S} \tag{9.1}$$

The expected variance of the noise is (Problem 9.1)

$$\begin{aligned} \sigma^2 &= E(|\mathbf{W}^{\mathbf{H}}\mathbf{N}|^2) = E[(\mathbf{W}^{\mathbf{H}}\mathbf{N})(\mathbf{W}^{\mathbf{H}}\mathbf{N})^*] = E(\mathbf{W}^{\mathbf{H}}\mathbf{N}\mathbf{N}^{\mathbf{H}}\mathbf{W}) = \mathbf{W}^{\mathbf{H}}E(\mathbf{N}\mathbf{N}^{\mathbf{H}})\mathbf{W} \\ &= \mathbf{W}^{\mathbf{H}}\mathbf{M}\mathbf{W} \end{aligned} \tag{9.2}$$

where **M** is the noise covariance matrix. **M** is hermitian: $\mathbf{M} = \mathbf{M}^{\mathbf{H}}$, and $\mathbf{W}^{\mathbf{H}}\mathbf{M}\mathbf{W}$ is a positive real number (Problem 9.2). We want to find **W** such that the SNR

$$SNR = \frac{|\mathbf{W}^{\mathbf{H}}\mathbf{S}|^2}{\mathbf{W}^{\mathbf{H}}\mathbf{M}\mathbf{W}} \tag{9.3}$$

is maximized.

Table 9.1
Types of Matrices

Real Matrices	**Complex Matrices**
Symmetric: $\mathbf{A}^{\mathbf{T}} = \mathbf{A}$ Orthogonal: $\mathbf{A}^{\mathbf{T}} = \mathbf{A}^{-1}$	Hermitian: $\mathbf{A}^{\mathbf{H}} = \mathbf{A}$ Unitary: $\mathbf{A}^{\mathbf{H}} = \mathbf{A}^{-1}$

Notes: $\mathbf{A}^{\mathbf{T}}$ = transpose of **A**; $\mathbf{A}^{-1}$ = inverse of **A**; $\mathbf{A}^{\mathbf{H}}$ = hermitian adjoint (conjugate transpose) of **A**. A real hermitian matrix is symmetric; a real unitary matrix is orthogonal.

Because **M** is positive definite [6] as well as hermitian, it has a unique positive definite hermitian square root [2], which we denote as $\mathbf{D} = \mathbf{M}^{1/2}$. Denoting $SNR \equiv \alpha^2$, we have

$$\alpha^2 \mathbf{W}^{\mathrm{H}}\mathbf{M}\mathbf{W} = |\mathbf{W}^{\mathrm{H}}\mathbf{S}|^2 = |\mathbf{W}^{\mathrm{H}}\mathbf{D}\mathbf{D}^{-1}\mathbf{S}|^2 = \left|\sum_{j=1}^{n} a_j b_j\right|^2 \tag{9.4}$$

The inverse $\mathbf{D}^{-1}$ is also hermitian (Problem 9.3). Then

$$\begin{aligned} a_j &= \sum_{k=1}^{n} w_k^* D_{kj} \\ b_j &= \sum_{k=1}^{n} D_{jk}^{-1} S_k \end{aligned} \tag{9.5}$$

We now apply the Schwarz inequality [6, p. 147]:

$$\left|\sum_{j=1}^{n} a_j b_j\right|^2 \le \left(\sum_{j=1}^{n} |a_j|^2\right)\left(\sum_{j=1}^{n} |b_j|^2\right) \tag{9.6}$$

Thus,

$$\begin{aligned} \alpha^2(\mathbf{W}^{\mathrm{H}}\mathbf{M}\mathbf{W}) &\le (\mathbf{W}^{\mathrm{H}}\mathbf{D})(\mathbf{D}\mathbf{W})(\mathbf{S}^{\mathrm{H}}\mathbf{D}^{-1})(\mathbf{D}^{-1}\mathbf{S}) \\ &= (\mathbf{W}^{\mathrm{H}}\mathbf{M}\mathbf{W})(\mathbf{S}^{\mathrm{H}}\mathbf{M}^{-1}\mathbf{S}) \end{aligned} \tag{9.7}$$

and

$$\alpha^2 \le \mathbf{S}^{\mathrm{H}}\mathbf{M}^{-1}\mathbf{S} \tag{9.8}$$

The solution for **W** is

$$\mathbf{W} = \beta \mathbf{M}^{-1}\mathbf{S} \tag{9.9}$$

For that value of **W**, the inequality (9.8) becomes an equality, and the *SNR* is maximized. From (9.3)

$$\alpha^2 = \frac{|\mathbf{S}^{\mathrm{H}}\mathbf{M}^{-1}\mathbf{S}|^2}{(\mathbf{S}^{\mathrm{H}}\mathbf{M}^{-1})\mathbf{M}(\mathbf{M}^{-1}\mathbf{S})} = \frac{(\mathbf{S}^{\mathrm{H}}\mathbf{M}^{-1}\mathbf{S})^2}{\mathbf{S}^{\mathrm{H}}\mathbf{M}^{-1}\mathbf{S}} = \mathbf{S}^{\mathrm{H}}\mathbf{M}^{-1}\mathbf{S} \tag{9.10}$$

which is the maximum value of α^2. We can replace $|\mathbf{S}^\mathbf{H}\mathbf{M}^{-1}\mathbf{S}|$ with $\mathbf{S}^\mathbf{H}\mathbf{M}^{-1}\mathbf{S}$ because the latter quantity is a positive real number (per Problems 9.2 and 9.3).

The test can be normalized to have a CFAR property [5]:

$$\frac{|\mathbf{S}^\mathbf{H}\hat{\mathbf{M}}^{-1}\mathbf{Z}|^2}{\mathbf{S}^\mathbf{H}\hat{\mathbf{M}}^{-1}\mathbf{S}} \underset{H_0}{\overset{H_1}{\gtrless}} T \tag{9.11}$$

where T is a chosen threshold and $\hat{\mathbf{M}}$ signifies an estimate of the noise covariance. The numerator is the output of the AMF, tuned to the target of interest, analogous to a $(\text{SNR})^2$ at the region of interest. The denominator is analogous to a SNR evaluated using the signal from the target of interest and the local noise value. That is the usual form of the AMF CFAR test [5]. The noise covariance matrix must somehow be estimated from other data in the same data set or from other, similar data sets.

Consider a complex SAR image that may or may not contain targets of interest. If (a) the expected target signal **S** could somehow be precisely known (where the vector components represent image pixels), (b) the I and Q values of the interference—noise plus clutter—were Gaussian (and the magnitude were therefore Rayleigh; see Section 4.1), (c) the interference covariance matrix could somehow be adequately estimated, and (d) the processor could perform the requisite matrix inversions and other iterations in a reasonable time, then ATR could, in principle, be performed with the AMF. Different regions of the image would be evaluated for the presence or absence of a target. The local noise covariance would be estimated from the surrounding pixels, and the denominator of (9.11) would be computed. For each candidate image chip, the pixel values would constitute **Z**; the ratio in (9.11) would be computed and compared with a threshold in a CFAR procedure. However, we do not, in general, know the target orientation, whether it is articulated, whether it is partially obscured by foliage, nets, or weather, or whether it is dirty. Therefore, this procedure is beyond realistic computation, even for individual images. The next section discusses actual methods used by ATR researchers.

9.3 Automatic Target Recognition in SAR Imagery

Many ATR techniques for SAR have been developed since about 1980. This section focuses on the techniques that were developed by the Massachusetts Institute of Technology (MIT) Lincoln Laboratory, Lexington, Massachusetts, under the leadership of Dr. Leslie Novak.

Novak and his colleagues have published extensively concerning ATR [7–12 and references therein]. They developed an algorithm suite that has been effective in tests using actual complicated cultural targets. The Lincoln SAR ATR system is described in [10] and [12], which are used as the basis for the following discussion. The Lincoln system comprises three general steps: prescreening, discrimination, and classification.

9.3.1 Prescreening

The relatively simple prescreener [9] (sometimes called the detector) uses a two-parameter CFAR algorithm. At each pixel, the algorithm compares a test statistic with a threshold and makes a preliminary target-declaration decision:

$$\frac{x - \mu_c}{\sigma_c} \underset{H_0}{\overset{H_1}{\gtrless}} T \tag{9.12}$$

where x is the intensity (power) of the pixel under consideration (test pixel), μ_c and σ_c are estimates of the local clutter mean and standard deviation, respectively, and T is a threshold. The estimates of the mean and standard deviation are computed from a square annular region surrounding the pixel of interest but far enough away from it to preclude the possibility of a target of interest occupying both the test pixel and a portion of the annulus. For Rayleigh clutter, the procedure will be CFAR [9,13]. However, even when that condition is not met, the procedure can still be useful and is still generally termed CFAR. The result is a set of pixels identified as potentially part of targets. In many cases, an individual extended target will result in several selected pixels. The selected pixels are clustered to form a region of interest (ROI) approximately the size of a target. The result of the prescreener step is a set of ROIs in the image, each of which may or may not represent a target.

9.3.2 Discrimination

The discrimination step computes a series of features for each ROI. The values of those features are then compared with the expected corresponding values for an actual target, and a decision is made as to whether the ROI is or is not similar enough to a target to be further considered as such.

A highly effective discriminator, described in 1998, is the quadratic gamma discriminator [11]. A stencil based on two-dimensional gamma functions is used to analyze the characteristics of the ROI and to compare its characteristics with those of real targets. Another discrimination algorithm (described in 1997) involves a multiresolution approach, based on variations in ROI characteristics as image resolution is varied from coarse to fine [9].

Earlier discrimination algorithms [9] involved several of the following features based on a principal-object region *P*, essentially the bright section near the ROI center. Discrimination is performed by setting up a multidimensional feature space and measuring the distance, in that space, between the feature vectors corresponding to the observed ROI and to an image of the target-type of interest.

9.3.2.1 Size Features

- *Mass* is the number of pixels in *P*.
- *Diameter* is the length of the diagonal of the smallest rectangle (either horizontally or vertically oriented) that encloses *P*.
- *Rotational inertia* is the normalized second mechanical moment about the center of mass of *P* (analogous to the moment of inertia of a massive object).

9.3.2.2 Contrast Features

- *Peak CFAR* is the maximum value of the pixels within *P*.
- *Mean CFAR* is the average value of the pixels within *P*.
- *Percent bright CFAR* is the percentage of pixels in *P* that exceed a certain value.

9.3.2.3 Textural Features (Based on a Target-Sized Region *T* in the ROI)

- *Standard deviation* is the standard deviation of the pixel values in *T*.
- *Ranked fill ratio* is the percentage of power contained in the brightest 5% of the pixels in *T*.
- *Fractal dimension* is discussed in detail next.

Fractal dimension is defined as follows [14]. Consider a set of square boxes of dimension ϵ by ϵ. We define M_ϵ as the minimum number of such boxes required to cover an image region S, and d as a characteristic distance across S, typically the length of the greatest line segment that can be drawn in the region. Then the fractal dimension of S is

$$\dim(S) = \lim_{\epsilon \to 0} \frac{\log(M_\epsilon)}{\log\left(\dfrac{d}{\epsilon}\right)} \tag{9.13}$$

Three examples follow.

- S is a single point. In this case, one box always covers S, and $M_\epsilon = 1$. We require that d cannot be smaller than a single pixel; thus, $\dim(S) = 0$.
- S is a line segment of length L, either straight or "not too curved." Then (Problem 9.4)

$$\dim(S) = \underset{\epsilon \to 0}{Lim} \frac{\log(L/\epsilon)}{\log(d/\epsilon)} = \underset{\epsilon \to 0}{Lim} \left(1 + \frac{\log(L/d)}{\log(d/\epsilon)}\right) = 1 \quad (9.14)$$

- S is an extended, simply connected area without "too many" wiggles in its boundary, of area L^2. Then (Problem 9.5)

$$\dim(S) = \underset{\epsilon \to 0}{Lim} \frac{\log[(L/\epsilon)^2]}{\log(d/\epsilon)} = \underset{\epsilon \to 0}{Lim} \left[2 + 2\frac{\log(L/d)}{\log(d/\epsilon)}\right] = 2 \quad (9.15)$$

In summary, fractal dimensions of some simple shapes are: dim(point) = 0, dim(line) = 1, dim(area) = 2. If the region has gaps, "peninsulas," complicated "wiggles" in its boundaries, and so on, the fractal dimension may increase beyond 2 or assume a noninteger value [15, 16].

9.3.2.4 Polarimetric Features

Novak et al. have also considered ATR using polarimetric SAR imagery [7, 8]. In addition to the nine features already described, they use three additional features (originally developed by Dr. Stuart DeGraaf of ERIM, Ann Arbor, Michigan) [8] appropriate to the additional polarimetric information. These features are based on the facts that the PSM is quite different for dihedrals and trihedrals (Section 3.1.4), and, for frequencies above 1 GHz, few dihedrals exist in natural clutter. Thus, if the PSM of a patch of clutter corresponds closely to a dihedral, then the clutter patch is relatively likely to represent a cultural object.

9.3.3 Classification

The discrimination step results in a set of relatively small image regions, called chips, which are relatively likely to contain targets. Lincoln Laboratory has developed two classifiers, shown in Figure 9.1. One tests the entire chip at fine resolution (0.5m by 0.5m) using a classifier based on the HDVI superresolution technique (Section 8.2.5) followed by a computation of the mean square error

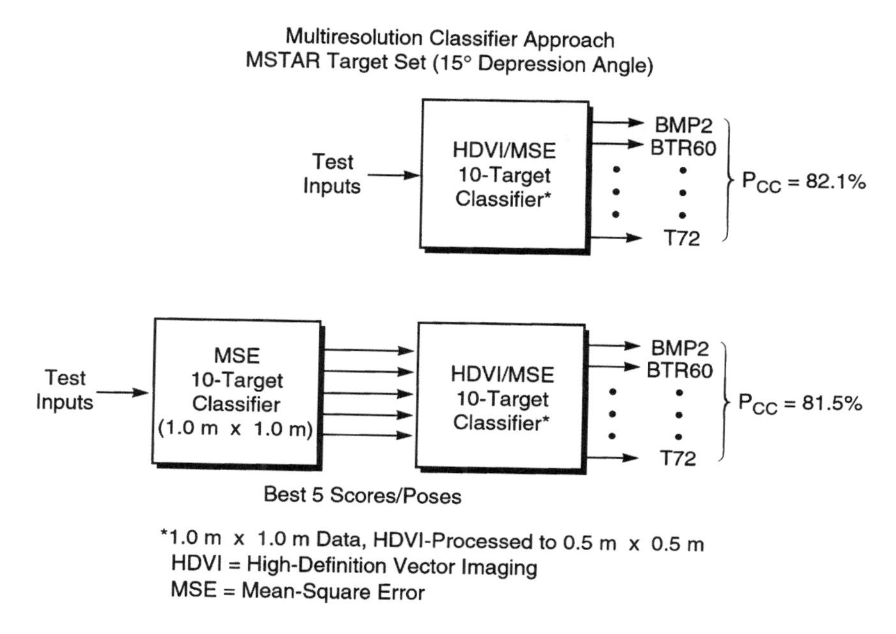

Figure 9.1 Two HDVI-MSE classifiers. (After [12] copyright 1999 IEEE.)

(MSE) between the image chip and a preobtained image of the target being sought. The algorithm tests for targets at every possible orientation (typically every 5 degrees of aspect angle). The other classifier, less optimum but much faster, uses an MSE stage on a reduced-resolution chip (1.0m by 1.0m) to determine likely target orientations, then an HDVI-MSE stage on the fine-resolution image (0.5m by 0.5m), limiting the search space by testing at the reduced set of orientations.

Another classification procedure is to construct a many-dimensional feature space. The features of a particular target cause the target to be associated with a particular location in feature space. The space is divided into regions such that, if a target location is in a particular region, then the probability is relatively high that the target is of a particular type. One can then formulate a classification tree, a series of yes-no questions such that their answers specify the region of feature space and therefore, with high probability, the target type [17].

The several probabilities of misclassification are enumerated in terms of a matrix called (unfortunately, in this author's opinion) a confusion matrix. The rows correspond to the true target, and the columns correspond to the declaration by the algorithm suite. Tables 9.2 and 9.3 show (for the SAIP program see Section 9.4) the results for conventionally processed (i.e., no

Table 9.2
SAIP Classifier Performance Results (Confusion Matrix)
HH Data, 0.3m-by-0.3m Resolution

Specific Vehicle \ Vehicle Type	Number of Targets Classified as										
	BMP2	**BTR60**	**BTR70**	**M109**	**M110**	**M113**	**M1**	**M2**	**M548**	**T72**	**Unknown**
BMP2#1	255										
BMP2#2	251							1			3
BMP2#3	251							1		2	2
BTR60		256									
BTR70			256								
M109				256							
M110					256						
M113						256					
M1							256				
M2#1								256			
M2#2								251			4
M2#3								252			3
M548									256		
T72#1										256	
T72#2										216	39
T72#3										247	4
HMMWV	12	6				37		2			187
M35									4		251

Source: Reproduced from [12] copyright 1999 IEEE.

Table 9.3
SAIP Classifier Performance Results (Confusion Matrix)
HH Data, 1.0m-by-1.0m Resolution

Specific Vehicle \ Vehicle Type	Number of Targets Classified as										
	BMP2	**BTR60**	**BTR70**	**M109**	**M110**	**M113**	**M1**	**M2**	**M548**	**T72**	**Unknown**
BMP2#1	255							1			
BMP2#2	129	15	21	8	1	39	3	15		18	6
BMP2#3	135	7	10	5	2	43		25		18	11
BTR60		255				1					
BTR70			256								
M109				256							
M110					256						
M113	1					253					2
M1				1		1	253				
M2#1						1		255			
M2#2	14	3	2	37	14	36	5	121		16	7
M2#3	21	6		18	6	47	9	118		27	4
M548									256		
T72#1						1				255	
T72#2	7	12	3	19	9	23	17	32	1	94	38
T72#3	19	8	6	23	5	12	14	19		127	18
HMMWV	48	7	4	4		145		13	1		22
M35	8			12	9	29	4	4	13	1	175

HDVI) data of resolution (i.e., 3-dB IPR width) 0.3m and 1.0m; finer resolution, of course, leads to better classification. Tables 9.4 and 9.5 show the improvement due to addition of the HDVI step. Table 9.6 shows the results from a multiresolution classifier for 1.0m imagery; although the probability of correct classification drops slightly (see Figure 9.1), the computational speed increases substantially, implying that multiresolution is a valuable algorithm when large quantities of imagery must be evaluated.

9.4 Semiautomated Image Intelligence Processing

Because radar waves readily penetrate clouds, SAR is an extremely important technique for practical reconnaissance. For example, the Global Hawk UAV "is projected to collect in one day enough data sampled at a resolution of 1.0m by 1.0m to cover 140,000 km^2 (roughly the size of North Korea)" [10]. Current attempts to perform ATR on such volumes of imagery involve a combination of less than theoretically optimum automated techniques plus the admittedly splendid processing capabilities of the human eye and brain. Such techniques are termed semiautomated. Automated and semiautomated techniques were developed during the 1980s and 1990s. The present discussion focuses on the semiautomated image intelligence (IMINT) processing (SAIP) effort [10]. The program combines many algorithms developed by a number of contractors.

Figure 9.2 summarizes the SAIP ATR system. Image data are first analyzed by the CFAR prescreener to produce candidate targets (ROIs). Several operations then occur. A terrain delimitation step compares the known region of the image with a priori knowledge about the terrain and eliminates all ROIs that occur in regions where targets of interest (e.g., tanks) could not plausibly be located (e.g., lakes or high mountains). An object-level change detection step compares the image with previous images taken of the same region and gives greater weight to ROIs that occur at the site of previously detected targets. A cultural clutter identification step eliminates ROIs that are coincident with known cultural (i.e., human-made) objects that have signatures similar to targets. Spatial clustering looks for object grouping and gives greater weight to objects in groups that are similar in structure to groups of targets of interest (e.g., military vehicles typically travel and are parked in formations).

The next steps involve feature extraction, detection characterization, discrimination, thresholding, and priority ordering of candidate ROIs. The discrimination is based on some combination of the techniques discussed in Section 9.3.2. At this stage, the process makes use of SAR image chips of targets of interest, taken under controlled conditions. For the research described

Table 9.4
SAIP Classifier Performance Results (Confusion Matrix)
HDVI Processed, HH Data, 0.3m-by-0.3m Resolution

Specific Vehicle \ Vehicle Type	Number of Targets Classified as										
	BMP2	**BTR60**	**BTR70**	**M109**	**M110**	**M113**	**M1**	**M2**	**M548**	**T72**	**Unknown**
BMP2#1	256										
BMP2#2	202										3
BMP2#3	255							1			
BTR60		256									
BTR70			256								
M109				256							
M110					256						
M113						256					
M1							255				
M2#1								256			
M2#2								251	2		2
M2#3								248	3		3
M548									256		
T72#1										256	
T72#2										232	23
T72#3										245	6
HMMWV	4	1				13		3			223
M35								1	6		248

Source: Reproduced from [12] copyright 1999 IEEE.

Table 9.5
SAIP Classifier Performance Results (Confusion Matrix)
HDVI Processed, HH Data, 1.0m-by-1.0m Resolution

Specific Vehicle \ Vehicle Type	Number of Targets Classified as									
	BMP2	**BTR60**	**BTR70**	**M109**	**M110**	**M113**	**M1**	**M2**	**M548**	**T72**	**Unknown**
BMP2#1	256										
BMP2#2	202	4	2			1		17		10	19
BMP2#3	195	2	4	2		4		22		17	10
BTR60		254									2
BTR70			256								
M109				256							
M110					256						
M113						255					1
M1							255				
M2#1								256			
M2#2	4		2	13	11	2	1	188		24	10
M2#3	8	1	2	16	5	3	1	180		27	13
M548									256		
T72#1										256	
T72#2	3			5		1	7	10		190	39
T72#3	5	1		3	4		3	10		211	14
HMMWV	25	8	8	1		101		8			93
M35	1			3	1	6			14		230

Source: Reproduced from [12] copyright 1999 IEEE.

Table 9.6
SAIP Multiresolution Classifier Performance Results (Confusion Matrix)
HH Data, 1.0m-by-1.0m Resolution

Specific Vehicle \ Vehicle Type	Number of Targets Classified as									
	BMP2	BTR60	BTR70	M109	M110	M113	M1	M2	M548	T72	Unknown
BMP2#1	169	3	3	1				5		5	9
BMP2#2	131	2	8			4		16		21	14
BMP2#3	122	3	3	1	1	2		25		22	17
BTR60		171	4							1	19
BTR70		5	185					1		3	2
M109				185				2			9
M110				1	181		1	3			8
M113						162					34
M1	1						183			1	12
M2#1				4	4	4		172		1	5
M2#2	4			13	10	2	1	133		26	21
M2#3		1		9	6	2		145		15	14
M548						1			189		6
T72#1	3	1		1			2	2		180	7
T72#2	3	1		1	1	1	5	7		133	43
T72#3		1	1	2	2		1	9		152	23
HMMWV	15	9	7	3		71		5			84
M35				1		2			8		185

Source: Reproduced from [12] copyright 1999 IEEE.

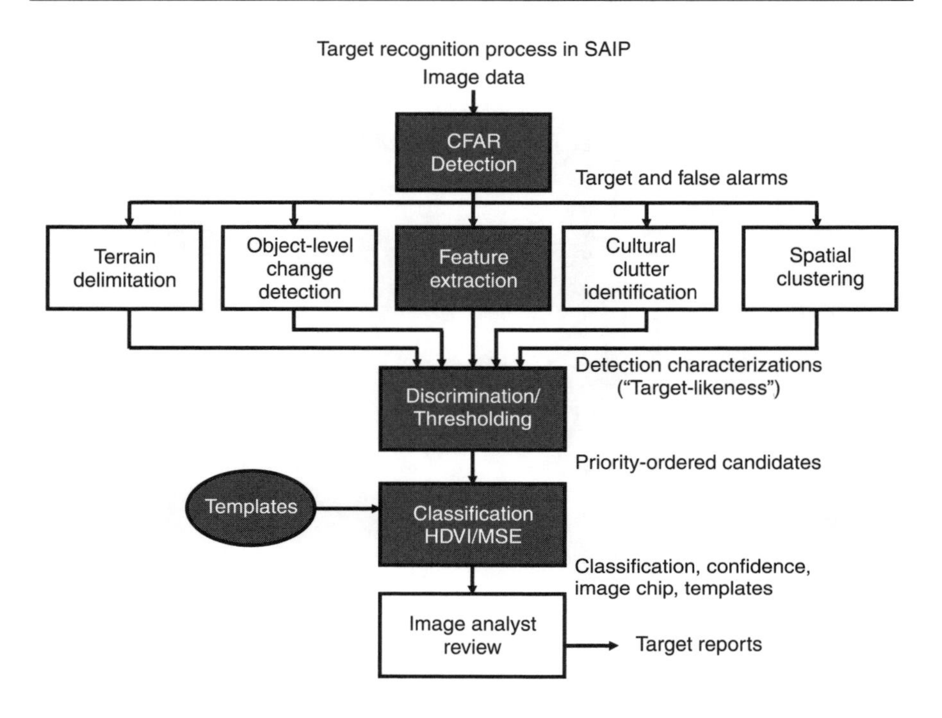

Figure 9.2 Simplified block diagram of elements of SAIP system; dark boxes represent Lincoln Laboratory ATR algorithm. (After [12] copyright 1999 IEEE.)

in [12], the images were from the Moving and Stationary Target Acquisition and Recognition (MSTAR) program, taken by the Sandia X-band HH-polarized spotlight SAR. They include images of non-obscured military targets taken over 360 degrees of aspect angle, plus about 30 km^2 of natural and cultural clutter stripmap imagery, for false-alarm measurement. For each target, 72 templates were created, that is, over every 5 degrees of aspect. The classifier (as described in Section 9.3.3) was developed using some images as training data, then tested using other images as test data.

Figure 9.3 shows a receiver operating characteristic (ROC) curve illustrating P_D for three stages of the overall SAIP process. The abscissa is expressed in false alarms per square kilometer, a more useful measure than P_{FA}. The two quantities are related by

$$FA/\text{km}^2 = P_{FA} \frac{(1{,}000\text{m})^2}{\delta_r \delta_{cr}} \tag{9.16}$$

where δ_r and δ_{cr} are the range and crossrange pixel sizes, respectively.

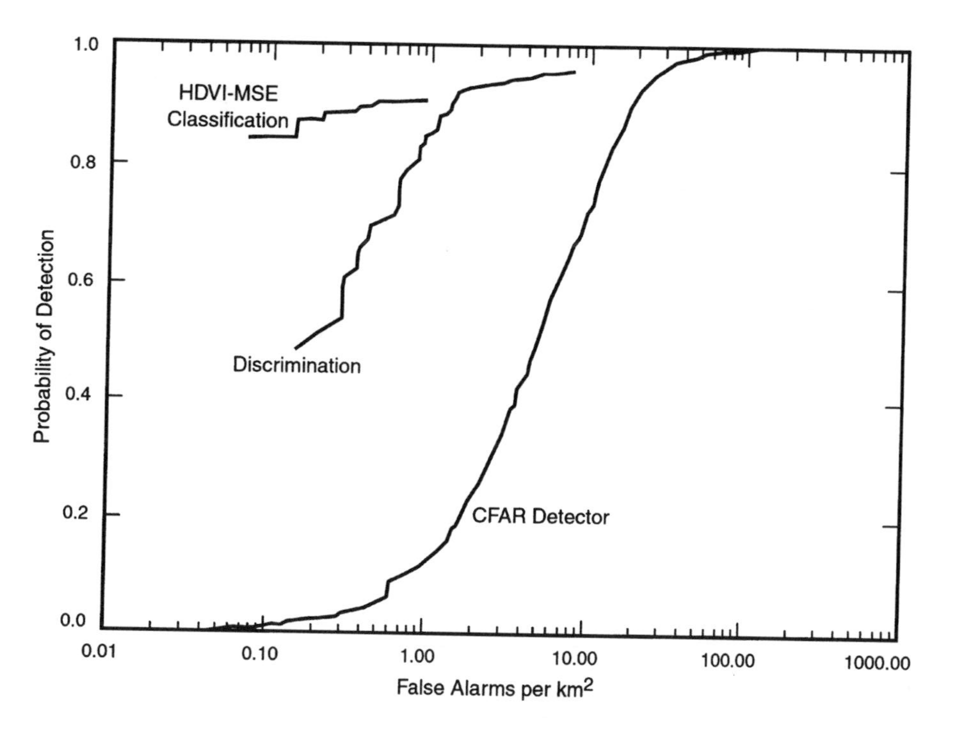

Figure 9.3 ROC performance curves for each element of the end-to-end ATR system. The HDVI-MSE classifier provided a probability of detection of greater than 0.8 with 0.1 false alarm/km^2. (After [12] copyright 1999 IEEE.)

As a final step (see Figure 9.2), the imagery is annotated with target cues and sent to image analysts (IAs), who make the final decision whether or not to consider the ROI a target. The information is then passed to the user.

9.5 Other Target-Recognition Algorithms

Rihaczek and Hershkowitz [18] have published a detailed description of a procedure for identifying fixed or moving targets in SAR or ISAR images, emphasizing the fact that in radar images, as opposed to optical images, the phase, as well as the magnitude, is available for each pixel. The authors call their technique complex-image analysis (CIA) and state (p. 14) that they "use the acronym CIA despite the fact that it is better known under a different meaning." They discuss in detail the expected complex response from single- and multiple-point targets, dihedrals, trihedrals, and curved surfaces; multipath; and human-made targets such as ground vehicles and aircraft. They also discuss the advantages and disadvantages of CIA versus more conventional target-recognition methods.

Borden [19] presents a mathematically oriented discussion of radar imaging of airborne targets for identification. He discusses such topics as improving one-dimensional downrange high-range-resolution (HRR) profiles by means of prolate-spheroidal wave expansion, Bayesian methods, and the CLEAN algorithm for sidelobe reduction [20; 21, Chap. 7]; formation of two-dimensional (ISAR) images via the projection-slice theorem and convolution-backprojection algorithm [22, 23]; and three-dimensional imaging (within some maximum range) by angle-of-arrival imaging using azimuth-elevation monopulse radar, considering the effects of glint (see Chapter 5). He also discusses the effects of radar wave polarization, multiple bounces on the target, and jet-engine modulation (JEM) [24].

References

[1] Minkler, G., and J. Minkler, *CFAR*, Baltimore: Magellan, 1990.

[2] Brennan, L. E., and I. S. Reed, "Theory of Adaptive Radar," *IEEE Trans. Aerospace and Electronic Systems*, Vol. AES-9, No. 2, March 1973, pp. 237–252.

[3] Reed, I. S., J. D. Mallett, and L. E. Brennan, "Rapid Convergence Rate in Adaptive Arrays," *IEEE Trans. Aerospace and Electronic Systems*, Vol. AES-10, No. 6, Nov. 1974, pp. 853–863.

[4] Robey, F. C., et al., "A CFAR Adaptive Matched Filter Detector," *IEEE Trans. Aerospace and Electronic Systems*, Vol. 28, No. 1, Jan. 1992, pp. 208–216.

[5] Goldstein, J. S., and I. S. Reed, "Theory of Partially Adaptive Radar," *IEEE Trans. Aerospace and Electronic Systems*, Vol. 33, No. 4, Oct. 1997, pp. 1309–1325.

[6] Strang, G. S., *Linear Algebra and Its Applications*, 3rd ed., San Diego: Harcourt Brace Jovanovich, 1988.

[7] Novak, L. M., M. C. Burl, and W. W. Irving, "Optimal Polarimetric Processing for Enhanced Target Detection," *IEEE Trans. Aerospace and Electronic Systems*, Vol. 29, No. 1, Jan. 1993, pp. 234–243.

[8] Novak, L. M., et al., "Effects of Polarization and Resolution on SAR ATR," *IEEE Trans. Aerospace and Electronic Systems*, Vol. 33, No. 1, Jan. 1997, pp. 102–116.

[9] Irving, W. W., L. M. Novak, and A. S. Willsky, "A Multiresolution Approach to Discrimination in SAR Imagery," *IEEE Trans. Aerospace and Electronic Systems*, Vol. 33, No. 4, Oct. 1997, pp. 1157–1169.

[10] Novak, L. M., et al., "The Automatic Target Recognition System in SAIP," *Lincoln Laboratory J.*, Vol. 10, No. 2, 1997 (Special Issue on Superresolution), pp. 187–202.

[11] Principe, J. C., et al., "Target Prescreening Based on a Quadratic Gamma Discriminator," *IEEE Trans. Aerospace and Electronic Systems*, Vol. 34, No. 3, July 1998, pp. 706–715.

[12] Novak. L. M., G. J. Owirka, and A. L. Weaver, "Automatic Target Recognition Using Enhanced Resolution SAR Data," *IEEE Trans. Aerospace and Electronic Systems*, Vol. 35, No. 1, Jan. 1999, pp. 157–175.

[13] Goldstein, G. B., "False Alarm Regulation in Log-Normal and Weibull Clutter," *IEEE Trans. Aerospace and Electronic Systems*, Vol. AES-9, No. 1, Jan. 1973, pp. 84–92.

[14] Burl, M. C., G. J. Owirka, and L. M. Novak, "Texture Discrimination in Synthetic Aperture Radar Imagery," *23rd Asilomar Conf. Signals, Systems, and Computers*, Pacific Grove, CA, 1989, pp. 399–404.

[15] Mandelbrot, B. B., *The Fractal Geometry of Nature*, San Francisco: W. H. Freeman, 1982.

[16] Barnsley, M., *Fractals Everywhere*, New York: Academic, 1988.

[17] Breiman, L., et al., *Classification and Regression Trees*, Belmont, CA: Wadsworth International Group, 1984.

[18] Rihaczek, A. W., and S. J. Hershkowitz, *Radar Resolution and Complex-Image Analysis*, Norwood, MA: Artech House, 1996.

[19] Borden, B., *Radar Imaging of Airborne Targets*, Philadelphia: Institute of Physics Publishing, 1999.

[20] Tsao, J., and B. D. Steinberg, "Reduction of Sidelobe and Speckle Artifacts in Microwave Imaging: The CLEAN Technique," *IEEE Trans. Antennas and Propagation*, Vol. 36, No. 4, 1988, pp. 543–556.

[21] Steinberg, B., and H. M. Subbaram, *Microwave Imaging Techniques*, New York: Wiley, 1991.

[22] Munson, D. C., J. D. O'Brien, and W. K. Jenkins, "A Tomographic Formulation of Spotlight-Mode Synthetic Aperture Radar," *Proc. IEEE*, Vol. 71, No. 8, 1983, pp. 917–925.

[23] Jakowatz, C. V., Jr., et al., *Spotlight-Mode SAR: A Signal-Processing Approach*, Boston: Kluwer Academic Publishers, 1996.

[24] Bell, M. R., and R. A. Grubbs, "JEM Modeling and Measurement for Radar Target Identification," *IEEE Trans. Aerospace and Electronic Systems*, Vol. 29, No. 1, Jan. 1993, pp. 73–87.

Problems

Problem 9.1

Show the following (see [6, p. 294]):

a. For a complex vector $\mathbf{V}$, $|\mathbf{V}|^2 = \mathbf{V}^{\mathbf{H}}\mathbf{V}$

b. For two complex matrices $\mathbf{A}$ and $\mathbf{B}$, $(\mathbf{AB})^{\mathbf{H}} = \mathbf{B}^{\mathbf{H}}\mathbf{A}^{\mathbf{H}}$

Problem 9.2

In (9.2), show that the expected value of the noise, $\mathbf{W}^{\mathbf{H}}\mathbf{M}\mathbf{W}$, is a positive real number (see [6, p. 294]).

Problem 9.3

Show that, if $\mathbf{M}$ is hermitian ($\mathbf{M} = \mathbf{M}^{\mathbf{H}}$), then $\mathbf{M}^{-1}$ is also hermitian.

Problem 9.4

Verify (9.14): For a line-segment target, the fractal dimension is

$$\dim(S) = \underset{\epsilon\to 0}{Lim}\frac{\log(L/\epsilon)}{\log(d/\epsilon)} = \underset{\epsilon\to 0}{Lim}\left[1 + \frac{\log(L/d)}{\log(d/\epsilon)}\right] = 1$$

Problem 9.5

Verify (9.15): For an area target, the fractal dimension is

$$\dim(S) = \underset{\epsilon\to 0}{Lim}\frac{\log[(L/\epsilon)^2]}{\log(d/\epsilon)} = \underset{\epsilon\to 0}{Lim}\left[2 + 2\frac{\log(L/d)}{\log(d/\epsilon)}\right] = 2$$

Part III:
Pulse-Doppler and MTI Radar

10

Pulse-Doppler Radar

A doppler radar is defined by [1] as "a radar which uses the doppler effect to determine the radial component of relative radar target velocity or to select targets having particular radial velocities." A pulse-doppler (or pulsed-doppler) (PD) radar is defined [1] as "a doppler radar that uses pulsed transmissions." PD radar is used extensively for detecting and characterizing moving targets.

10.1 Doppler Spectrum of a Sequence of Pulses

A modern PD radar typically observes the scene in a particular direction during a CPI long enough to collect a number of pulses, digitally records the I and Q of each pulse (providing amplitude and phase information), and performs an FT (see Section 2.1) of the complex returns, yielding the (doppler) frequency spectrum of the echoes, from which the LOS velocity spectrum can be determined.

This chapter considers monochromatic pulses, that is, those pulses formed by multiplying a single-frequency tone with a rectangular envelope of duration τ, with the resulting bandwidth $B \approx 1/\tau$. If modulated pulses and pulse compression (PC) are used, the PC is performed on each pulse as soon as it is received, and the doppler processing is performed for each range bin of interest.

We assume that the radar receives a sequence of pulses scattered from a point target. The pulses are assumed to have been transmitted with a constant PRF f_R and therefore constant PRI t_R (a single CPI). Such a simple waveform is nevertheless characterized by no fewer than four time scales:

- The period of the carrier signal, $t_0 = 1/f_0$, where f_0 is the carrier frequency;
- The pulse width τ;
- The time between pulses t_R (PRI);
- The total length of the CPI, or dwell time, T;

Morris and Harkness [2, pp. 46–49], summarize the Fourier spectrum of such a waveform, as it appears after doppler processing; their figures are reproduced here. Figure 10.1 illustrates the simple tone at frequency f_0, with the Fourier transform (FT) characterized by two delta functions at $\pm f_0$. Figure 10.2 shows the rectangular envelope of width τ; its FT is a sinc function with peak-to-first-null of $f_{pn} = 1/\tau$. Because a single pulse (two time scales) is formed by the product of those two functions, its FT (Figure 10.3) is described by the convolution of their individual FTs, consisting of the sinc function replicated at the positions of the delta functions.

Figure 10.4 illustrates an infinitely long sequence of pulses (three time scales). In the time domain, that may be considered as a comb function—an infinite sequence of delta functions—convolved with the signal of a single pulse. The FT of a comb of period t_R is another comb of period $1/t_R$ [3, p. 21]. Therefore, in the frequency domain the FT is another comb function multiplied by the FT of a single pulse (Figure 10.3). In the frequency domain, the lines of the comb are called PRF lines, because they are separated by a frequency equal to the PRF. In fact, we recognize the interval between two of those lines (the PRF) as corresponding to the unambiguous interval of doppler frequency (Section 1.15 and Section 4.2.4).

Finally, Figure 10.5 shows the fourth time scale, the dwell time T. Because that multiplies the previous time-domain signal, it appears as a convolution in the frequency domain. Its FT, $\text{sinc}(\pi t/T)$, is convolved with the delta functions in Figure 10.4, resulting in a broadening of each line to produce a set of peaks of finite width.

10.2 Phase Noise

Of particular concern in a PD radar is phase noise, noise due to the phase instability of the primary oscillator, which then propagates through the rest of the radar hardware. Phase noise effectively raises the level of interference that may reduce detectability of a moving target. It is discussed in detail in a number of references [4–9]. A brief overview is presented here.

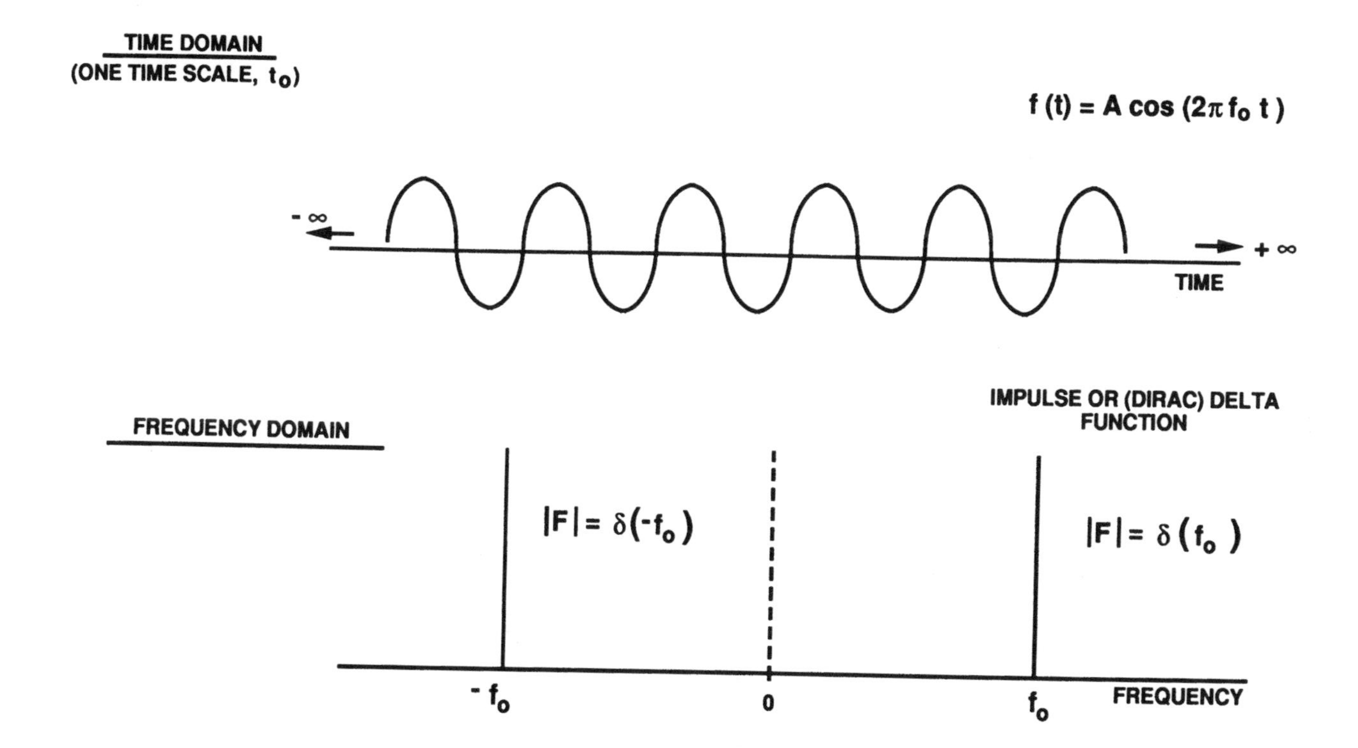

Figure 10.1 Waveform with one time scale (carrier period t_o). (Reproduced from [1], Figure 3.2, with permission of Artech House.)

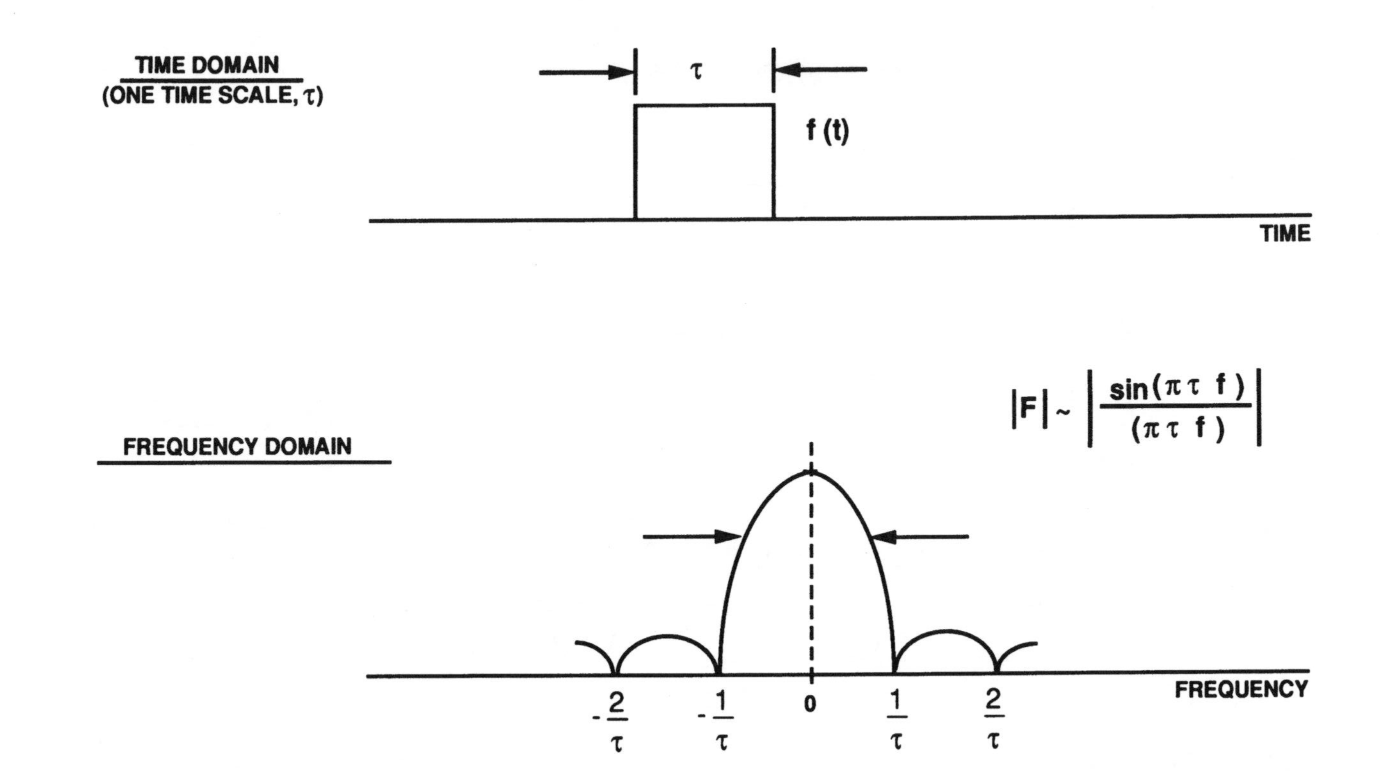

Figure 10.2 Waveform with one time scale (pulse width τ). (Reproduced from [1], Figure 3.3, with permission of Artech House.)

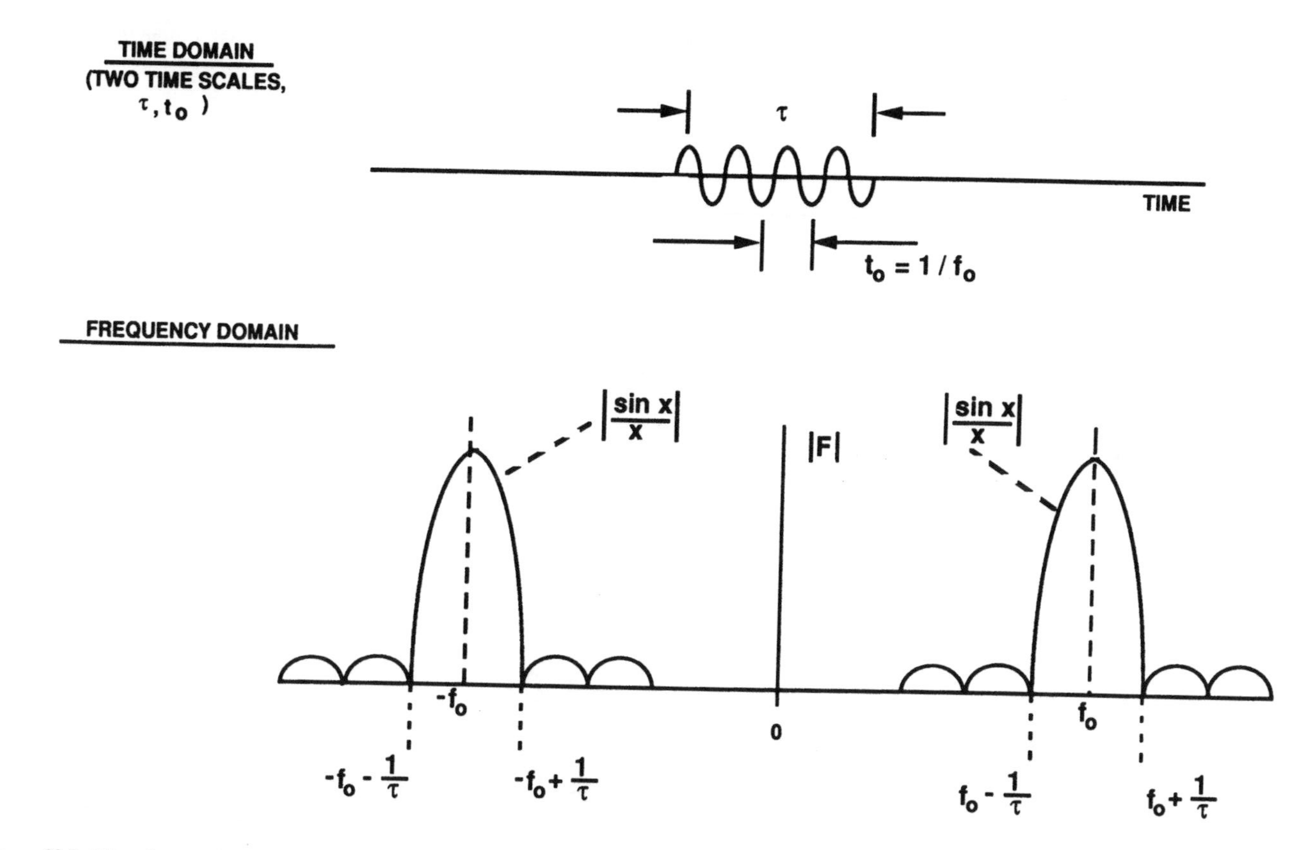

Figure 10.3 Waveform with two time scales (t_o and τ). (Reproduced from [1], Figure 3.4, with permission of Artech House.)

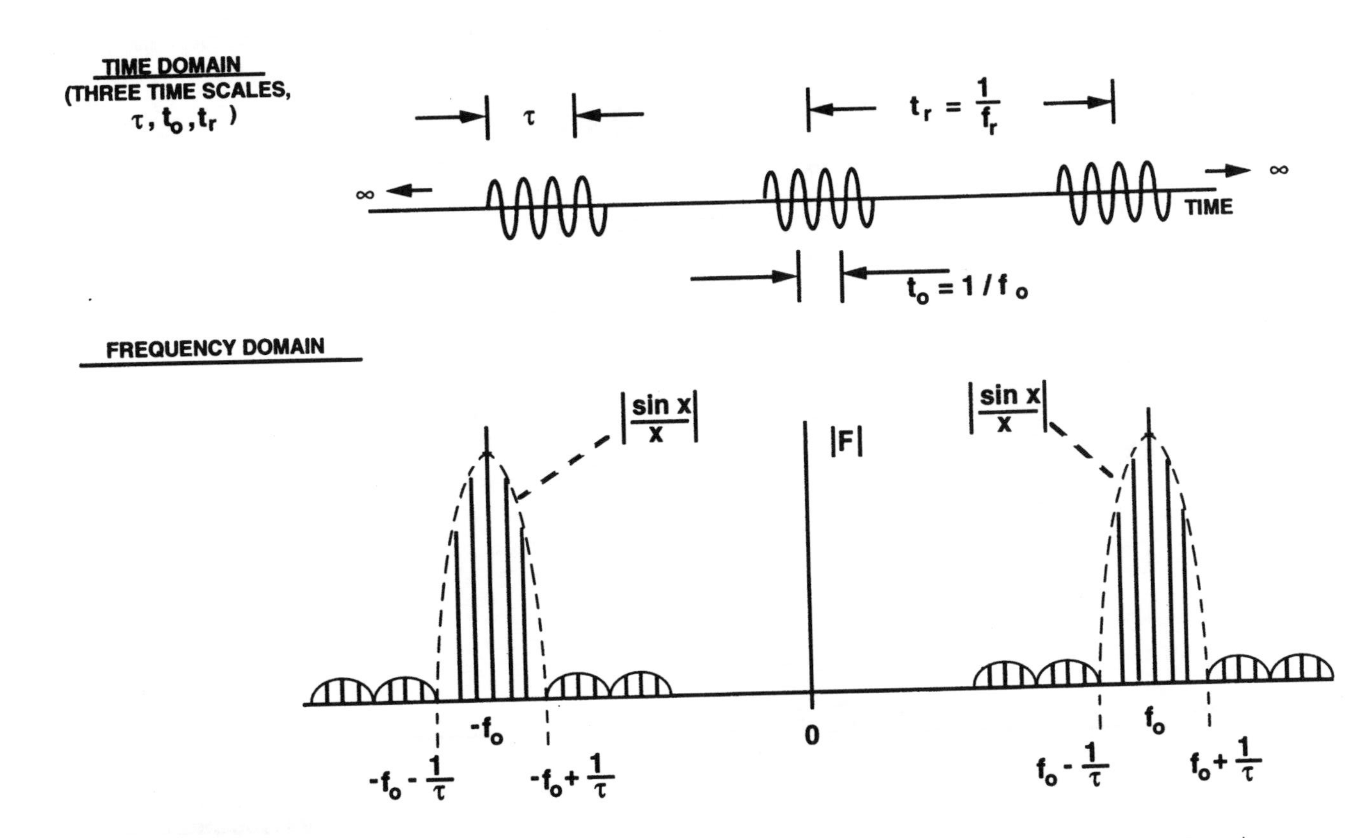

Figure 10.4 Waveform with three time scales (t_o, τ, and t_R). (Reproduced from [1], Figure 3.5, with permission of Artech House.)

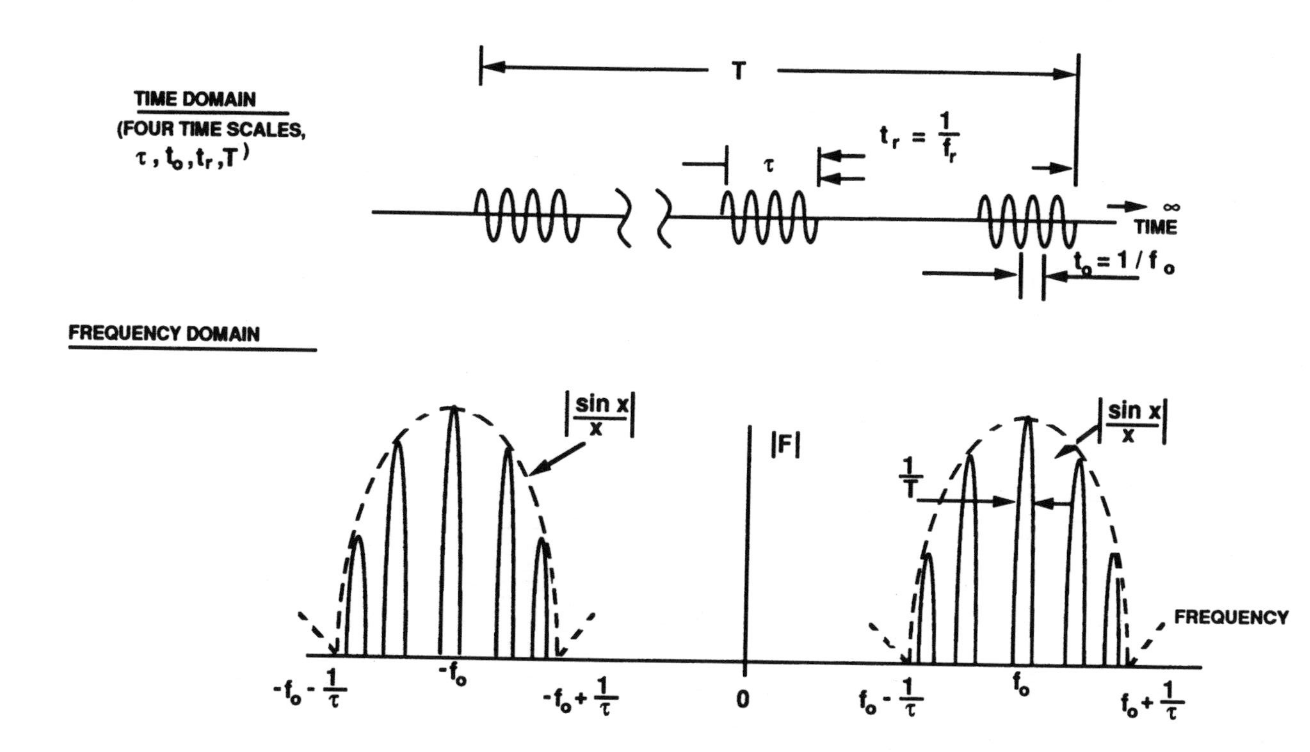

Figure 10.5 Waveform with four time scales (t_o, τ, t_R, T_{dwell}). (After [1], Figure 3.6, with permission of Artech House.)

We can describe phase noise in either time-domain or frequency-domain nomenclature; this text uses frequency-domain terminology only. Phase noise is typically discussed using the formalism that treats all frequencies as positive (see Section 4.2.1). Expressed in that way, for an oscillator of nominal frequency f_0, the frequency is not exactly constant but has components in a range of frequencies around f_0.

As mentioned in Section 2.2.1, an oscillator ideally produces a single-frequency tone, the FT of which is a delta function. The FT of a real oscillator exhibits a peak at f_0 with "skirts" that fall away on either side, as shown in Figure 10.6. The units of the ordinate are decibels relative to the carrier per hertz (dBc/Hz). Thus, the integral over a small region of frequency is the fraction, measured in decibels, of the total power in the interval, and the overall integral is unity.

Figure 10.6(a) shows the phase noise-generated frequencies both above and below the carrier f_0, a double-sideband (DSB) representation. Typically, that curve is quite symmetric. Usually, the phase noise is plotted versus the absolute value of the difference between the frequency and f_0. That is equivalent to "folding over" the lower sideband onto the upper sideband, producing a single-sideband (SSB) representation, thus, in practice, doubling the value of the power density at any particular frequency, as shown in Figure 10.6(b). Scheer and Kurtz [4] illustrate an SSB phase noise plot for a typical oscillator (Figure 10.7). This spectrum includes some spurious peaks, or "spurs" (Section 2.2.1).

As discussed in Section 10.1, for a series of pulses the transmitted spectrum is replicated about each PRF line. Figure 10.8 shows the result, illustrated in DSB format. In the interval between 0 and f_R appears (1) the original phase noise curve corresponding to this interval, (2) phase noise corresponding to the interval between 0 and $-f_R$, replicated from the PRF line at $+f_R$, (3) phase noise corresponding to the intervals between $\pm f_R$ and $\pm 2f_R$, replicated from the PRF lines at $-f_R$ and $+2f_R$, and so forth. Thus all of the phase noise power appears in the interval 0 to f_R.

We can group phase noise into several different types, as illustrated in Figure 10.9 (the origin of the terminology is described in [6]), decreasing according to different power laws as the abscissa value increases:

- Random walk frequency: $\sim f^{-4}$
- Flicker frequency: $\sim f^{-3}$
- White frequency: $\sim f^{-2}$
- Flicker phase: $\sim f^{-1}$
- White phase: $\sim f^{0}$ = constant

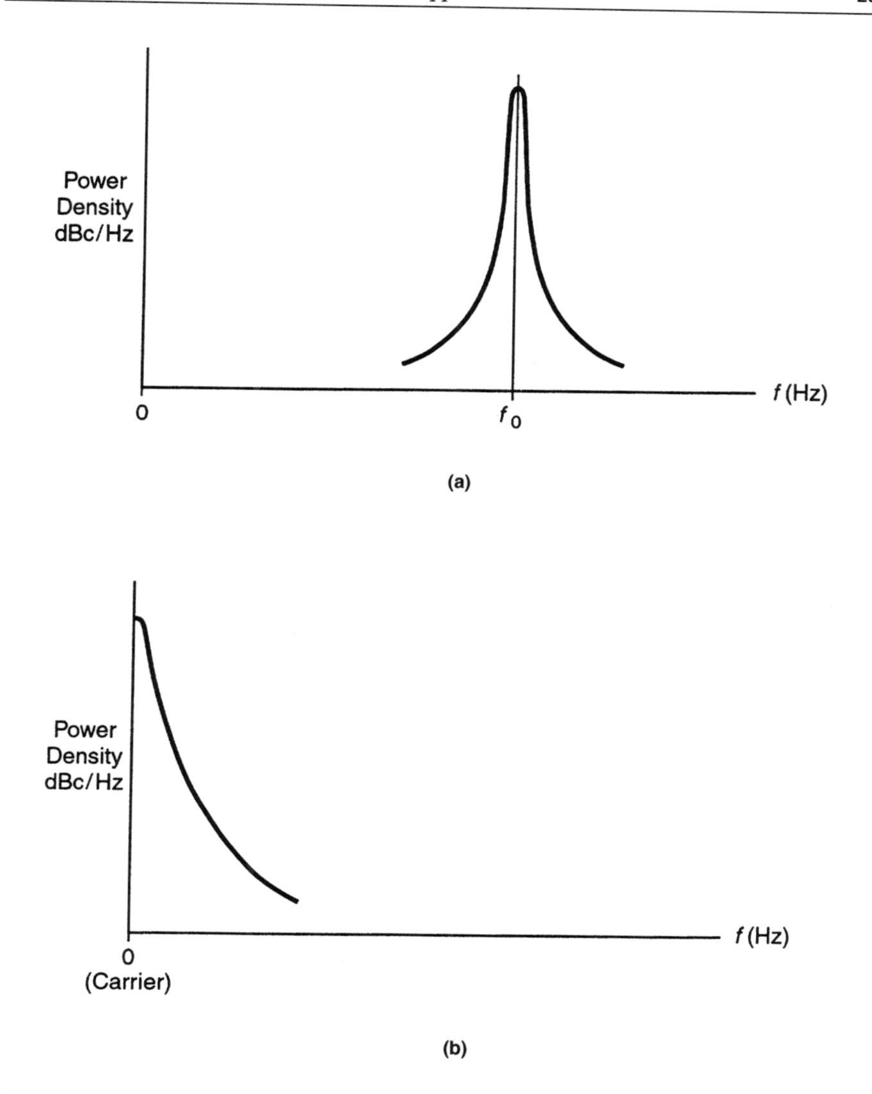

Figure 10.6 Phase noise spectrum (schematic): (a) double-sideband; (b) single sideband.

The phase noise curve for a radar results from the curve for the primary oscillator. Each time the frequency is multiplied by M, the phase noise is multiplied by M^2 [5, p. 78]. For example, the oscillator in Figure 10.7 has a nominal frequency of 106 MHz and a floor of about −160 dBc/Hz between 10^4 and 10^6 Hz. If the carrier frequency is upconverted by a factor of 100 to 10.6 GHz, then the phase noise floor will increase by a factor of at least 10,000 (40 dB) to at least −120 dBc/Hz, perhaps more due to additive noise contributions from the other RF components.

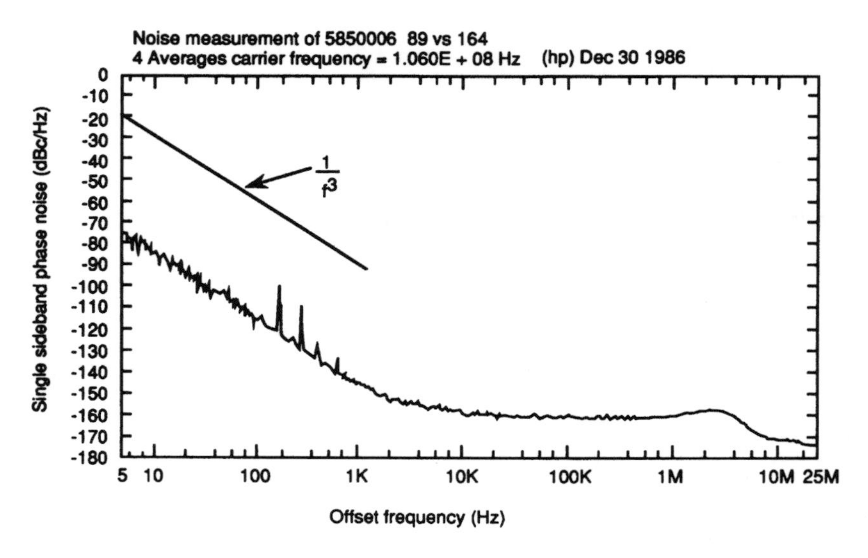

Figure 10.7 Example of phase noise. (Reproduced from [4], Figure 1.11, with permission of Artech House.)

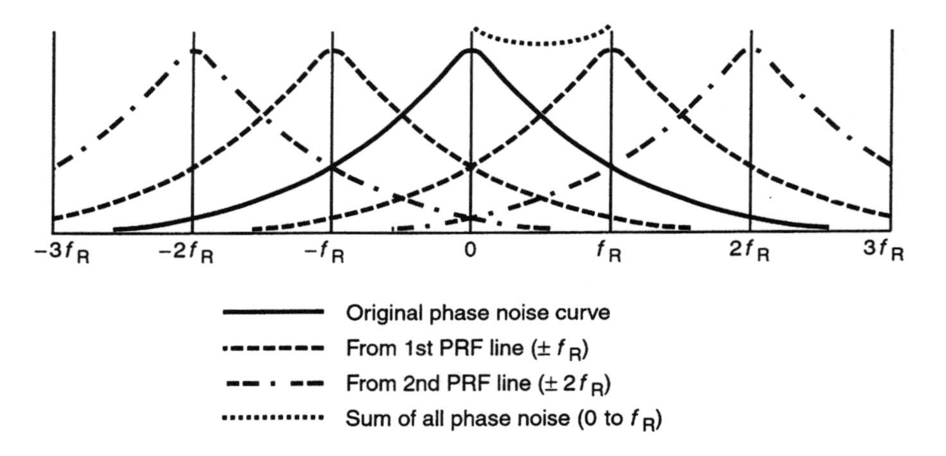

Figure 10.8 Convolution of phase noise with PRF lines.

In designing a PD radar, the designer must ensure that the expected phase noise spectrum is included in the interference against which any target must compete. Furthermore, if the primary oscillator will be subject to significant vibration, the phase noise will increase. Thus, if the radar is mounted in an aircraft, care must be taken to shock mount the exciter or otherwise mitigate the vibration-induced phase noise.

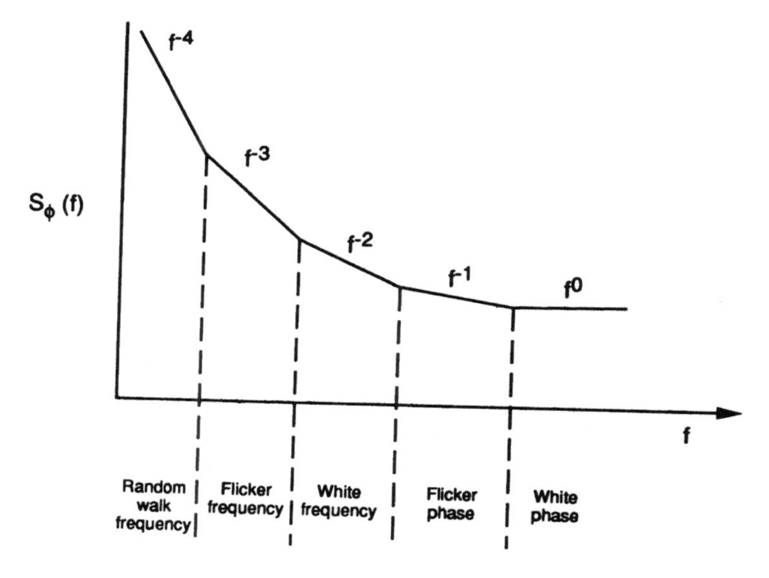

Figure 10.9 Types of phase noise. (Reproduced from [4], Figure 2.3, with permission of Artech House.)

10.3 Range and Velocity Ambiguities

Chapter 1 showed that for a waveform consisting of a sequence of identical monochromatic pulses with PRF of f_R and PRI of $t_R = 1/f_R$, the intervals of unambiguous range and radial velocity are given by

$$R_u = \frac{c}{2f_R} \qquad (10.1)$$

$$v_u = \frac{f_R \lambda}{2}$$

(The interval of unambiguous velocity could be $\pm f_R \lambda/4$, or 0 to $f_R \lambda/2$, or some other convenient interval.) We also saw that, for many cases of interest, it is not possible to avoid ambiguity in both range and velocity. Figure 10.10 presents a quantitative summary of that fact, and Figure 10.11 schematically illustrates the ambiguity function (see Section 4.2) of a sequence of monochromatic pulses, showing the ambiguities summarized by (10.1).

Chapter 1 defined the following terms (usually used in reference to an X-band radar):

- Low PRF: Unambiguous range, ambiguous velocity (typically < 4 kHz);

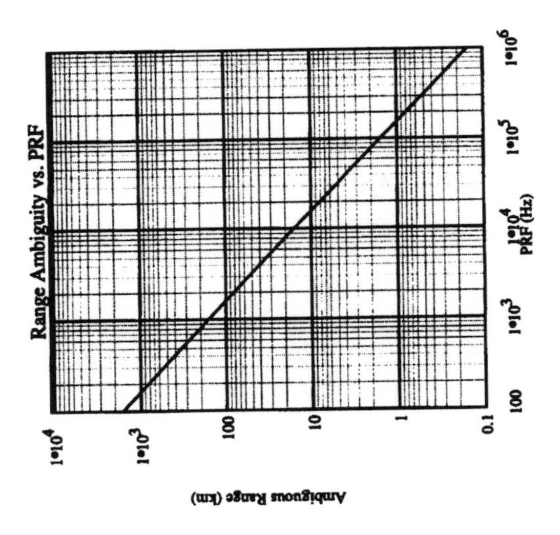

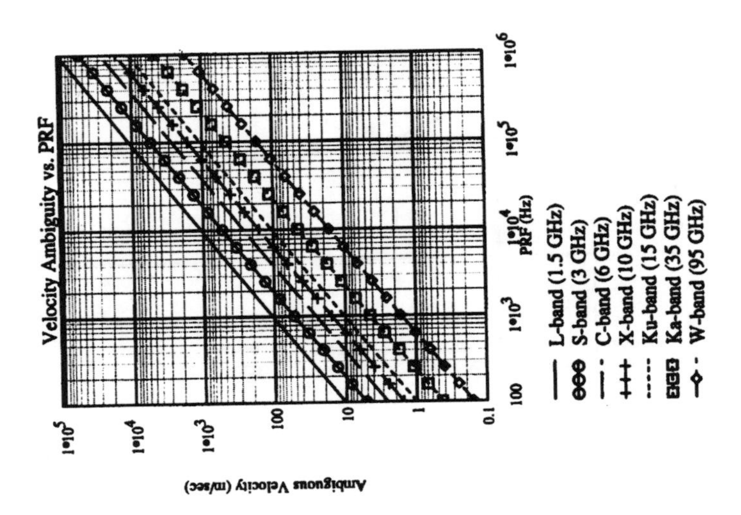

Figure 10.10 Ambiguity in range and velocity.

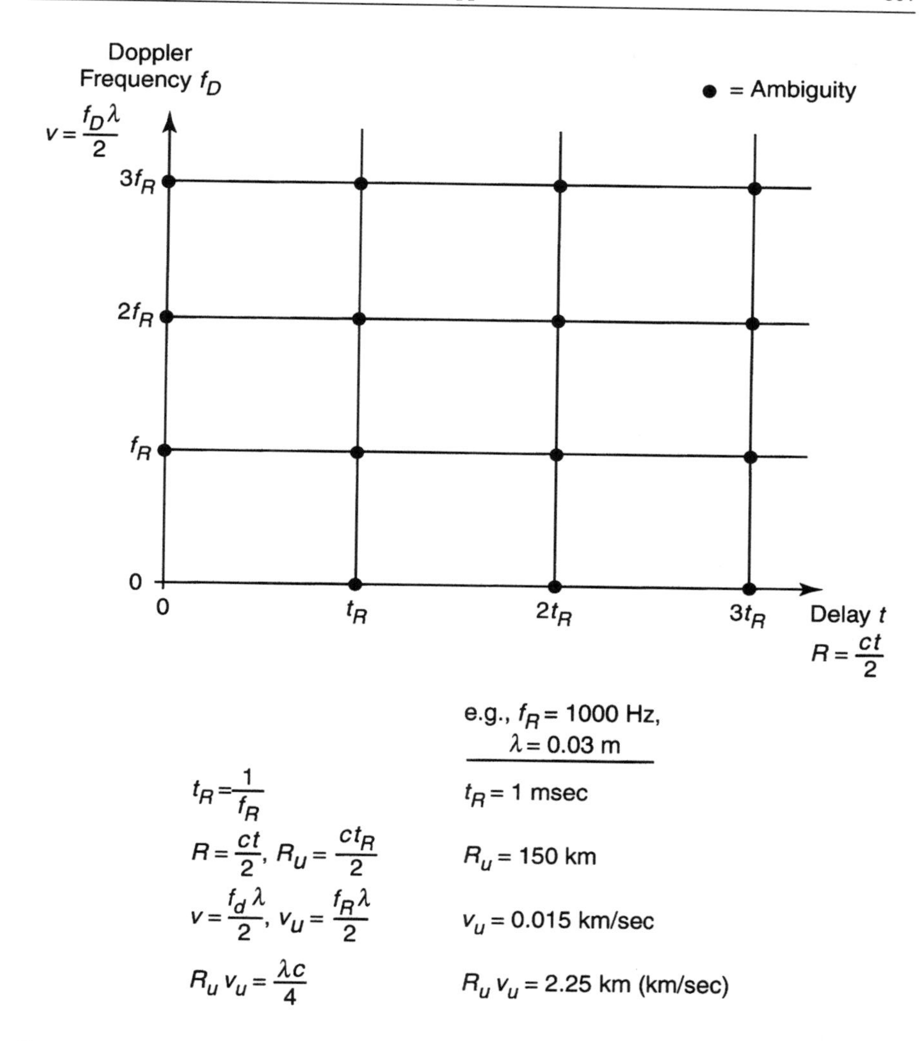

Figure 10.11 Ambiguity function for sequence of monochromatic pulses.

- High PRF: Unambiguous velocity, ambiguous range (typically > 100 kHz);
- Medium PRF: Ambiguous range and velocity (typically 4–100 kHz).

Furthermore, a monostatic radar usually cannot transmit and receive simultaneously, because it is almost impossible (or expensive) to detect the very low received power in the presence of the very high transmitted power present in the hardware. Thus, we typically assume that a radar cannot receive while it is transmitting, a phenomenon known as eclipsing. While a pulse of width τ is being transmitted by a radar, there is a time interval of width τ, corresponding to a sequence of range swaths of length $c\tau/2$, spaced at intervals of $ct_R/2$, which are eclipsed. Such swaths are often known as *blind zones* in range.

Figure 10.12 illustrates the example of a point target at range $R = (2.7)R_u$. If the radar is transmitting a single-PRF waveform, a received pulse begins $(0.7)t_R$ after the beginning of each transmitted pulse. The time interval from the start of a transmitted pulse to the start of the next received pulse corresponds to the apparent range:

$$R_{\text{app}} = R - NR_u = R\,\text{mod}(R_u) \tag{10.2}$$

$$N = \left[\frac{R}{R_u}\right]$$

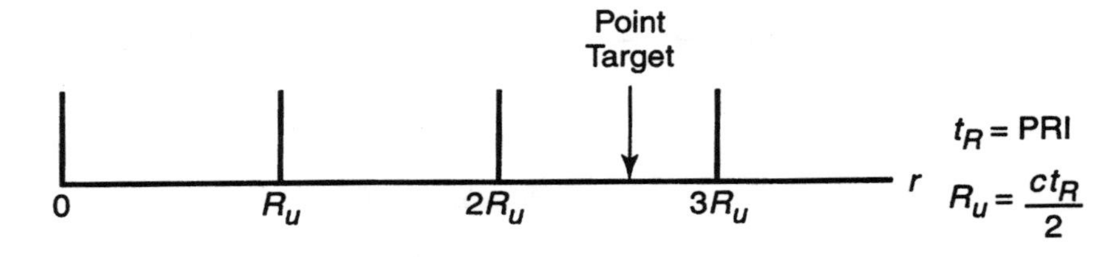

Suppose target is at range $R_t = 2.7\ R_u$

$t_{\text{delay}} = 2R/c = (2.7)\left(\frac{2R_u}{c}\right) = (2.7)\ t_R$

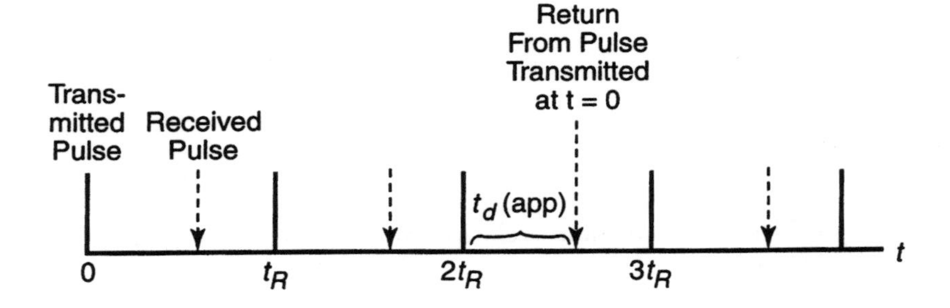

$R_{\text{app}} = \frac{ct_d\,(\text{app})}{2} = R\,\text{mod}\,(R_u) = R - NR_u$

$N = [R/R_u]$ = largest integer in R/R_u

Figure 10.12 Range ambiguity: one PRF.

where $[a]$ represents the largest integer in a and mod(a) is an abbreviation for "modulo (a)." (If the SNR were very high, it might be possible to measure the true range by noting that the first received pulse does not occur until a time $2R/c$ after the first transmitted pulse. However, we usually need the benefit of coherent integration of many pulses to achieve satisfactory SNR; thus, we usually measure only R_{app}.)

A completely analogous situation occurs for the measurement of the LOS velocity V of a point target:

$$\begin{aligned} v_{app} &= v - Mv_u = v \bmod (v_u) \\ M &= \left[\frac{v}{v_u}\right] \end{aligned} \tag{10.3}$$

When $v_{app} = 0$ and $M > 0$, we say that v is a *blind speed.*

When $N > 0$ (for range ambiguity) or when $M > 0$ (for velocity ambiguity) and we measure only a remainder, we often say that the return is aliased from its correct position (or velocity) to the apparent position (or velocity).

10.4 Pulse Tagging

In principle, one could mitigate the ambiguities described above by a procedure known as pulse tagging [10, p. 156]. Successive pulses are made different by means of RF variation, PM, FM, pulse-width modulation, or AM. In practice, AM would appear to be almost completely impractical, because any type of temporary signal attenuation due to a fluctuation in the target orientation or intervening atmospheric condition could cause a modulation in the received amplitude. Similarly, pulse-width modulation would be almost undetectable if each transmitted pulse resulted in an extended return echo due to clutter. Any of the methods involves considerable complexity in the exciter/transmitter hardware and the receiver and the processor software. Furthermore, doppler processing on the full set of returned pulses cannot be accomplished unless the pulse-tagging modulation is first removed, which is also challenging.

Pulse tagging, specifically PM, has been extensively implemented only in radars on RCS measurement ranges. With continuing improvements in hardware and processing, however, the technique may be used in a variety of radars in the future.

10.5 PRF Switching

A commonly used and less complex (i.e., less expensive) method to alleviate ambiguity is PRF switching. The radar beam observes a particular region of space during a coherent dwell or CPI (see Section 4.2.8), using a single PRF; it then observes the same region using a second PRF, then perhaps a third, and so forth. (If the radar simply switches between two PRFs to see how the apparent range changes, the technique is known as PRF jitter [10, p. 155].) For the ith PRF, the unambiguous range is R_{ui}.

We first consider a case involving two PRFs, with unambiguous ranges R_{u1} and R_{u2}. We assume that

$$\begin{aligned} R_{u1} &= m\Delta r \\ R_{u2} &= n\Delta r \\ \frac{R_{u1}}{R_{u2}} &= \frac{m}{n} \\ R_{u2} &> R_{u1} \\ n &> m \end{aligned} \tag{10.4}$$

where m and n are integers with no common factor, that is, the fraction m/n is reduced to lowest terms. The two integers are therefore noncommensurate. For a point target at range R greater than R_{u2}, the radar measures

$$\begin{aligned} R_{\text{app1}} &= R \bmod(R_{u1}) \\ R_{\text{app2}} &= R \bmod(R_{u2}) \end{aligned} \tag{10.5}$$

and we investigate whether we can determine R from R_{u1}, R_{u2}, R_{app1}, and R_{app2}.

Clearly, when $R = R_0 < R_{u1}$, then $R_{\text{app1}} = R_{\text{app2}} = R_0$. Furthermore, when $R = mn\Delta r + R_0$, then again $R_{\text{app1}} = R_{\text{app2}} = R_0$. Evidently, the overall unambiguous range R_u cannot be greater than $mn\Delta r$. By means of a proof by contradiction, we can also show that R_u cannot be less than $mn\Delta r$ and therefore that the overall unambiguous range is $R_u = mn\Delta r$.

> Assume that there exists a target at some range R less than $mn\Delta r$ but greater than R_{u1} for which $R_{\text{app1}} = R_{\text{app2}} = R_0$. Then for a target at range $R - R_0$, $R_{\text{app1}} = R_{\text{app2}} = 0$. In the latter case there exist integers $m' < m$ and $n' < n$ such that
>
> $$\begin{aligned} R_{u1} &= m'\Delta r \\ R_{u2} &= n'\Delta r \\ \frac{R_{u1}}{R_{u2}} &= \frac{m}{n} = \frac{m'}{n'} \end{aligned} \tag{10.6}$$

This is a contradiction, because it was assumed that m/n is reduced to lowest terms.

We often choose $m - n = 1$. In that case,

$$R_u = mn\Delta r = \frac{mn}{m - n}\Delta r = \frac{m\Delta r \cdot n\Delta r}{m\Delta r - n\Delta r} = \frac{R_{u1}R_{u2}}{R_{u1} - R_{u2}} \tag{10.7}$$

as stated in [10, p. 160].

We can also show that the two apparent ranges, R_{app1} and R_{app2}, along with R_{u1} and R_{u2}, uniquely determine R. Again, we prove that by contradiction.

Assume there are two point targets at ranges R_a and R_b, where $R_a < R_b < R_u$ and where the measured values of the ordered pair $(R_{\text{app1}}, R_{\text{app2}})$ are the same. (We also assume that the target at R_a is the shortest range target to produce those values.) For a target at range $R_a - R_a = 0$, the apparent ranges are clearly (0,0). Similarly, for a target at range $R_b - R_a$, the apparent ranges must be (0,0). That, however, is a contradiction, because we have assumed that the minimum unambiguous range is $R_u > R_b$.

Figure 10.13 illustrates an example of two PRFs producing an unambiguous range of $mn\Delta r$. Figures 10.13(a) and 10.13(b) show the apparent ranges as a function of actual range. Figure 10.13(c) shows that for each integer value of $R/\Delta r$ (where Δr is an arbitrary range increment), the value of the ordered pair $(R_{\text{app1}}, R_{\text{app2}})$ is unique. Figure 10.13(d) illustrates the procedure for finding R given R_{app1}, R_{app2}, R_{u1}, and R_{u2}. R is equal to $1/\sqrt{2}$ times the length of the solid line beginning at (0, 0) and proceeding by means of zero-length "jumps," indicated by dashed lines, until the appropriate value of $(R_{\text{app1}}, R_{\text{app2}})$ is reached. (A mathematical purist might note that in principle, if R_{u1}/R_{u2} were irrational, then R_u would be infinite. However, that consideration is made moot by finite measurement error and eclipsing.)

In similar fashion (Problem 10.1), we can show that for N PRFs, each producing an unambiguous range R_{ui}, $i = 1, \ldots, N$, where $R_{ui} = m_i\Delta r$ and no two m_i have a common factor, in principle (neglecting eclipsing and finite measurement errors) the unambiguous range becomes

$$R_u = \Delta r \cdot \prod_{i=1}^{N} m_i \tag{10.8}$$

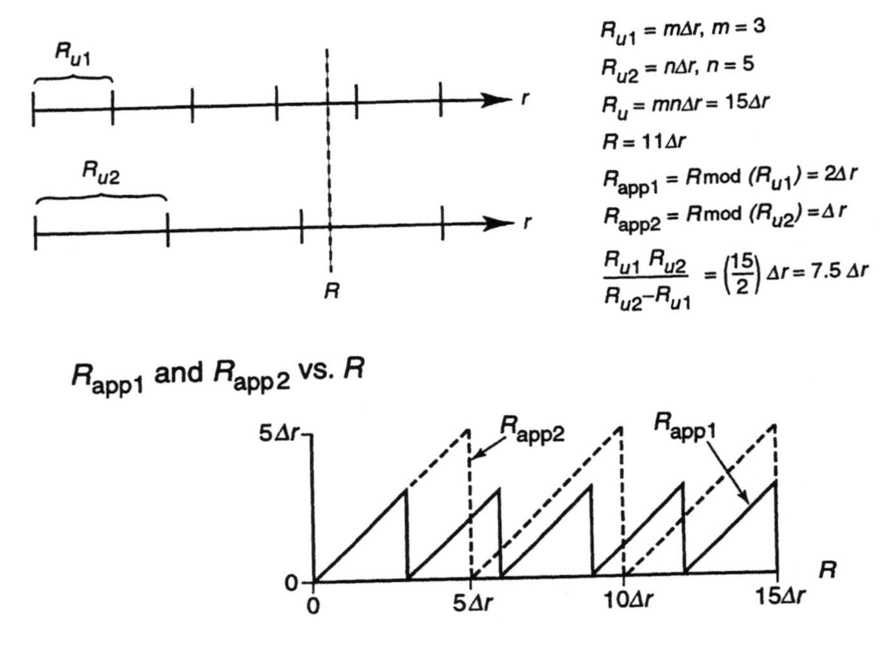

Figure 10.13 Range ambiguity: two PRFs: (a) geometry; (b) R_{app1} and R_{app2} versus R; (c) Uniqueness of (R_{app1}, R_{app2}) if $R < R_u$; (d) R versus R_{app1}, R_{app2}.

Using MathCad®, Figure 10.14 shows the use of three noncommensurate PRFs to measure the range to a point target. Only at the correct range do the three echoes "line up." Figure 10.14 also illustrates the fraction of possible target ranges that cannot be measured due to eclipsing. The larger the number of PRFs used in the waveform, the greater the eclipsing loss becomes, putting a practical limit on the potential number of PRFs.

10.6 Chinese Remainder Theorem

A general procedure for determining R for N PRFs, given the unambiguous and apparent ranges (and neglecting eclipsing time and measurement errors), is found in the Chinese Remainder Theorem [11, 12]. Stated in radar context, that theorem is as follows [11, p. 17.22; 12, pp. 246–249].

If:

for $i = 1$ to N, $R_{ui} = m_i$ (noncommensurate integers) and $R_{appi} = R \bmod (R_{ui})$

Then:

Uniqueness of
(R_{app1}, R_{app2}) if $R < R_u$

$R/\Delta r$	$R_{app1}/\Delta r$	$R_{app2}/\Delta r$
0	0	0
1	1	1
2	2	2
3	0	3
4	1	4
5	2	0
6	0	1
7	1	2
8	2	3
9	0	4
10	1	0
11	2	1
12	0	2
13	1	3
14	2	4
15	0	0

R vs. R_{app1}, R_{app2}

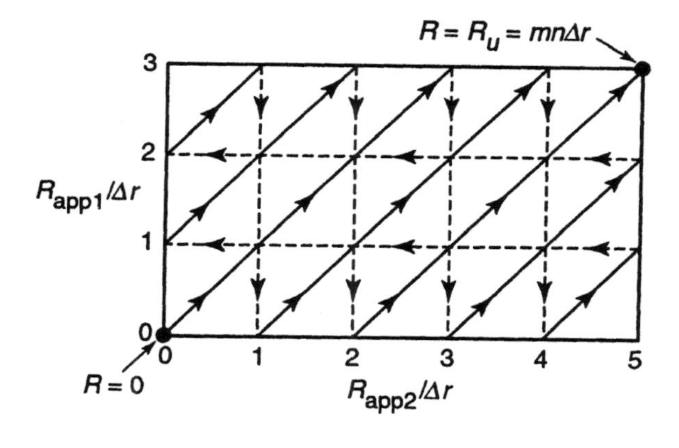

Figure 10.13 (continued).

Table 10.1
PRF Switching Calculations

Time in microseconds, distance in km speed of light: c: = 0.3 km/microsec

Three PRIs: $\Delta t := 50 \quad m := \begin{bmatrix} 4 \\ 5 \\ 7 \end{bmatrix} \quad k := 0, 1 \,..\, 2 \quad t_{R_k} := m_k \cdot \Delta t$

$t_R = \begin{bmatrix} 200 \\ 250 \\ 350 \end{bmatrix}$ microsec

Three Ru's (km) and PRFs (KHz): $Ru_k := \dfrac{c \cdot t_{R_k}}{2} \quad Ru = \begin{bmatrix} 30 \\ 37.5 \\ 52.5 \end{bmatrix} \quad f_{R_k} := \dfrac{1000}{t_{R_k}}$

$f_R = \begin{bmatrix} 5 \\ 4 \\ 2.857 \end{bmatrix}$

Overall Unambiguous Delay: $t_u := \left(\prod_k m_k\right) \cdot \Delta t \quad t_u = 7000$

Pulse Width (τ) and Actual Delay (T): $\tau := 20 \quad T := 5050 \quad R := \dfrac{c \cdot T}{2} \quad R = 757.5$

Functions Representing Transmitted and Received Pulses:

$t_{Rmin} := \min(t_R) \quad t_{Rmin} = 200 \quad N := \mathrm{ceil}\left(\dfrac{t_u}{t_{Rmin}}\right) \quad N = 35 \quad i := 0, 1 \,..\, t_u \quad t_i := i$

$F(i, \tau, t_R, T) := \begin{vmatrix} 0 \\ 1 \text{ if } \mathrm{mod}(i - T + N \cdot t_R, t_R) < \tau \end{vmatrix} \quad F_{Tx_{i,k}} := F(i, \tau, t_{R_k}, 0)$

$F_{Rx_{i,k}} := (F(i, \tau, t_{R_k}, T)) \cdot (0.5 \quad F_{Rx_{i,k}} := \text{if } (F_{Tx_{i,k}} > .9, 0, F_{Rx_{i,k}})$

$$R = \left(\sum_i C_i R_{appi}\right) mod\left(\prod_i m_i\right)$$

where

$$C_i = b_i \prod_{j \neq i} m_j$$

and b_i = least positive integer (lpi) such that

$$\frac{b}{m_i}\left(\prod_{j \neq i} m_j\right) - \left[\frac{b}{m_i}\left(\prod_{j \neq i} m_j\right)\right] = \frac{1}{m_i}$$

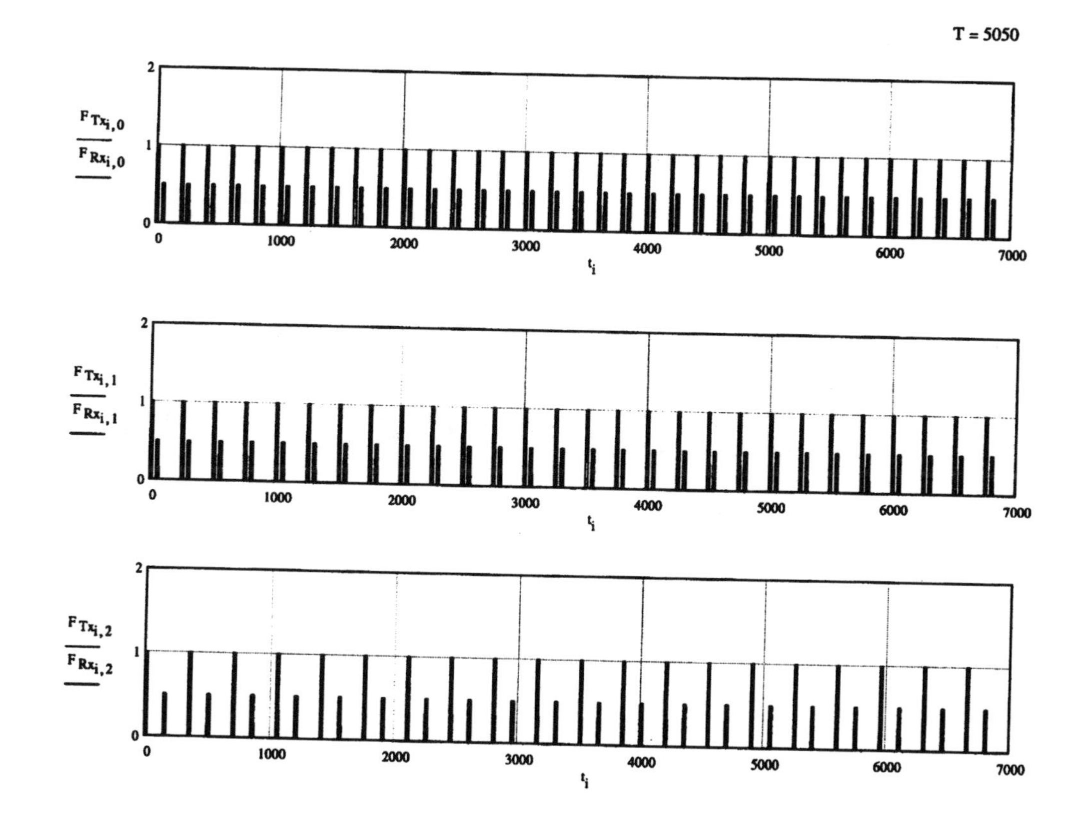

Figure 10.14 PRF switching: (a) transmitted (amplitude 1.0) and received (amplitude 0.5) pulses, versus time, for three PRFs; (b) determination of fraction (of collection time) eclipsed; (c) determination of actual target range (indicated by G_{Rx} value of 1.5).

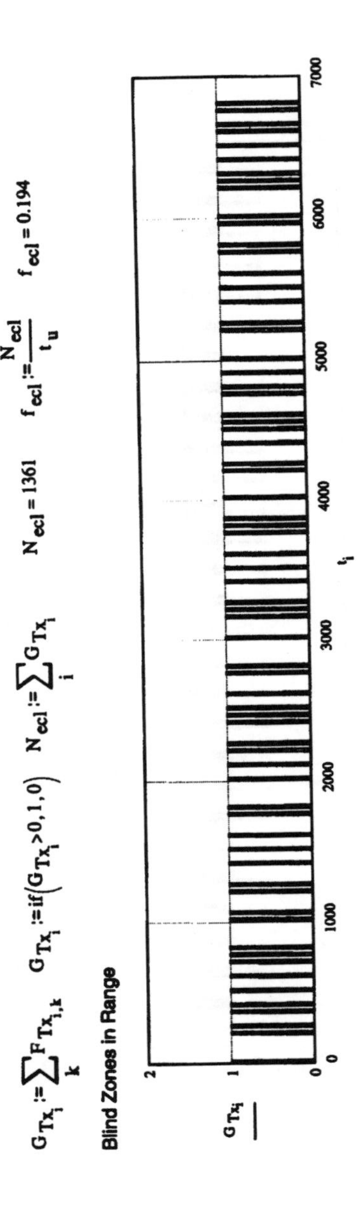

Figure 10.14 (continued).

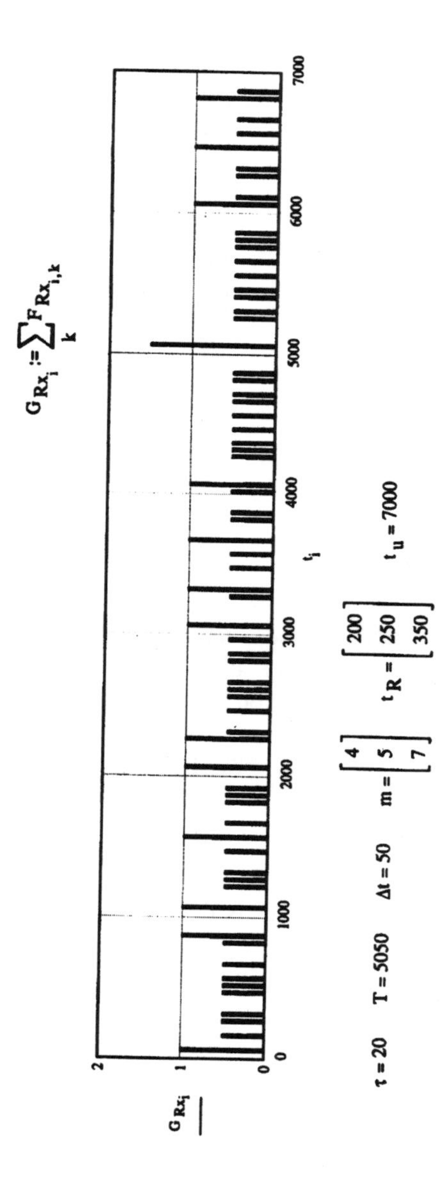

Figure 10.14 (continued).

An example follows (see Figure 10.13):
If: $N = 2,\ m_1 = 3,\ m_2 = 5$
$R_{app1} = 2,\ R_{app2} = 1$
Then:

$$b_1 = \text{lpi such that } \frac{b}{3} \cdot 5 - \left[\frac{b}{3} \cdot 5\right] = \frac{1}{3};\ b_1 = 2$$

$$b_2 = \text{lpi such that } \frac{b}{5} \cdot 3 - \left[\frac{b}{5} \cdot 3\right] = \frac{1}{5};\ b_2 = 2$$

$C_1 = b_1 m_2 = 2 \cdot 5 = 10$
$C_2 = b_2 m_1 = 2 \cdot 3 = 6$
$R = (10 \cdot 2 + 6 \cdot 1) \bmod (15)$
$= 26 \bmod 15 = 11$

10.7 PRF Selection Procedure

For some radars, several PRFs spaced fairly widely apart, called majors, are chosen to mitigate doppler (velocity) ambiguity. Surrounding each major, several PRFs spaced relatively closely together, called minors, are provided to mitigate delay (range) ambiguity [2, pp. 273–277]. Another strategy is the *M:N* method. As described in [2, pp. 277–278]:

1. A set of *N* PRFs is selected to ensure that the target will be in a clear doppler region (i.e., not eclipsed) for at least *M* . . . PRFs.
2. Detection on at least *M* PRFs is required. . . . Usually, $M = 3$.
3. Any three of the PRFs may be used for range resolving; that is, none is designated as major or minor.

In practice, one typically uses a computer to produce blind zone charts in range and velocity to form the basis for the choice of PRFs. Further details of PRF selection are provided in [2, Chap. 12].

As each new PRF is initiated, a time of $\sim 2R/c$ passes before the corresponding pulses are received. Here, R is the greatest range from which detectable echoes are expected, and it may extend beyond the range of an expected target or of the first range ambiguity, perhaps to a horizon. Because the transmitter and the receiver typically are controlled by the same timing, the time to "fill the space with pulses," or fill time (1 ms for R = 150 km), may be lost to the detection procedure.

During each CPI, the pulses are processed coherently. The phase of the combined result from the CPI is often discarded, a procedure sometimes called detection [1]. Using the methods described in Section 4.1, noncoherent combination of the magnitudes resulting from different CPIs is then referred to as postdetection integration (PDI) [10, p. 504].

References

[1] Kurpis, G. P., and C. J. Booth (eds.), *The New IEEE Standard Dictionary of Electrical and Electronics Terms*, 5th ed., New York: Institute of Electrical and Electronics Engineers, 1993.

[2] Morris, G., and L. Harkness (eds.), *Airborne Pulsed Doppler Radar*, 2nd ed., Norwood, MA: Artech House, 1996.

[3] Brigham, E. O., *The Fast Fourier Transform and Its Applications*, Englewood Cliffs, NJ: Prentice Hall, 1988.

[4] Scheer, J. A., and J. L. Kurtz, *Coherent Radar Performance Estimation*, Norwood, MA: Artech House, 1993.

[5] Robins, W. P., *Phase Noise in Signal Sources*, London: Peregrinus, 1984.

[6] Shoaf, J. H., *Frequency Stability Specification and Measurement: High Frequency and Microwave Signals*, Boulder, CO: National Bureau of Standards, COM-73-50238, Jan. 1973.

[7] Barnes, J. A., et al., *Characterization of Frequency Stability*, Boulder, CO: National Bureau of Standards, Tech. Note 394, Oct. 1970.

[8] Howe, D. A., D. W. Allan, and J. A. Barnes, "Properties of Signal Sources and Measurement Methods," *Proc. 35th Annual Frequency Control Symp.*, Washington, DC: Electronic Industries Association, May 1981, pp. A1–A47.

[9] Cutler, L. S., and C. L. Searle, "Some Aspects of the Theory and Measurement of Frequency Fluctuations in Frequency Standards," *Proc. IEEE*, Vol. 54, No. 2, Feb. 1966, pp. 136–154.

[10] Stimson, G. W., *Introduction to Airborne Radar*, 2nd ed., Mendham, NJ: SciTech, 1998.

[11] Long, W. H., D. H. Mooney, and W. A. Skillman, "Pulse Doppler Radar," Chap. 17 in M. Skolnik (ed.), *Radar Handbook*, 2nd ed., New York: McGraw-Hill, 1990.

[12] Ore, O., *Number Theory and Its History*, New York: McGraw-Hill, 1948.

Problems

Problem 10.1

Show that for N PRFs, each producing an unambiguous range R_{ui}, $i = 1, \ldots, N$, where $R_{ui} = m_i \Delta r$ and no two m_i have a common factor, in principle (neglecting eclipsing and finite measurement errors) the unambiguous range becomes

$$R_u = \Delta r \cdot \prod_{i=1}^{N} m_i$$

11

Observation of Moving Targets by an Airborne Radar

We begin this chapter with some definitions [1].

- *Doppler radar* is "a radar which uses the Doppler effect to determine the radial component of relative radar target velocity or to select targets having particular radial velocities."
- *Pulse doppler (PD) radar* (or pulsed-doppler radar) is "a Doppler radar that uses pulsed transmissions."
- *Moving-target indication (MTI)* "is a technique that enhances the detection and display of moving radar targets by suppressing fixed targets. *Note:* Doppler processing is one method of implementation."

Skolnik [2, p. 101] draws the following distinction between PD and MTI:

> A pulse radar that utilizes the Doppler frequency shift as a means for discriminating moving from fixed targets is called an MTI . . . or a pulse Doppler radar. The two are based on the same physical principle, but in practice there are generally recognizable differences between them. . . . The MTI radar, for instance, usually operates with ambiguous Doppler measurement (so-called *blind speeds*) but with unambiguous range measurement. . . . The opposite is generally the case for a pulse Doppler radar. Its pulse repetition frequency is usually high enough to operate with unambiguous Doppler (no blind speeds) but at the expense of range ambiguities.

Many users (e.g., Entzminger, Fowler, and Kenneally [3]) use the term *ground-moving-target indication* (GMTI) to mean detection of ground-moving targets from an airborne or spaceborne radar. The term describing detection of airborne targets is particularly confusing. *IEEE Standard Radar Definitions* [4] states:

> *Airborne Moving-Target Indication (AMTI) radar:* An MTI radar flown in an aircraft or other moving platform with corrections applied for the effects of platform motion, which include the changing clutter Doppler frequency and the spread of the clutter Doppler spectrum.

That would imply that the adjective *airborne* refers to the platform, not the target. However, often the term airborne-moving-target indication" (note the different use of hyphens) is used, with *airborne* referring to the target, not the platform. Furthermore an MTI radar is often characterized as having a stopband for clutter and a passband for targets, while a PD radar has a large set of narrowband filters, some of which may be assigned to clutter rejection.

To avoid confusion, this book does not use the term *AMTI* (either meaning). Rather, it uses *GMTI* to refer to observation of a ground-moving target from an airborne (or spaceborne) platform.

11.1 Low, Medium, and High PRFs

Assume a radar on an aircraft that is moving horizontally at speed V at altitude H above a flat ground. The radar will observe a return signal from the ground clutter that exhibits definite characteristics. This chapter investigates that definitive clutter pattern. The aircraft carrying the radar is referred to as the *ownship*.

Assume a single PRF (f_R) and investigate the resulting ambiguities (see Sections 1.11 and 1.15). From Section 1.15, the unambiguous velocity interval is $v_u = f_R\lambda/2$. If the radar is pointing in the forward direction at a low depression angle, the ground will have an apparent speed $-V$ relative to the aircraft; similarly, if the radar is pointing directly aft at a low depression angle, the apparent ground speed will be $+V$. The expected range of apparent speeds of stationary ground clutter, therefore, is 2V. If we want unambiguously to measure the speeds of objects on the ground and of other aircraft approaching the ownship (*closing*) or receding from it (*opening*), we need an unambiguous velocity interval v_u greater than $2V + 2|V_{tmax}|$, that is, a PRF at least $(2/\lambda) \cdot (2V + 2|V_{tmax}|)$, where $\pm V_{tmax}$ are the limits of radial target velocity. On the other hand, we also typically want to receive target echoes out to some unambiguous range $R_u = c/2f_R$, which requires that f_R be less than $c/2R_u$. Typically, it is not possible to meet those two conditions simultaneously, and we must distinguish among low PRF (LPRF) (unambiguous range), high PRF (HPRF) (unambiguous radial velocity), and medium PRF (MPRF) (both range and radial velocity are ambiguous).

An HPRF waveform will have a set of ambiguous ranges, each corresponding to a sphere in space (with the radar at the center) and a circle where the sphere intersects the (flat) ground.

An LPRF waveform will have a set of ambiguous velocities. If a target is stationary, its apparent velocity will be

$$v_{\text{app}} = -V\cos\theta,\ \theta = \cos^{-1}\left(\frac{-v_{\text{app}}}{V}\right) \tag{11.1}$$

where θ is the angle between the ownship velocity vector and the vector from the ownship to the target. The locus of all possible positions of targets with the apparent velocity v_{app} is a right-circular cone with generating angle θ and axis parallel to the ownship velocity vector (see Section 7.1.2). The set of ambiguous velocities will correspond to a set of such cones in space, and their intersections with the flat ground will be a set of hyperbolas (isodops). If the ground is in the xy plane and the ownship is at $(0, 0, H)$ flying in the $+x$ direction, then the isodop hyperbola is given by (Problem 11.1) [10]

$$\frac{x^2}{(H\cot\theta)^2} - \frac{y^2}{H^2} = 1 \tag{11.2}$$

Figure 11.1 illustrates examples of ground clutter patterns. The portions of the ground that produce sidelobe clutter are

- For HPRF, in Figure 11.1(a), a set of circles (it is assumed that $f_R > 2V/\lambda$; thus, no hyperbolas are formed);
- For MPRF, in Figure 11.1(c), sets of circles and hyperbolas;
- For LPRF, in Figure 11.1(b), a set of isodop hyperbolas.

Figure 11.1 shows the circles and hyperbolas that are ambiguous with zero range and zero velocity. For any particular target range and velocity, the ambiguous ranges and velocities would be determined by adding or subtracting multiples of R_u or v_u.

Most modern airborne radars use LPRF for ground targets and HPRF or MPRF for airborne targets. Consider an airborne radar that samples its

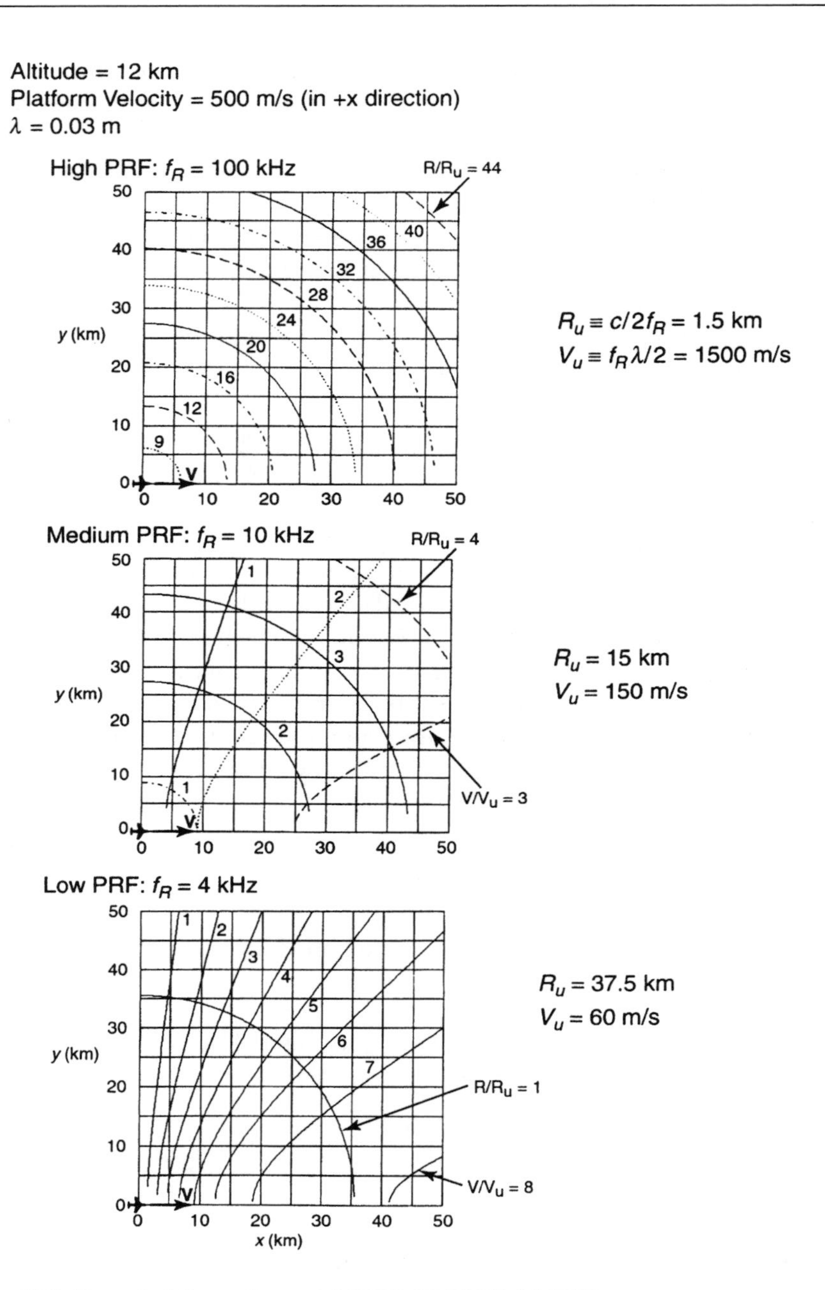

Figure 11.1 Ground clutter patterns: (a) HPRF; (b) MPRF; (c) LPRF.

returned signal so as to observe a particular range gate and doppler processes the signal to observe a particular velocity. Competing clutter will be caused by the echoes from the intersections on the ground of the corresponding range-ambiguous circles and velocity-ambiguous hyperbolas, multiplied by the two-way power pattern of the antenna.

- For an LPRF waveform used to observe a stationary or slowly moving ground target (e.g., SAR) at a range r less than R_u, the echoes resulting from $r + nR_u$ ($n > 1$) are usually negligible because of the $1/R^4$ factor and/or the horizon. The first velocity-ambiguous hyperbola should be kept outside the main beam power pattern.
- For an HPRF waveform used to observe an airborne target, if the PRF is sufficiently high that clutter results from essentially all ranges, then a clutter-free echo results if the target LOS velocity V_{tgt} is such that $V < |V - V_{tgt}| < \frac{f_R \lambda}{2} - V$, in which case there is no velocity-ambiguous ground echo, as discussed in more detail in Section 11.2.
- For an MPRF waveform or for an HPRF waveform when the above condition is not satisfied, the clutter is typically significant and must be determined by the sum of the echoes from the circle-hyperbola intersections, multiplied by the two-way antenna power pattern.

11.2 High-PRF Clutter Pattern

This section assumes that the PRF is very high and that the radar is looking in the forward direction (zero azimuth) at a slight depression angle to illuminate the ground relatively far ahead of the aircraft. In this case, the apparent speed of the mainlobe clutter will be $+V_g$, slightly less than $+V$. This case applies to, for example, a high-PRF mode of a typical fighter radar.

11.2.1 Sidelobe Clutter

From Chapter 2, recall that antenna sidelobes generally will be higher near the main beam. Thus, we expect sidelobe ground clutter to be present at all doppler frequencies between $+2V/\lambda$ and $-2V/\lambda$ falling off in intensity as angular distance from the main beam increases. This pattern is illustrated in Figure 11.2(a).

11.2.2 Mainlobe Clutter

Mainlobe clutter will be much stronger than sidelobe clutter and will peak at frequency $+2V_g/\lambda$, as shown in Figure 11.2(b). If the mainlobe is pointed

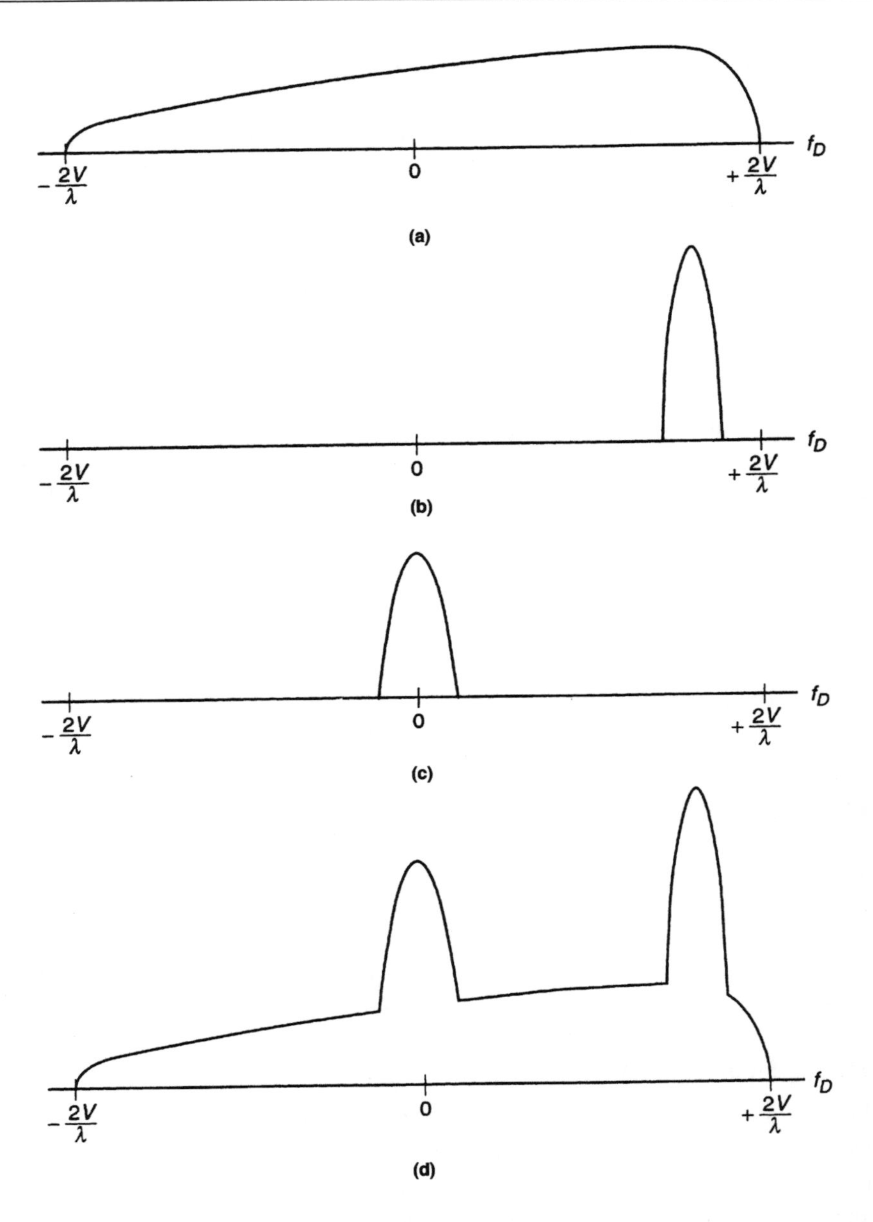

Figure 11.2 Typical ground clutter return for HPRF airborne radar: (a) sidelobe clutter; (b) mainlobe clutter; (c) altitude return; (d) overall ground clutter return.

away from the forward direction, then the mainlobe clutter peak will appear at a lower speed and will be broadened (Problem 11.2).

11.2.3 Altitude Return

From Figure 11.3, we see that for a range gate between slant ranges R and $R + \Delta R$ ($\Delta R << R$), the annular ground area covered is $A \approx 2\pi R \Delta R$. (The annular area $\approx 2\pi R_g \Delta R_g$, where R_g and ΔR_g are the parameters in the ground plane corresponding to R and ΔR, respectively; $R_g = R\cos(\psi)$ and $\Delta R_g \approx \Delta R/\cos(\psi)$.) The annular area increases linearly with R. However, the signal returned from each portion of that annulus decreases as $1/R^4$; thus, even if σ^0 were independent of grazing angle, the overall return from the annulus would decrease as $1/R^3$. Furthermore (Figure 3.15), σ^0 is considerably stronger at vertical incidence (specular reflection) than at other grazing angles. For those reasons, the return from the ground directly underneath the ownship, called the *altitude return*, is especially strong (Problem 11.3).

Figure 11.2(c) indicates the altitude return, and Figure 11.2(d) summarizes the overall ground clutter return. Figure 11.4 generalizes Figure 11.2(d) to illustrate the typical ground clutter return as a function of both velocity and range.

11.3 Air-to-Air Radar

This section considers the case of an airborne radar observing an airborne target.

11.3.1 HPRF Doppler Spectra

This section assumes that the PRF is high enough to avoid any velocity ambiguity. Except as specified, we assume that the target is directly ahead of or behind the ownship, and moving at V_{tgt} either directly toward or away from the ownship. The target apparent velocity is then $V_{tgt} - V$, and its doppler frequency is

$$f_D = -\frac{2}{\lambda}(V_{tgt} - V) = \frac{2}{\lambda}(V - V_{tgt}) \tag{11.3}$$

Some specific cases follow and are summarized in Figure 11.5.

- A *zero-closing-rate target* ($V_{tgt} = V$) could include an aircraft directly ahead of or directly behind the ownship and flying in the same direc-

$$R^2 = H^2 + R_g^2,\ A_1 = \pi R_g^2 = \pi(R^2 - H^2)$$

$$\begin{aligned} A_2 &= \pi[(R + \Delta R)^2 - H^2)] \\ &= \pi(R^2 + 2R\Delta R + \Delta R^2 - H^2) \\ &\cong A_1 + 2\pi R\Delta R \end{aligned}$$

$$A_{\text{annulus}} = A_2 - A_1 \cong 2\pi R\Delta R$$

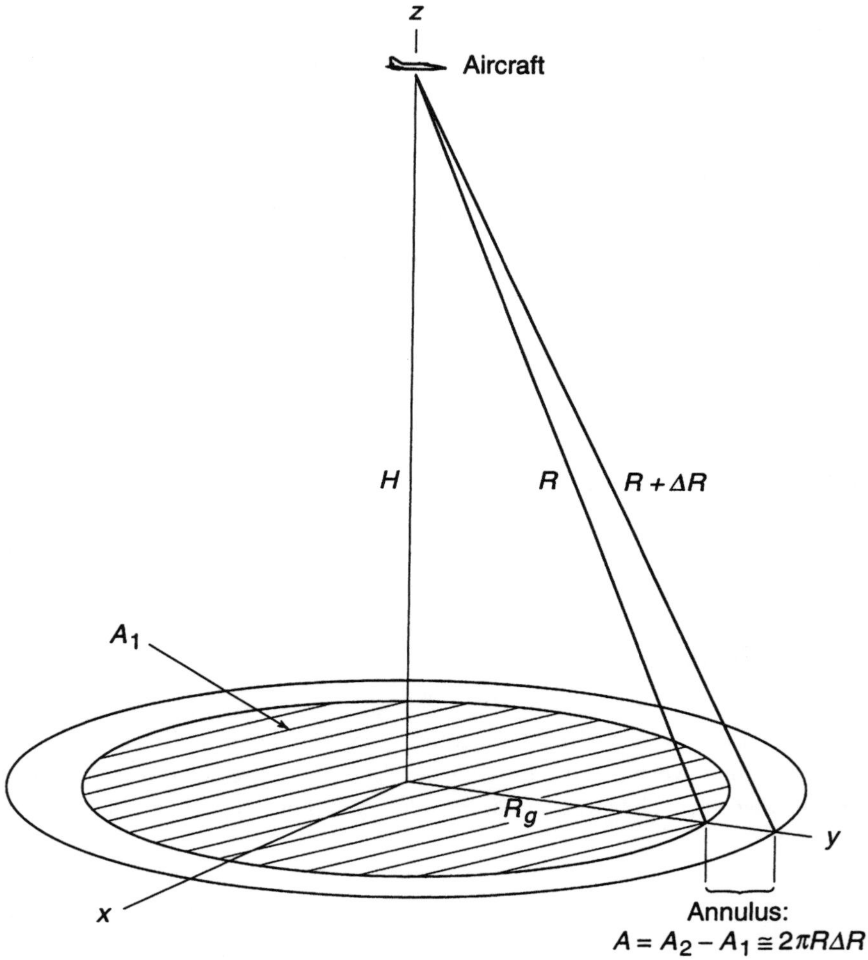

Figure 11.3 HPRF airborne radar: altitude return.

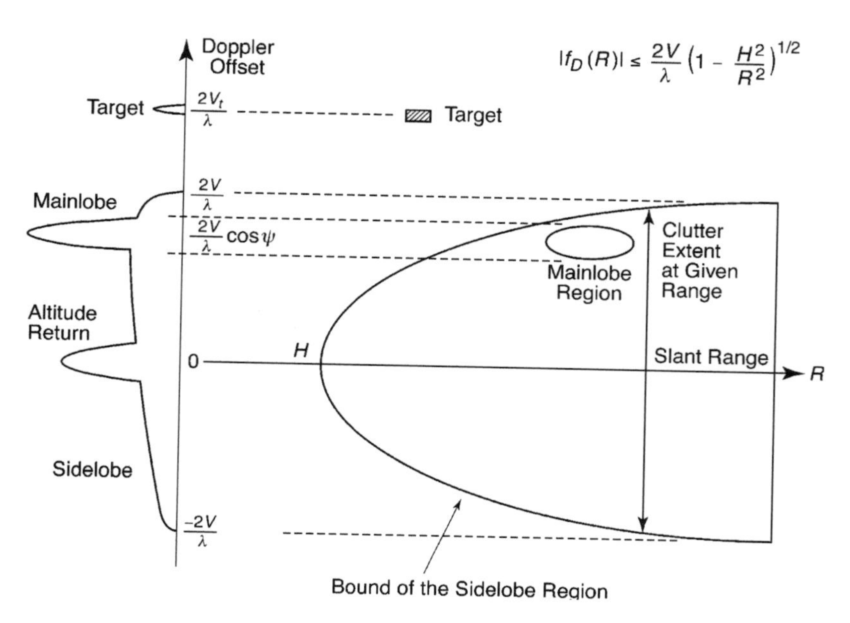

Figure 11.4 HPRF airborne radar: ground clutter versus velocity and range. (After [5], Figure 2.5 with permission of Artech House.)

tion, as illustrated in Figure 11.5(a), a situation known as *tail chase.* The return from the target will be "buried" in the altitude return.

- A *low-closing-rate target* could include an aircraft ahead of the ownship moving in the same direction, but more slowly, or one behind the ownship moving in the same direction and at a slightly faster speed, as illustrated in Figure 11.5(b).
- A *high-closing-rate target* could include an aircraft ahead of the ownship and flying toward it at low or high speed; in either case, it would appear to be closing more quickly than the ground clutter with greatest doppler frequency, as illustrated in Figure 11.5(c). It could also indicate an aircraft directly aft of the ownship flying toward it at ground speed greater than $2V$.
- A *low-opening-rate target* could include an aircraft directly ahead of the ownship flying in the same direction at a speed such that $V < V_{tgt} < 2V$, as illustrated in Figure 11.5(d), or one directly behind the ownship flying in the same direction and at a lower speed ($V_{tgt} < V$).
- A *high-opening-rate* target could include an aircraft directly aft of the ownship flying in the opposite direction at any speed or an aircraft

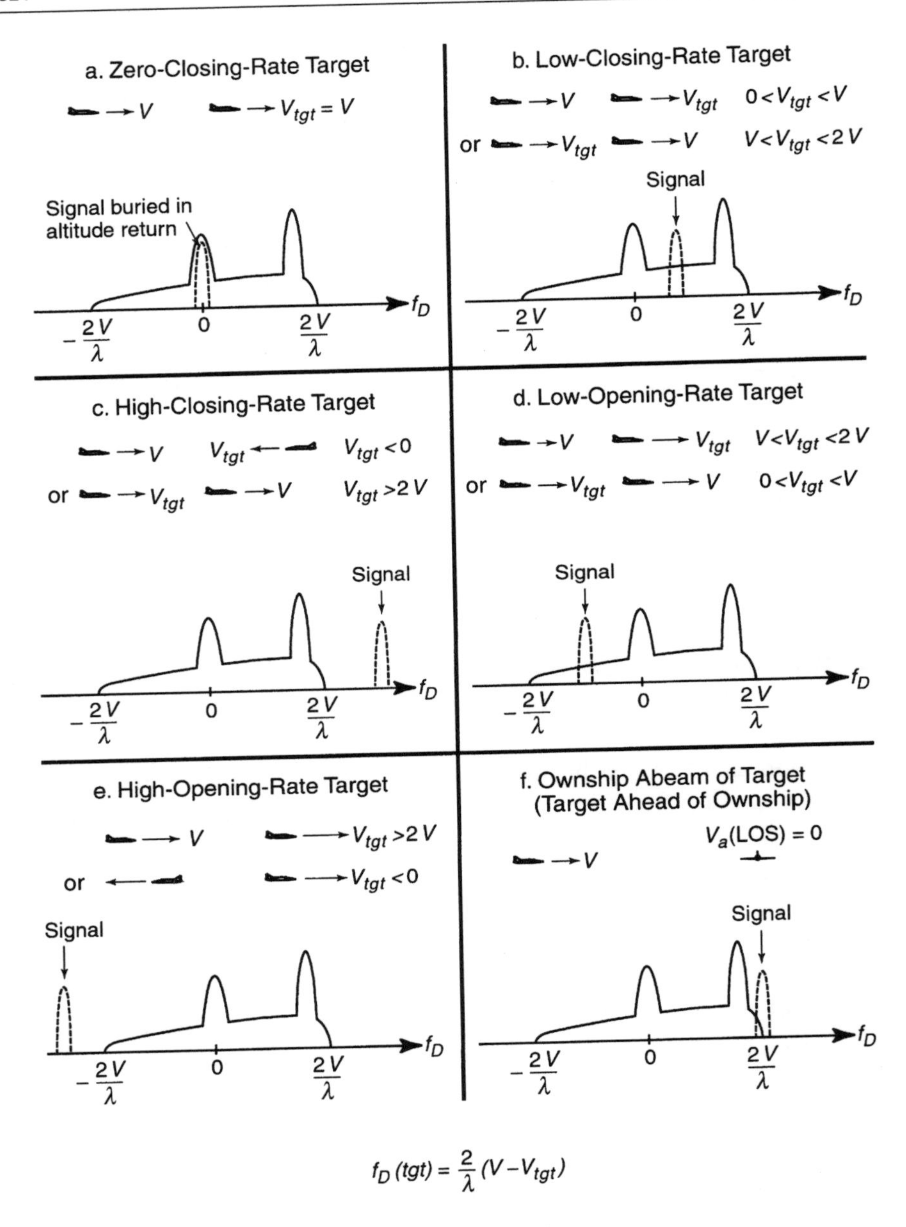

Figure 11.5 Typical return from aircraft using HPRF airborne radar: (a) zero-closing-rate target; (b) low-closing-rate target; (c) high-closing-rate target; (d) low-opening-rate target; (e) high-opening-rate target; (f) ownship abeam of target (target ahead of ownship).

ahead of the ownship flying in the same direction but at a much higher speed ($V_{tgt} > 2V$), as illustrated in Figure 11.5(e).

- For a target flying directly ahead of the ownship in a direction perpendicular to the ownship direction (ownship is abeam of the target), the apparent closing velocity is just V and the target return coincides with the maximum-speed sidelobe clutter return, as illustrated in Figure 11.5(f).

Table 11.1, from [5, Chaps. 4–6], summarizes the advantages and disadvantages of the different types of PRF.

11.3.2 Air-to-Air Radar Techniques

Sensitivity-time control (STC) is frequently used in LPRF air-to-air modes. The receiver amplification factor increases as the return from more distant ranges is received. The amplification tends to mitigate the effect of space loss (loss of signal due the factor $1/R^4$) and to attenuate the return from nearby clutter.

A guard antenna helps separate mainlobe targets from targets or jammers in the sidelobes [5, p. 119]. The guard antenna, typically a small horn, is located somewhere on the aperture of the main antenna. The guard has a very wide mainlobe and low gain. If a very high RCS target, such as a metal building, or a high-power jammer, is in the sidelobes of the main antenna, it can sometimes produce a return as large as a typical target in the mainlobe. Whereas a real mainlobe target will have a strong return in the main channel and a very low return in the guard channel, a sidelobe target or jammer will have a strong return in both channels (mainlobe of guard antenna, sidelobe of main antenna). The processing logic deletes the return whenever the guard channel signal is greater than a specific threshold. The concept is illustrated in Figure 11.6.

11.3.3 Air-to-Air Radar Modes

We define the following terms.

- A *waveform* describes a sequence of similar pulses, usually at the same beam position.
- A *mode* corresponds to a sequence of waveforms processed together to achieve a particular objective (e.g., PRF switching); PRF and beam position may vary.
- A *timeline* describes a sequence of modes.

Table 11.1
Benefits and Limitations of LPRF, MPRF, and HPRF

Benefits	**Limitations**
Low-PRF mode	
Precise range measurement	Low probability of detection or high false alarm rate in look-down missions
Ground maps	High peak power or pulse compression ratio
Simple method to achieve long unambiguous range	Highly ambiguous doppler
Range gating rejects sidelobe clutter	
Simple data processing	
Medium-PRF mode	
Better low-altitude tail-chase mode than high PRF	No velocities clear of sidelobe clutter
Accurate ranging	Many PRFs and pulse widths; complex processing needed to cope with range and doppler ambiguities
Good nose-aspect performance	Rejection of large targets in sidelobes
High-PRF mode	
For a fixed peak power, higher average power and thus longer detection range achieved as result of the higher duty factor	Highly ambiguous range
Unambiguous doppler, no blind zones except zero doppler and near mainlobe clutter	Eclipsing of target return interferes with initial detection and tracking
Good look-down nose-aspect detection (targets in clutter-free region)	Ranging methods more complicated, often less accurate
Illumination for semiactive missiles	Sidelobe clutter reduces tail-aspect detection sensitivity

Source: Reproduced from [5], pp. 64, 83, and 103, with permission of Artech House.

Stimson [6, p. 379] describes a potential search mode for a radar in the nose of an aircraft. In frame 1, the antenna scans over four bars:

- Bar 1: azimuth sweep left-to-right (L to R), constant elevation, HPRF;
- Bar 2: azimuth sweep, R to L, lower elevation, MPRF;
- Bar 3: azimuth sweep, L to R, lower elevation, HPRF;
- Bar 4: azimuth sweep, R to L, lower elevation, MPRF.

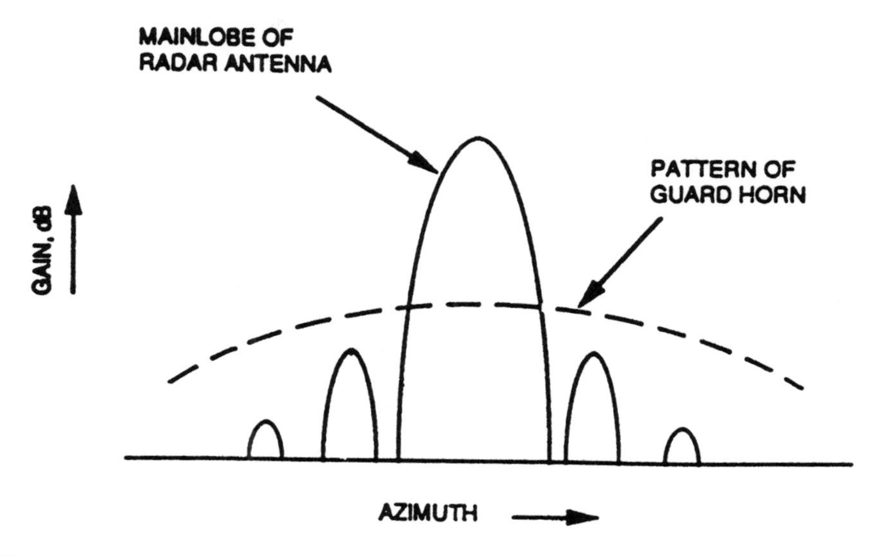

Figure 11.6 Guard channel. (Reproduced from [5], p. 119, with permission of Artech House.)

In frame 2, the scans are repeated with HPRF and MPRF reversed. In that way, each portion of the search volume (the region of solid angle over which the main beam may point) is scanned with both HPRF and MPRF, accruing the advantages of each mode. Stimson [6, p. 550] states that the future U.S. F-22 fighter aircraft will employ an active ESA, thus making the angular search even more flexible.

A modern fighter aircraft has a multimode radar, and pilots can switch between modes at their discretion. For example, Stimson's first edition [7] gave the modes for the APG-63 radar (F-15 aircraft) as follows: HPRF Range While Search, HPRF Velocity Search, HPRF Acquisition, HPRF Track, MPRF Search and Acquisition, MPRF Track, LPRF Air-to-Air, LPRF Ground Map (SAR), LPRF Beacon, LPRF Air-to-Ground Ranging, LPRF Track. Further details are provided in [7].

11.4 Air-to-Ground Radar

Air-to ground radar is typically LPRF to avoid range ambiguities. This section considers an LPRF air-to-ground radar designed to observe ground-moving targets.

11.4.1 MTI Radar

If a stationary LPRF PD radar is observing a stationary scene (clutter) for a dwell time T and coherently receives and doppler processes the returns, the spectrum obtained will be of the type shown in Section 10.1 (Figure 10.5). We now investigate whether it is possible to observe a moving target in addition to the stationary clutter. We mix the received signal to baseband and combine the positive and negative frequency results. Figure 11.7 illustrates a schematic "magnification" of the region between the peaks at 0 and at f_R. We recognize the second peak as the first doppler ambiguity. (The radar return is assumed to be from the range gate of interest; other returns, such as the sidelobe altitude return, are assumed to be range-gated out.)

If the radar and the clutter are both stationary, if the noise is negligible, and if the FFT is zero-padded (see Section 8.1.3) to produce a large number of bins, then the broadening of the PRF line will be due solely to the finite dwell time and will have $f_{pn} = 1/T$. If the clutter has intrinsic motion, then the PRF line will be broadened further (Section 11.4.3).

Several authors have presented a thorough treatment of MTI radar [2, 5, 8–13]. Before extensive digital processing was developed, MTI was generally based on a pulse-cancelation procedure. The return from a particular pulse is coherently subtracted (canceled) from the return from the previous pulse. If the scene contains no moving targets, that cancelation should produce

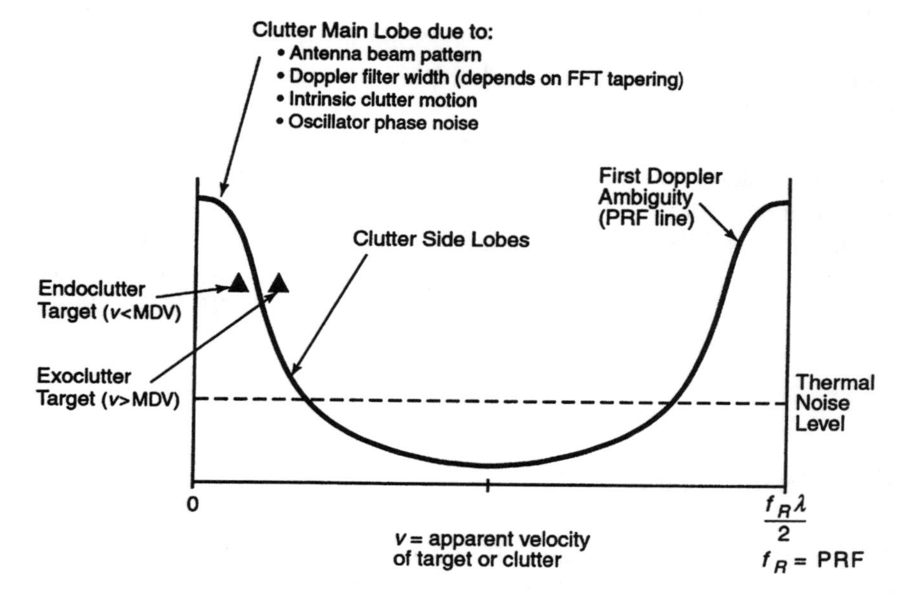

Figure 11.7 LPRF airborne radar: clutter and moving-target signatures.

a zero result; a moving target will produce a nonzero cancelation result (if its doppler frequency shift is not a multiple of the PRF). Additional pulses with appropriate weighting can be used to improve the response of the cancelation filter. This section does not attempt to reproduce that discussion; rather, it is confined to a presentation of a simple model of an airborne MTI searching for ground targets (a GMTI).

11.4.2 Detection of Ground-Moving Targets

As shown in Figure 11.8, we consider an airborne (or spaceborne) radar observing the ground at *squint angle* θ_{sq} to the broadside direction, and at grazing angle $\psi << \pi/2$. We denote $\theta_{sq} - \theta = \Delta\theta$. The finite radar beam will receive echoes from portions of the ground moving at different apparent LOS speeds relative to the radar. ψ is assumed to be constant over the range of interest of $\Delta\theta$. The apparent LOS speed of any portion of the ground is

$$
\begin{aligned}
v &= -V \sin\theta \cos\psi \\
\frac{dv}{d\theta} &= -V \cos\theta \cos\psi
\end{aligned}
\tag{11.4}
$$

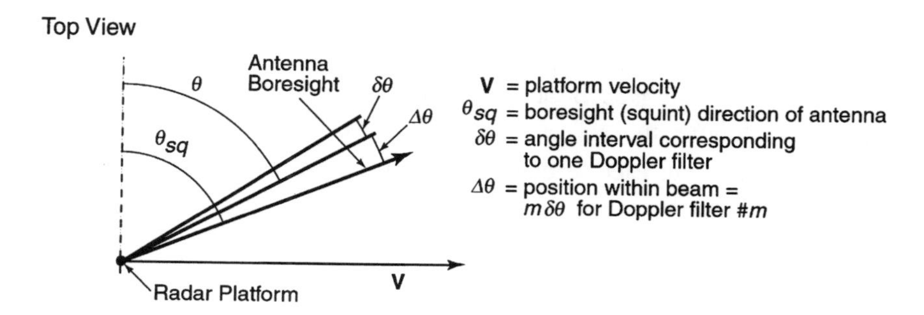

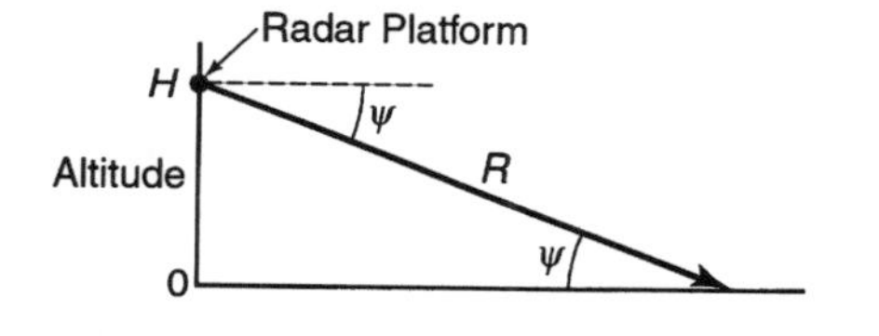

Figure 11.8 Airborne GMTI geometry: (a) top view; (b) side view, normal to boresight direction.

We normalize the apparent speed of the clutter at the beam center to zero. Then, for small $\Delta\theta$,

$$v = -V\Delta\theta\cos\theta\cos\psi \approx -V\Delta\theta\cos\theta_{sq}\cos\psi \tag{11.5}$$

The principles illustrated in Figure 11.8 are summarized here:

- From the horizontally moving airborne platform, the real-beam radar stares at squint angle θ_{sq} and grazing angle ψ, at a patch of ground (clutter), and transmits N pulses in a dwell time T. Thus, the PRF is N/T, and the time between pulses (PRI) is $\delta t = T/N$. The results are interpreted as a plot of clutter return (voltage) versus time.
- The processor performs an FT and produces a plot of clutter return (voltage) as a function of frequency; this is magnitude-squared and normalized to produce clutter power versus frequency.
- The abscissa is multiplied by $\lambda/2$ to yield clutter power versus clutter LOS velocity. The origin is shifted so the clutter LOS velocity is zero in the antenna boresight direction (we refer to this as the *apparent LOS clutter velocity*).
- Because apparent clutter velocity is directly proportional to $\Delta\theta$, then for stationary clutter, the clutter power versus velocity function has the shape of the beam pattern of the antenna in the directions of the annular range gate. Because the grazing angle ψ is small, the clutter shape follows that of the azimuth beam pattern.
- The clutter in each velocity bin can be expressed as the RCS of the ground clutter σ_c (dBsm) in the corresponding range bin and azimuth angle bin.
- The interference is the combination of clutter plus jamming (if any); we then speak of the *signal to interference-plus-noise ratio* (SINR). Given a probability of false alarm, P_{FA}, per range-doppler bin, the probability of target detection, P_D, can be determined from the PDFs of the clutter and the noise. To a good approximation, P_D can be estimated from the SINR. The minimum SINR required for target detection is denoted as $SINR_{\min}$.
- If a moving target with RCS σ_{tgt} at actual squint angle θ_{sq} is present in a particular velocity bin, corresponding to its LOS velocity, then
  - If σ_{tgt}(dBsm) $>$ σ_c(dBsm) + $SINR_{\min}$(dB), we describe the target as an exoclutter target (from the Latin *exo*, meaning "outside of").
  - If σ_{tgt}(dBsm) $<$ σ_c(dBsm) + $SINR_{\min}$(dB), we describe the target as an endoclutter target (from the Latin *endo*, "inside of").

- This section discusses exoclutter GMTI, that is, detection of exoclutter targets. Endoclutter GMTI, which can be performed using a multiple-subaperture antenna, is discussed in Section 11.4.5.
- From Figure 11.8, as the apparent clutter velocity approaches zero, the clutter RCS increases. Thus, except for very high-RCS targets, there is some velocity such that, for lower velocities, σ_{tgt} (dBsm) < σ_c (dBsm) + $SINR_{\min}$ (dB), and the target is "buried" in the clutter. This velocity is called the minimum detectable velocity (MDV). The MDV is a function of a number of variables, including target RCS, clutter σ^0, platform speed, squint angle, grazing angle, wavelength, antenna length, target range, and range bin width.

Assume that the number of doppler filters is equal to the number of pulses, $n_p = f_R T$. The width of a frequency bin is $1/T$, and that of a velocity bin is $\delta V = \lambda/2T$. Thus, from (11.5), the angular interval corresponding to one doppler bin is

$$|\delta\theta| = \frac{|\delta v|}{V\cos\theta_{sq}\cos\psi} = \frac{\lambda}{2VT\cos\theta_{sq}\cos\psi} \tag{11.6}$$

We assume that the antenna is gimballed so that its face is perpendicular to the boresight direction (if the antenna is a phased array oriented parallel to the direction of flight, then L becomes $L\cos\theta_{sq}$ in the equations that follow). From Section 2.3.2, for a uniformly weighted aperture, the one-way antenna azimuth beam power pattern (assume that the elevation pattern is constant) can be represented as

$$f(x) = \left(\frac{\sin x}{x}\right)^2, \; x = \frac{\pi L \sin(\Delta\theta)}{\lambda} \approx \frac{\pi L \Delta\theta}{\lambda} \tag{11.7}$$

For the (more useful) tapered aperture, the one-way power pattern can be denoted as

$$f_B\left(\frac{\pi L \Delta\theta}{\lambda}\right) = f_B\left(\frac{\pi L m \delta\theta}{\lambda}\right) \equiv f_B(U_m) \tag{11.8}$$

The normalized clutter RCS in doppler bin m is

$$\delta\sigma_m = \sigma^0 \frac{\delta R}{\cos\psi} R\delta\theta \cdot f_B^2(U_m) \tag{11.9}$$

where δR is the range bin width and R is the target range. The distribution is normalized so that, in the boresight direction (maximum gain), the clutter has its true value; in off-boresight directions, it is reduced according to the antenna gain relative to the maximum. Thus, a stationary target on boresight is in the same doppler bin as the peak main-beam clutter return. However, a moving target on boresight is in a bin with clutter reduced by the corresponding beam-pattern factor. If the LOS velocity of such a target, with RCS of σ_{tgt}, is

$$v_{tgt} = \frac{m\lambda}{2T} \tag{11.10}$$

then the signal-to-clutter ratio is

$$SCR_m = \frac{\sigma_{tgt}}{\delta\sigma_m} \tag{11.11}$$

To find the MDV, we obtain M, the corresponding value of m, as follows. (Because we are finding the velocity such that (in dB units) $\sigma_{tgt} = \sigma_c + SINR_{\min}$, we assume that the clutter at this value of m is much greater than the noise; and we then neglect the latter. That assumption may be easily relaxed, and the equations become somewhat more cumbersome (Problem 11.5)).

$$\begin{aligned} SINR_{\min} &\cong \frac{\sigma_{tgt}}{\delta\sigma_M} = \frac{\sigma_{tgt}\cos\psi}{\sigma^0\delta R \cdot R|\delta\theta| \cdot f_B^2(U_M)} \\ f_B^2(U_M) &= \frac{\sigma_{tgt}\cos\psi}{\sigma^0\delta R \cdot R|\delta\theta| \cdot SINR_{\min}} \end{aligned} \tag{11.12}$$

and from (11.6)

$$f_B^2(U_M) = \frac{2\sigma_{tgt}VT(\cos\psi)^2(\cos\theta_{sq})}{\sigma^0\delta R \cdot R\lambda \cdot SINR_{\min}} \tag{11.13}$$

It is presumed that the user knows all the factors on the right side of (11.12) and (11.13).

This transcendental equation, (11.13), can be solved graphically for $|U_M|$. Using that result plus (11.6) and (11.8), we can obtain MDV:

$$|U_M| = \frac{\pi L|M|\delta\theta}{\lambda} = \frac{\pi L|M|}{2VT\cos\theta_{sq}\cos\psi}$$
$$|M| = \frac{2VT(\cos\theta_{sq}\cos\psi)|U_M|}{\pi L} \quad (11.14)$$
$$MDV = |M|\delta v = |M|\frac{\lambda}{2T} = \frac{V\lambda(\cos\theta_{sq}\cos\psi)|U_M|}{\pi L}$$

The number of doppler filters is proportional to T; and M, the number of the filter such that the target is just detectable, also is proportional to T. Thus $|U_M|$ is a function of L, V, λ, θ_{sq}, ψ, σ_{tgt}, σ_0, δR, R, and $SINR_{\min}$, but not M or T; and for a given V, λ, θ_{sq}, ψ, σ_{tgt}, σ_0, δR, R, and $SINR_{\min}$, MDV is inversely proportional to antenna length.

Tables 11.2–11.5 and Figure 11.9 give example estimates of those parameters.

Two parameters sometimes used to characterize GMTI systems are MDV and false alarms per unit area, that is, false alarm density (FAD). The expected number of false alarms in the dwell is the number of doppler bins (equal to the number of pulses) multiplied by the probability of false alarm per bin, P_{FA}:

$$\hat{n}_{FA} = n_p P_{FA} \quad (11.15)$$

The area under scrutiny is

$$A = \delta R \cdot R\delta\theta \quad (11.16)$$

and FAD is therefore

$$FAD = \frac{n_p P_{FA}}{\delta R \cdot R\delta\theta} \quad (11.17)$$

Entzminger et al. [3] provide an interesting history and summary of some GMTI radars, including those on the Joint Strategic Target Attack Radar System (Joint STARS) and the Global Hawk UAV. The aircraft carrying those radars fly a narrow "racetrack" pattern, continually scanning a sector of a circle $\pm\theta$ off broadside. Thus, the region of the ground scanned is the overlap of the sectors scanned from the two ends of the racetrack, which has the area (Figure 11.10)

$$A = R^2(\theta - \sin^{-1}\alpha - \alpha(1 - \alpha^2)^{1/2} + \alpha^2\cot\theta) \quad (11.18)$$

Table 11.2
Example of Airborne GMTI—1

(All units MKS, except as noted) deg = 0.017 rad := deg^{-1} rad = 57.296

GEOMETRY: Slant Range = R := 100000
Assume $\cos\psi \sim 1$

BEAMWIDTH: $c := 3 \cdot 10^8$ $f := 10 \cdot 10^9$ $\lambda := \frac{c}{f}$ $\lambda = 0.03$

Antenna Height = $L_v := 0.4$

Antenna Length = $L_h := 1.4$ Antenna Area = $A := L_v \cdot L_h$ $A = 0.56$

Horizontal Beamwidth = $\theta := \frac{1.2 \cdot \lambda}{L_h} \cdot \text{rad}$ $\theta = 1.473$ degrees

Azimuth Field of Regard = $\Theta := 90$ deg No. Beams = $n_B := \text{floor}\left(\frac{\Theta}{\theta}\right)$ $n_B = 61$

Gain = $G := \frac{4 \cdot \pi \cdot A}{\lambda^2} \cdot .5$ $10 \cdot \log(G) = 35.921$ dB

PRF; DOPPLER:

PRF = $f_R := 2500$ Unambiguous Velocity Interval = $\Delta V := f_R \cdot \frac{\lambda}{2}$ $\Delta V = 37.5$

Unambiguous Range = $R_u := \frac{c}{2 \cdot f_R}$ $R_u = 6 \cdot 10^4$ Range Bin = $\delta R := 15$

Scan Time = T := 20 Coherent Processing Intervals (CPIs) Per Beam = $N_d := 4$ (to disambiguate range, velocity)

CPI = $\delta T := \frac{T}{n_B \cdot N_d}$ $\delta T = 0.082$ Pulses Per CPI = $n_p := f_R \cdot \delta T$ $n_p = 204.918$

Assume No. Doppler Filters = No. Pulses Per CPI $n_d := \text{floor}(n_p)$ $n_d = 204$

where R is the maximum ground range, $2d$ is the racetrack length and $\alpha = d/R$.

11.4.3 Intrinsic Clutter Motion

To minimize MDV, it is desirable to keep the width of the clutter peak (Figure 11.7) as narrow as possible. As previously discussed, the primary source of peak width is the antenna pattern. The antenna illumination function is typically tapered to reduce sidelobes, resulting in beam mainlobe broadening (Section 2.3.4) relative to a uniformly illuminated aperture. Two other key sources of

Table 11.3
Example of Airborne GMTI—2

VELOCITY vs. ANGLE: $\delta v := \frac{\lambda}{2} \cdot \frac{f_R}{n_d}$ $\delta v = 0.184$

Velocity Corresponding to m'th Doppler Filter: $m := 1, 2 .. n_d$ $v_m := m \cdot \delta v$

Squint Angle = $\theta_{sq} := 45 \cdot \mathrm{deg}$ $\cos(\theta_{sq}) = 0.707$

Platform Velocity = $V := 180$

Angle Corresponding to m'th Doppler Filter: $\delta\theta := \frac{|\delta v|}{V \cdot \cos(\theta_{sq})}$

$\delta\theta \cdot \mathrm{rad} = 0.083$ degrees $\Delta\theta_m := m \cdot \delta\theta$

ANTENNA PATTERN: $a_m := \frac{\pi \cdot \Delta\theta_m \cdot L_h}{\lambda}$

Assume $(\mathrm{sinc}(0.265)x)^4$ one-way antenna power pattern (rough approximation to pattern for low sidelobe tapered aperture):

Power Pattern = $f_{B_m} := \left(\frac{\sin(0.265 \cdot a_m)}{0.265 \cdot a_m}\right)^4$ $Lf_{B_m} := 10 \cdot \log(f_{B_m})$ dB

$$\theta 3dB_{goal} := 1.2 \cdot \frac{\lambda}{L_h} \cdot \mathrm{rad}$$

$$\theta 3dB_{goal} = 1.473 \text{ degrees}$$

clutter peak broadening are STALO phase noise (Section 10.2) and intrinsic clutter motion.

If the clutter itself is moving (e.g., bushes or trees blown by the wind, typical sea-surface motion), the peak will clearly broaden. Summaries of expected RMS velocities of land and sea clutter are shown in Figure 11.11. Estimation of the amount of peak broadening depends on assumptions about the clutter velocity PDF and the antenna pattern. Within a doppler filter, the total clutter power is the sum of that from the antenna pattern at the appropriate angle and that from the intrinsic clutter motion at the appropriate clutter velocity.

11.4.4 Terrain Limitations on Range

Section 3.2.2 determined the following procedure for calculating the grazing angle ψ corresponding to a radar at altitude H above the earth of radius R_E observing a point on the ground at ground range R_g from the subradar point on the Earth's surface (see Figure 3.8):

Table 11.4
Example of Airborne GMTI—3

CLUTTER: Clutter Reflectivity = $\sigma_0 := 10^{-1.5}$

Clutter RCS per doppler bin = $\sigma_{c_m} := \sigma_0 \cdot \delta R \cdot R \cdot \delta\theta \cdot (f_{B_m})^2$

$L\sigma_{c_m} := 10 \cdot \log(\sigma_{c_m})\,dB$

NOISE: $kT := 10^{-20.4}$ Noise Figure = $F := 2.5$ Losses = $L := 20$ $P_{avg} := 1000$

Noise-equivalent RCS in one doppler bin = $\sigma_{n_m} := \dfrac{(4 \cdot \pi)^3 \cdot R^4 \cdot kT \cdot F \cdot L}{P_{avg} \cdot G^2 \cdot \lambda^2} \cdot \dfrac{1}{\delta T}$

$L\sigma_{n_m} := 10 \cdot \log(\sigma_{n_m})\,dB$

PROBABILITY OF DETECTION:

Three target sizes: $l := 1, 2 \,..\, 3$ $\sigma_{t_l} := 10 \cdot 10^{-(l-1)}$ $\sigma_t = \begin{bmatrix} 10 \\ 1 \\ 0.1 \end{bmatrix}$

Signal-to-interference-plus-Noise Ratio (Rayleigh Interference) =

$$SINR_{m,l} := \frac{\sigma_{t_l}}{\sigma_{n_m} + \sigma_{c_m}}$$

Assume Swerling-2 Target, and Rayleigh Interference; Per CPI: $p_{fa} := 10^{-2.2}$

Require M out of N_d (3 out of 4) detections: $N_d = 4$ $M := 3$

$$p_{d_{m,l}} := p_{fa}\left(\frac{1}{1 + SINR_{m,l}}\right)$$

$$P_{D_{m,l}} := \left[\sum_{J=M}^{N_d}\left[\frac{N_d!}{(N_d - J)! \cdot J!} \cdot (p_{d_{m,l}})^J \cdot (1 - p_{d_{m,l}})^{N_d - J}\right]\right]$$

$$P_{FA} := \left[\sum_{J=M}^{N_d}\left[\frac{N_d!}{(N_d - J)! \cdot J!} \cdot (p_{fa})^J \cdot (1 - p_{fa})^{N_d - J}\right]\right] \quad P_{FA} = 1 \cdot 10^{-6}$$

$$\alpha = \frac{R_g}{R_E}$$

$$R_S = [(R_E + H)^2 + R_E^2 - 2(R_E + H)R_E\cos\alpha]^{1/2} \tag{11.19}$$

$$B = \sin^{-1}\left(\frac{\sin\alpha}{R_S}\right)(R_E + H), \ \frac{\pi}{2} < B < \pi$$

$$\psi = B - \frac{\pi}{2}$$

Figure 11.12 summarizes ψ versus R_g and H for a 4/3 Earth (R_E = 8,495 km).

Table 11.5
Example of Airborne GMTI—5

FALSE ALARMS: $P_{FA} = 1 \cdot 10^{-6}$ Range Bin = $\delta R := 15$

Max. Range = $R_{max} := 110000$ Min. Range = $R_{min} := 90000$ $\Delta R := R_{max} - R_{min}$

$\Delta R = 2 \cdot 10^4$

Sector scanned = $A_{sector} := \pi \cdot (R_{max}^2 - R_{min}^2) \cdot \frac{\Theta}{360}$ $A_{sector} = 3.142 \cdot 10^9$

Sector area (sq km) = $Asqkm := A_{sector} \cdot 10^{-6}$ $A_{sqkm} = 3.142 \cdot 10^3$

False Alarms per sector = $N_{FA} := \frac{\Delta R}{\delta R} \cdot n_d \cdot P_{FA} \cdot n_B$ $N_{FA} = 16.592$

False Alarms per sq km = $\rho_{FA} := \frac{N_{FA}}{Asqkm}$ $\rho_{FA} = 5.281 \cdot 10^{-3}$

Scan Time = T = 20

False Alarms per Second = $r_{FA} := \frac{N_{FA}}{T}$ $r_{FA} = 0.83$

Burge and Lind [14] have provided a summary of minimum grazing angles for various terrain types. They obtained their data by visiting various types of terrain and measuring the minimum elevation angle at which they could obtain a clear LOS (CLOS) to the sky, averaged over azimuth. Table 11.6 summarizes those results. For example, for "fairly smooth desert," the median minimum elevation (grazing) angle is 2 degrees. A comparison of Table 11.6 with Figure 11.12 shows that for a radar at altitude of 10 km (33 Kft) observing fairly smooth desert the maximum CLOS range is effectively about 210 km.

11.4.5 Imaging a Ground-Moving Target of Unknown Constant Velocity

Having established that certain moving targets on the ground can be detected by airborne radar, we now inquire as to whether, by altering our radar waveform, we can acquire images of such a target. Section 6.12 discussed the possibility of using ISAR to image a target that is undergoing unknown variable velocity, including both unknown translation and rotation. Section 7.5.1 mentioned that a SAR could perhaps image a target that is moving on the ground in a straight line at unknown constant velocity.

Perry et al. [15] have developed a method for SAR imaging of moving targets that have unknown rectilinear motion. They process the dechirped phase history (see Section 7.2.1.2) with a keystone formatting procedure that eliminates the effects of linear range migration (Section 7.1.6) for all ground-moving targets regardless of their unknown velocity. The processing procedure

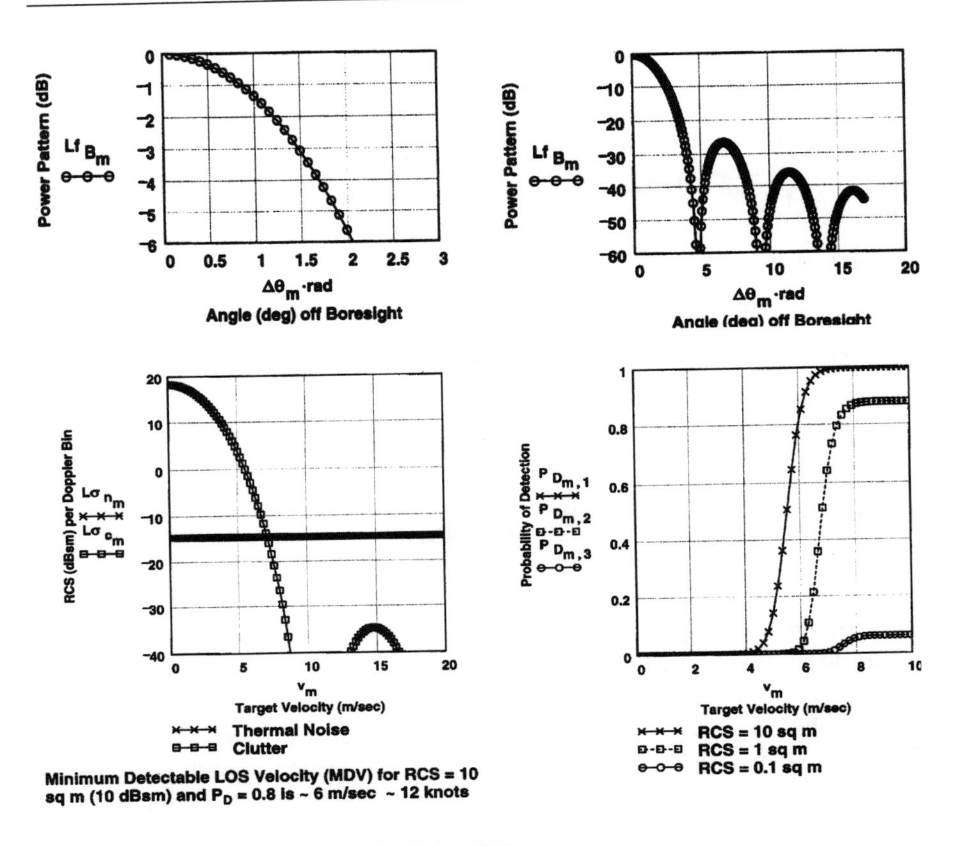

Figure 11.9 Example of airborne GMTI.

then automatically focuses the moving targets. Figure 11.13 illustrates the results. Figure 11.13(a) shows a conventionally processed SAR image containing three moving targets: an M813 truck, a tractor-trailer truck, and a surrogate (i.e., a full-size replica) of a missile transporter-erector-launcher (TEL). Figure 11.13(b) shows the focused image of the tractor-trailer resulting from the processing. The 2-foot resolution clearly shows the outline of the cab and trailer of the truck.

11.4.6 Displaced Phase Center Antenna

Section 11.4.1 stated that classical MTI is based on a pulse-cancelation procedure. The return from a particular pulse is subtracted (canceled) from the return from the previous pulse. If the scene contains no moving targets, that

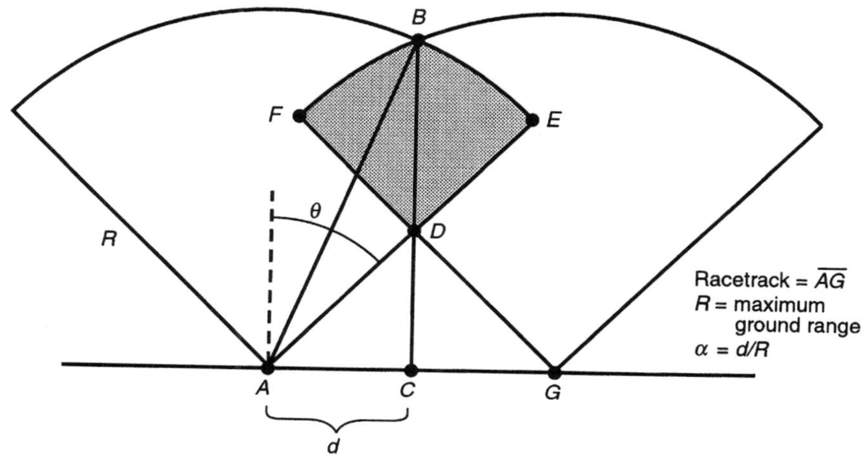

$$\text{Area }(ABE) = \tfrac{1}{2}R^2\,(\theta - \sin^{-1}\alpha)$$
$$\text{Area }(ABC) = \tfrac{1}{2}r \cdot [R^2 - r^2]^{\frac{1}{2}} = \tfrac{1}{2}\alpha R^2\,[1-\alpha^2]^{\frac{1}{2}}$$
$$\text{Area }(ADC) = \tfrac{1}{2}r^2 \cot\theta = \tfrac{1}{2}R^2\alpha^2 \cot\theta$$
$$\text{Area }(DBE) = \text{Area }(ABE) - [\text{Area }(ABC) - \text{Area }(ADC)]$$
$$= \tfrac{1}{2}R^2\,[\theta - \sin^{-1}\alpha - \alpha\,(1-\alpha^2)^{\frac{1}{2}} + \alpha^2\cot\theta]$$
$$\text{Area }(DFBE) = 2\text{ Area }(DBE)$$
$$= R^2\,[\theta - \sin^{-1}\alpha - \alpha\,(1-\alpha^2)^{\frac{1}{2}} + \alpha^2\cot\theta]$$
$$\alpha_{max} = \sin\theta$$
$$A\,(\alpha = \alpha_{max}) = 0$$
$$A\,(\alpha = 0) = R^2\theta$$

Figure 11.10 Ground region observed by GMTI radar.

cancelation should produce a zero result, whereas a moving target will produce a nonzero cancelation result (unless its doppler frequency shift is a multiple of the PRF).

If the radar antenna is moving normal to the LOS, the MTI procedure may be compromised. To retain good MTI performance, a displaced phase center antenna (DPCA) procedure was introduced [16]. In physical DPCA, the apertures of two side-looking antennas are aligned parallel to the aircraft flight direction. The PRF is adjusted so that, when pulse number $n + 1$ is transmitted, the second antenna is at the position that was occupied by the first antenna when pulse number n was transmitted. Thus, the two pulses can be canceled as if they were transmitted from a stationary aperture. Because that procedure is difficult to implement without significant errors, electronic DPCA was introduced. In electronic DPCA, the PRF need not be matched

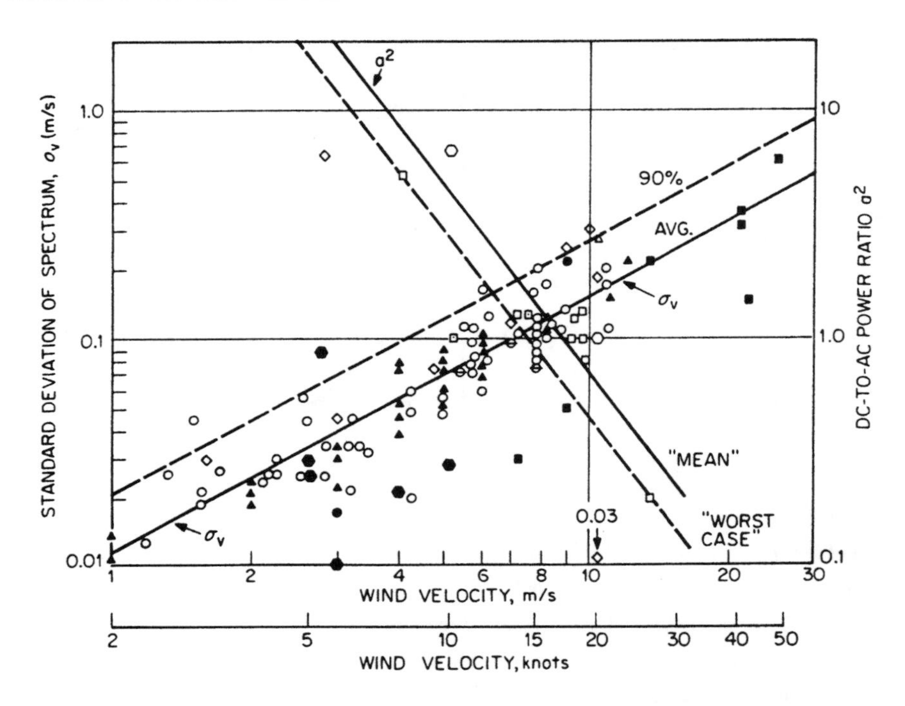

Figure 11.11 Expected clutter motion: (a) land clutter (wooded terrain); (b) sea clutter. (Reproduced from [11], Figures 7.32 and 7.5, with permission of McGraw-Hill.)

exactly to the antenna spacing and platform velocity, and phase corrections are applied. Furthermore, transmission is typically on the sum channel and reception is on the sum and difference channels (the monopulse technique [see Chapter 5]).

Section 11.4.2 described exoclutter GMTI. With DPCA, endoclutter GMTI can be achieved; certain targets that are buried in the clutter and nondetectable by the exoclutter method may be detectable by DPCA. Staudaher [16] points out that further improvement can be obtained with multiple apertures and use of adaptive processing (Chapters 8 and 9). That leads us to space-time adaptive processing, the subject of the next chapter.

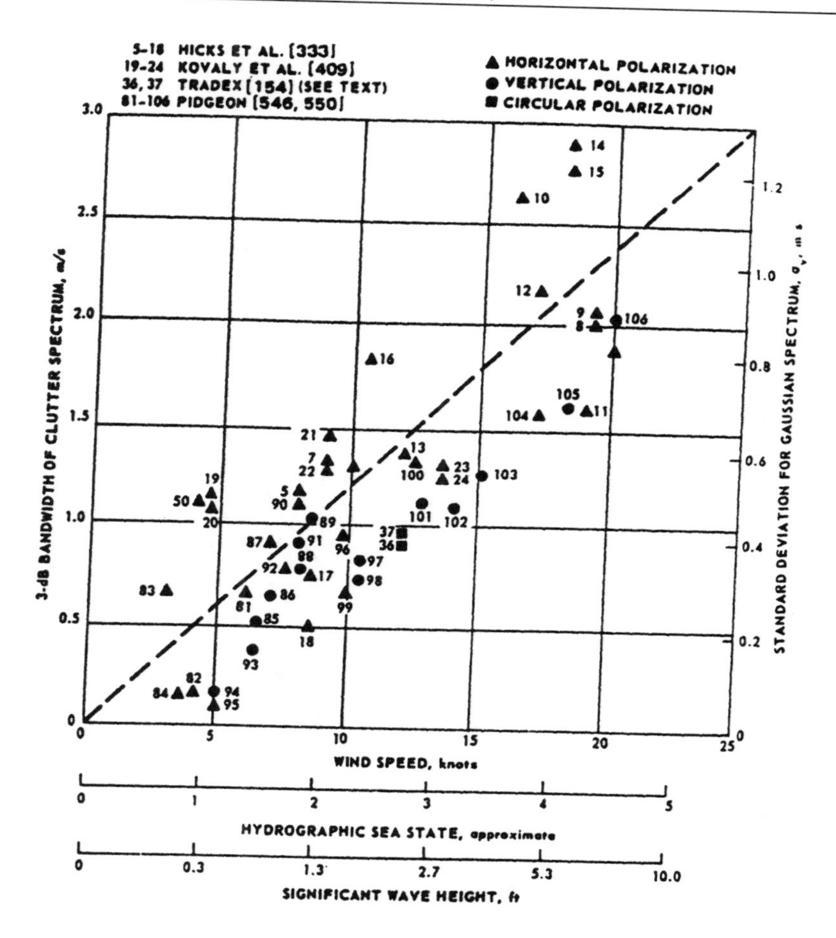

Figure 11.11 (continued).

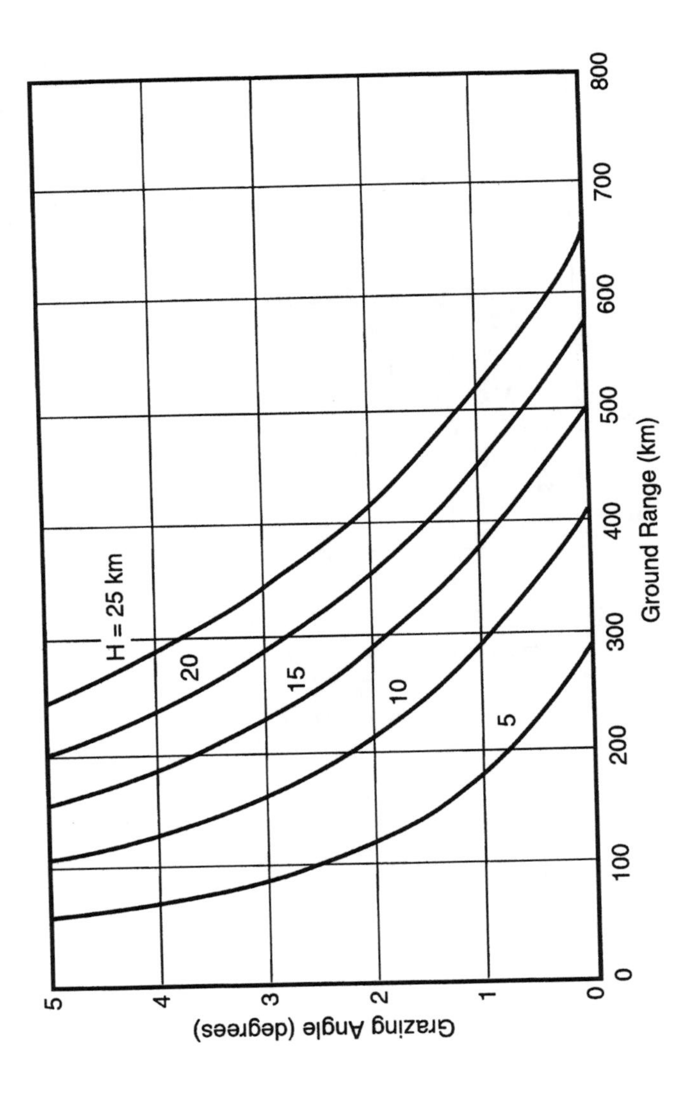

Figure 11.12 Grazing angle versus ground range and platform altitude.

Table 11.6
Minimum Grazing Angles for CLOS [14]

Terrain Designation	Median (Degrees)	97.5% (Degrees)
Flat farmland, forests in distance	1	3
Fairly smooth desert	2	4
Gently rolling farmland with smooth deserts	2	4
Moderate rough desert	4	11
Flat farmland with close forests	5	14
Gently rolling hills with scattered trees	6	12
Rough desert with little vegetation	9	18
Sharply rolling hills with thickly scattered trees	11	24

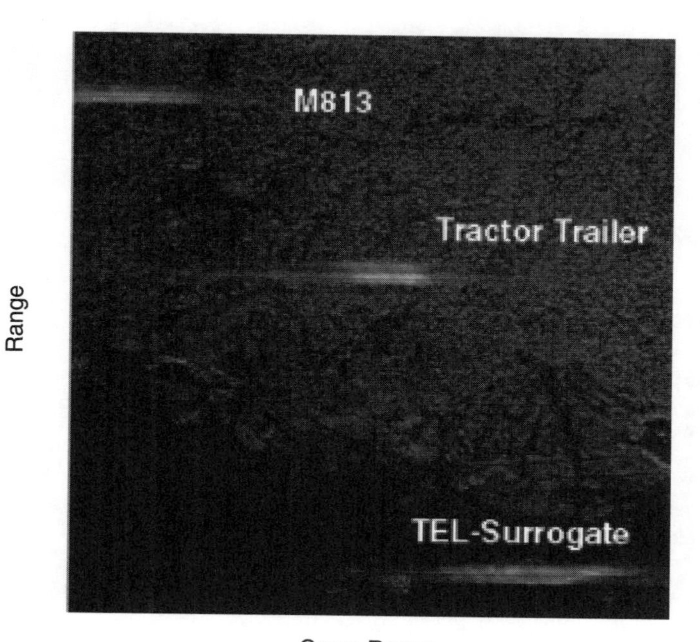

Figure 11.13 Examples of SAR imaging of target with unknown rectilinear velocity: (a) conventional SAR processing; (b) processing with methods in [15]. (Reproduced from [15], Figures 5 and 14, copyright 1999 IEEE.)

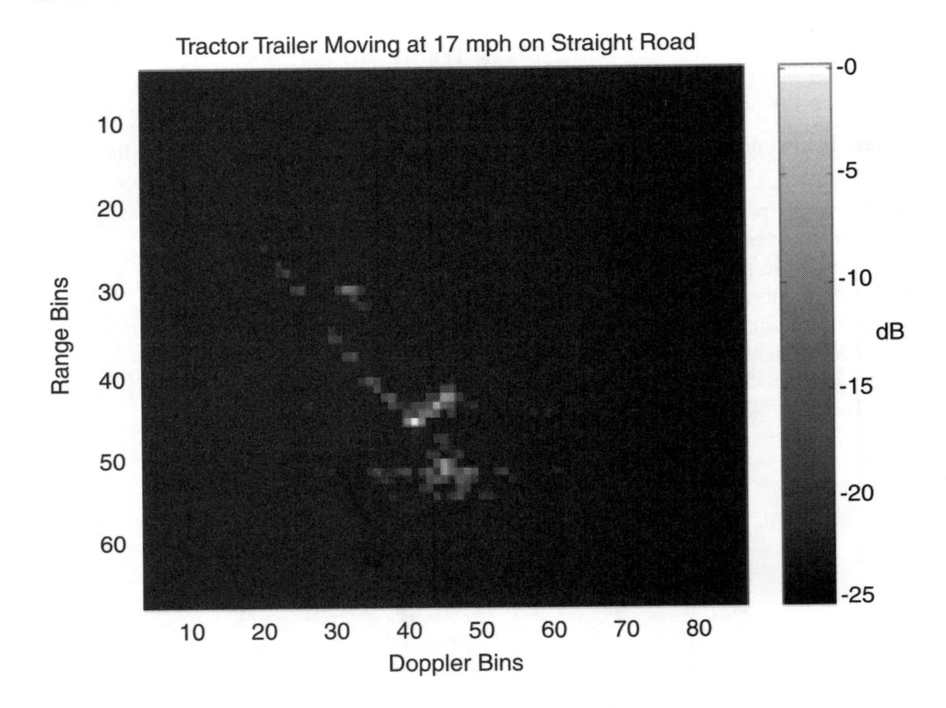

Figure 11.13 (continued).

References

[1] Kurpis, G. P., and C. J. Booth (eds.), *The New IEEE Standard Dictionary of Electrical and Electronics Terms*, 5th ed., New York: Institute of Electrical and Electronics Engineers, 1993.

[2] Skolnik, M. I., *Introduction to Radar Systems*, 2nd ed., New York: McGraw-Hill, 1980.

[3] Entzminger, J. N., C. A. Fowler, and W. J. Kenneally, "Joint-STARS and GMTI: Past, Present, and Future," *IEEE Trans. Aerospace and Electronic Systems*, Vol. 35, No. 2, April 1999, pp. 745–761.

[4] *IEEE Standard Radar Definitions* (Std 686-1997, rev. IEEE Std 686-1990), Piscataway, NJ: Institute of Electrical and Electronics Engineers, 1997.

[5] Morris, G., and L. Harkness (eds.), *Airborne Pulsed Doppler Radar*, 2nd ed., Norwood, MA: Artech House, 1996.

[6] Stimson, G. W., *Introduction to Airborne Radar*, 2nd ed., Mendham, NJ: SciTech, 1998.

[7] Stimson, G. W., *Introduction to Airborne Radar*, 1st ed., El Segundo, CA: Hughes Aircraft, 1983.

[8] Schleher, D. C., *MTI and Pulsed Doppler Radar*, Norwood, MA: Artech House, 1991.

[9] Barton, D. K., *Modern Radar System Analysis*, Norwood, MA: Artech House, 1988.

[10] Levanon, N., *Radar Principles*, New York: Wiley, 1988.

[11] Nathanson, F. E., J. P. Reilly, and M. N. Cohen, *Radar Design Principles*, 2nd ed., New York: McGraw-Hill, 1991.

[12] Skolnik, M. I., *Radar Handbook*, New York: McGraw-Hill, 1990.

[13] Mahafza, B. R., *Introduction to Radar Analysis*, New York: CRC Press, 1998.

[14] Burge, C. J., and Lind, J. H., *Line-of-Sight Handbook*, NWC-TP-5908, Naval Weapons Center, China Lake, CA, January 1977 (Defense Technical Information Center [DTIC] Number AD A038548).

[15] Perry, R. P., R. C. DiPietro, and R. L. Fante, "SAR Imaging of Moving Targets," *IEEE Trans. Aerospace and Electronic Systems,* Vol. 35, No. 1, Jan. 1999, pp. 188–200.

[16] Staudaher, F. M., "Airborne MTI," Chapter 16 of Skolnik, M., Ed., *Radar Handbook,* 2nd Edition, New York: McGraw-Hill, 1990.

Problems

Problem 11.1

For an airborne radar located at $(0, 0, H)$ and flying at speed V in the $+x$ direction, show that the isodop (hyperbola on the ground) corresponding to apparent speed $v_{\text{app}} < V$ is given by

$$\frac{x^2}{(H \cot \theta)^2} - \frac{y^2}{H^2} = 1, \; \theta = \cos^{-1}\left(\frac{-v_{\text{app}}}{V}\right)$$

Problem 11.2

If the mainlobe with 3-dB width $\Delta\theta$ is pointed at an angle θ away from broadside and at a negligible depression angle, show that the mainlobe clutter peak will appear at the lower speed $+V \sin \theta$ and will be broadened. Calculate the broadened 3-dB width in velocity space.

Problem 11.3

Compute the relative strength of the sidelobe ground clutter return from an annulus of width $\Delta R = 100$m and minimum slant range $R > H = 10$ km as a function of R. Assume that the beam pattern is isotropic and that σ^0 is given by the constant-γ approximation (Section 3.2.7). Repeat the calculation approximating σ^0 from Figure 3.15(b). Approximately how much stronger is the altitude return than the return from the annulus with $R = 100$ km? See also Problem 3.11.

Problem 11.4

Show that the maximum doppler frequency due to apparent ground motion at slant range R is

$$f_D \leq \frac{2V}{\lambda}\left(1 - \frac{H^2}{R^2}\right)^{1/2}$$

Problem 11.5

Generalize (11.11) to (11.13) to include the case for which the thermal noise is not negligible with respect to the clutter.

Part IV:
Special Radar Topics

12

Space-Time Adaptive Processing

Chapter 11 examined exoclutter airborne MTI (using a single-aperture radar) and the two-subaperture DPCA technique for clutter suppression (endoclutter MTI). This chapter considers a radar with an antenna consisting of several subapertures, or antenna elements; the echoes received by each subaperture are processed using the adaptive matched filter (see Section 9.2). Such a radar is capable of a certain degree of suppression of jammers, as well as (for an airborne radar) more effective endoclutter MTI.

A single-aperture radar receives a series of echoes, each characterized by the time at which it is received. A multiple-subaperture radar receives a set of echoes, each characterized not only by the reception time but also by the subaperture on which it is received. Because the subapertures are characterized by their location in space, the processing of the combined set of echoes is referred to as space-time processing. If spatial and temporal weights are adaptively calculated (i.e., using the adaptive matched filter), the technique is called space-time adaptive processing (STAP). This discussion of STAP follows that of Ward [1]; Klemm [2] has also developed a detailed exposition. This chapter summarizes the results; more complete derivations of equations can be found in [1].

Although the discussion here is in the context of an airborne radar, it can also be applied to a stationary surface-based radar by setting the platform velocity equal to zero. That is particularly applicable to jammer suppression.

12.1 Initial Assumptions

Assume that the antenna is an ESA characterized by a series of N subapertures (or antenna elements). During a CPI, M pulses are transmitted from the full aperture, and the echoes are received on each subaperture.

Figure 12.1 illustrates the assumed geometry. The antenna face is in the *xz* plane. The array consists of $N = N_x \times N_z$ subapertures, arranged in a rectangular array, with rows parallel to **x** and columns parallel to **z**. The array is at altitude H with its normal in the y direction, moving at velocity $\mathbf{V_a}$ parallel to the *xy* plane but not necessarily parallel to **x**. In other words, the direction of flight is parallel to the ground but not necessarily in the direction of the nose of the aircraft; aircraft "crab" is allowed. The radar transmits pulses from the full aperture with gain $G(\theta, \phi)$, where θ is the depression angle and ϕ is the angle in the *xy* plane between the beam and the normal to the antenna.

We typically perform pulse compression and sample in fast time to obtain the radar return in L range bins. We now have LMN complex radar echoes, denoted as x_{lmn}. Those echoes are visualized as a datacube (Figure 12.2), a rectangular solid array in three-dimensional space. We consider only one range bin and restrict our attention to the corresponding $M \times N$ two-dimensional matrix.

We now represent those MN values as an $MN \times 1$ column vector $\boldsymbol{\chi}$:

$$\boldsymbol{\chi} = (x_{l,0,0}; \ldots; x_{l,0,N-1}; \ldots; x_{l,M-1,0}; \ldots; x_{l,M-1,N-1}) \quad (12.1)$$

where the semicolons indicate a column vector (commas would indicate a row vector). $\boldsymbol{\chi}$ is referred to as the space-time snapshot.

The expected return in the absence of a target of interest (i.e., the return consists of only noise, clutter, and possibly jamming) is denoted as $\boldsymbol{\chi}_u$. That

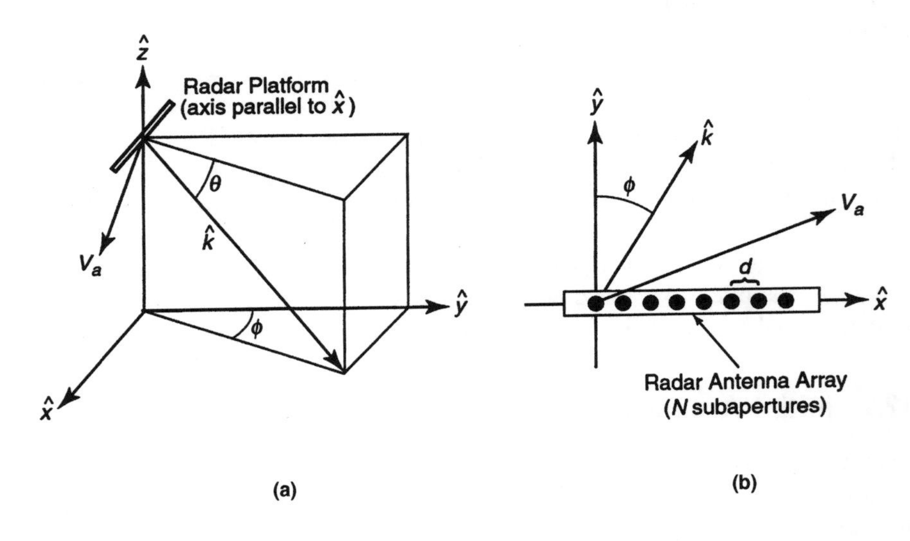

Figure 12.1 Assumed geometry: (a) perspective view; (b) top view. (After [1], Figure 2, with permission of MIT Lincoln Laboratory, Lexington, MA.)

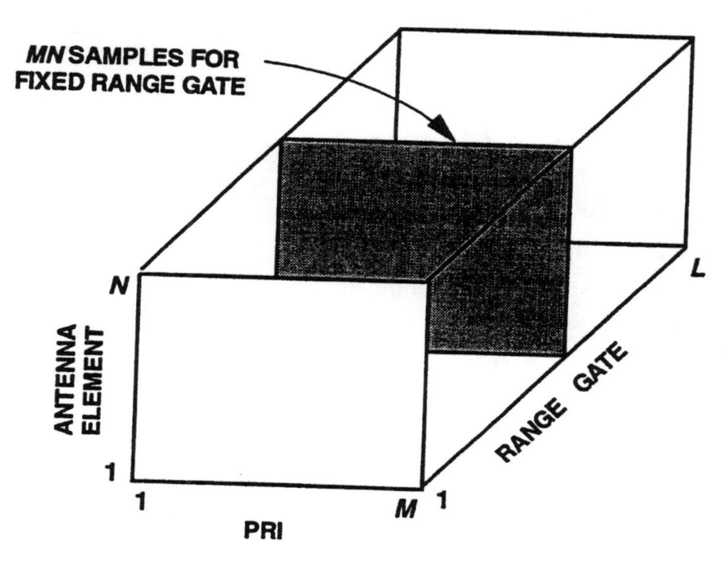

Figure 12.2 Datacube. (Reproduced from [1], Figure 4, with permission of MIT Lincoln Laboratory, Lexington, MA.)

corresponds to hypothesis H_0 (see Section 4.1.2). The response from a target of known type, position, and velocity is considered known except for an unknown scalar amplitude α_T, and is denoted by the $MN \times 1$ column vector $\alpha_{\mathbf{T}}\mathbf{v}_{\mathbf{T}}$. The return with target present (hypothesis H_1) is therefore

$$\boldsymbol{\chi} = \boldsymbol{\chi}_{\mathbf{u}} + \alpha_{\mathbf{T}}\mathbf{v}_{\mathbf{T}} \tag{12.2}$$

The theory of fully adaptive STAP is that of the adaptive matched filter (discussed in Section 9.2 in the context of detection of a fixed target in a SAR image). Accordingly, we need an estimate of the covariance matrix of $\boldsymbol{\chi}_u$. This $MN \times MN$ matrix is defined (Section 8.2.1) as

$$\mathbf{R}_{\mathbf{u}} = \mathbf{E}(\boldsymbol{\chi}_{\mathbf{u}}\boldsymbol{\chi}_{\mathbf{u}}^{\mathbf{H}}) \tag{12.3}$$

where E indicates the expected value and the superscript $\mathbf{H}$ indicates the hermitian adjoint. We assume that the noise, clutter, and jamming are mutually independent; thus, we can write

$$\mathbf{R}_{\mathbf{u}} = \mathbf{R}_{\mathbf{N}} + \mathbf{R}_{\mathbf{C}} + \mathbf{R}_{\mathbf{J}} \tag{12.4}$$

where the three terms on the right side represent the $MN \times MN$ covariance matrices for noise, clutter, and jamming, respectively.

12.2 Representation of Noise, Target, Jamming, and Clutter

We use the following notation for vectors and matrices. Assume that **a** and **b** are $N \times 1$ and $M \times 1$ column vectors, respectively.

- Vectors: lowercase boldface;
- Matrices: uppercase boldface;
- $\mathbf{A}^{\mathbf{T}}$ = transpose, $\mathbf{A}^*$ = conjugate, $\mathbf{A}^{\mathbf{H}}$ = hermitian adjoint (conjugate transpose);
- $[a_1; a_2; \ldots]$ = column vector;
- $[a_1, a_2, \ldots]$ = row vector;
- $[\mathbf{a};\mathbf{b}] = (M + N) \times 1$ column vector;
- $\mathbf{a} \otimes \mathbf{b}$ = Kronecker (vector) product ($MN \times 1$ vector);
- $\mathbf{a} \cdot \mathbf{b}$ = Hadamard (vector) product = element-by-element product vector ($M \times 1$ vector, $M = N$);
- $\mathbf{ab}$ = outer (matrix) product ($M \times N$ matrix);
- $\mathbf{a}\cdot\mathbf{b} = \mathbf{a}^{\mathbf{H}}\mathbf{b}$ = inner (scalar) product ($M = N$).

12.2.1 Noise

We assume that the noise consists entirely of thermal noise internal to the radar. We also assume that each subaperture has its own separate receiver channel with σ^2 = variance of the single-pulse received noise energy. The noise in any channel is uncorrelated with that in the other channels, and the off-diagonal elements of $\mathbf{R_N}$ are zero. Therefore, we have

$$\mathbf{R_N} = E\{\boldsymbol{\chi}_{\mathbf{N}}\boldsymbol{\chi}_{\mathbf{N}}^{\mathbf{H}}\} = \sigma^2\mathbf{I_N} \tag{12.5}$$

where $\mathbf{I_N}$ is the $\mathbf{N} \times \mathbf{N}$ identity matrix.

12.2.2 Target

We consider a moving target with LOS velocity $\mathbf{V_T}$ (magnitude V_T) and corresponding target doppler frequency $f_T = 2V_T/\lambda_c$. Here, λ_c is the wavelength corresponding to the center frequency of the transmitted waveform; the bandwidth B is assumed to be much less than the center frequency f_c. The normalized doppler frequency is defined to be

$$\omega_T = \left\{\frac{f_T}{f_R}\right\} \tag{12.6}$$

f_R is the PRF, which we assume to be range-unambiguous for the range bin being evaluated. The braces in (12.7) indicate the unambiguous representation of the argument:

$$\begin{aligned} \{z\} &\equiv (z + 0.5) \bmod 1 - 0.5 \\ -0.5 &\leq \{z\} \leq 0.5 \end{aligned} \tag{12.7}$$

We now assume that the antenna consists of a single horizontal row of equally spaced subapertures, that is, $N_z = 1$ (see Problem 12.1). θ_T and ϕ_T are the elevation and the azimuth of the (far-field) target with respect to the radar. We define

$$\Theta_T = \textit{spatial frequency} = \frac{d}{\lambda_c} \cos\theta_T \sin\phi_T \tag{12.8}$$

where d is the distance between subapertures. If Θ_T is > 0.5 or < −0.5, then $\mathbf{a}(\Theta_T)$ is ambiguous; a target in a certain direction could not be distinguished from a target in another direction. We assume $-0.5 < \Theta_T < 0.5$, that is, $\mathbf{a}(\Theta_T)$ is unambiguous.

We now define the following.

$$\mathbf{a}(\Theta_T) = [1;\ e^{j2\pi\Theta_T};\ \ldots;\ e^{j2\pi(N-1)\Theta_T}] \tag{12.9}$$

represents the target phasor signal (i.e., with amplitude normalized to 1) received at the N subapertures from the first pulse. $\mathbf{a}(\Theta_T)$, the spatial steering vector, is multiplied by the (scalar) target amplitude to produce the overall complex target signal corresponding to the first pulse.

$$\mathbf{b}(\omega_T) = [1;\ e^{j2\pi\omega_T};\ \ldots;\ e^{j2\pi(N-1)\omega_T}] \tag{12.10}$$

the temporal steering vector, represents the target phasor signal received from the M pulses at the first subaperture.

$$\begin{aligned} \mathbf{v}_T(\Theta_T, \omega_T) &= \mathbf{a}(\Theta_T) \otimes \mathbf{b}(\omega_T) \\ &= [\mathbf{a}(\Theta_T);\ e^{j2\pi\omega_T}\mathbf{a}(\Theta_T);\ \ldots;\ e^{j2\pi(M-1)\omega_T}\mathbf{a}(\Theta_T)] \end{aligned} \tag{12.11}$$

the $MN \times 1$ target steering vector, represents the target phasor signal received from the M pulses at the N subapertures.

12.2.3 Jamming

We assume that noise jamming radiation is being received from one or more fixed directions. Thus, the jamming is decorrelated in time and correlated in space.

We consider a single jammer at θ_J, ϕ_J, with (power) jammer-to-noise ratio equal to ξ_J. For the space representation of the jammer, we consider a jammer steering vector, completely analogous to the target steering vector:

$$\begin{aligned} \Theta_J &\equiv \frac{d}{\lambda_c}\cos\theta_J \sin\phi_J \\ \mathbf{a_J} &\equiv [1;\ e^{j2\pi\Theta_J};\ \ldots;\ e^{j2\pi(N-1)\Theta_J}] \end{aligned} \tag{12.12}$$

For the time representation, we assume that, on the first subaperture, the complex jammer signals received at the M times are

$$\mathbf{b_J} = [\alpha_0;\ \alpha_1;\ \ldots;\ \alpha_{M-1}] \tag{12.13}$$

and that they are decorrelated:

$$E(\mathbf{b_J b_J^H}) = \sigma^2 \xi_J \mathbf{I_M} \tag{12.14}$$

where $\mathbf{I_M}$ is the $M \times M$ identity matrix. The space-time snapshot and covariance matrix corresponding to the jammer are then

$$\begin{aligned} \boldsymbol{\chi}_\mathbf{J} &= \mathbf{a_J} \otimes \mathbf{b_J} \\ \mathbf{R_J} &= E(\boldsymbol{\chi}_\mathbf{J}\boldsymbol{\chi}_\mathbf{J}^\mathbf{H}) = \sigma^2 \xi_J \mathbf{I_M} \otimes \mathbf{a_J a_J^H} \end{aligned} \tag{12.15}$$

$\mathbf{R_J}$ is $MN \times MN$ and is block-diagonal, that is, it is much simpler than a general $MN \times MN$ matrix, because many of its off-diagonal elements are zero. Ward [1] extends the discussion to cover multiple jammers.

12.2.4 Clutter

We assume that all clutter is associated with the ground, which is smooth and diffuse with uniform γ (see Section 3.2.7). Between the radar and the horizon, N_r clutter-generating annuli are assumed to exist. Because the waveform is assumed to be range-unambiguous, the nearest of these is at some range R_0 less than the maximum unambiguous range R_{amb}, and the others are at $R_n = R_0 + iR_{amb}$, where i is an integer. The portion of ground illuminated in the appropriate range gate between R_i and $R_i + \Delta r$ ($\Delta r << r_0$) is an annulus, the inner circle of which is at slant range R_i from the radar. N_c clutter patches are distinguished within each annulus, where the azimuthal width of a clutter patch is $\Delta\phi = 2\pi/N_c$. We assume that the clutter return is decorrelated from one clutter patch to another, but that the return from a given clutter patch is correlated in time (pulse to pulse) and space (subaperture to subaperture).

The clutter spatial frequency for the ith range and kth azimuthal position is

$$\Theta_{i,k} \equiv \frac{d}{\lambda_c} \cos\theta_i \sin\phi_k \tag{12.16}$$

and the corresponding clutter spatial steering vectors are

$$\mathbf{a_{i,k}} \equiv [1;\ e^{j2\pi\Theta_{i,k}};\ \ldots;\ e^{j2\pi(N-1)\Theta_{i,k}}] \tag{12.17}$$

The clutter temporal steering vectors are

$$\mathbf{b_{i,k}} = [1;\ e^{j2\pi\omega_{i,k}};\ \ldots;\ e^{j(M-1)2\pi\omega_{i,k}}] \tag{12.18}$$

(We discuss later how to determine the $\omega_{i,k}$.) The overall clutter steering vectors are then

$$\mathbf{v_{i,k}} = \mathbf{a_{i,k}} \otimes \mathbf{b_{i,k}} \tag{12.19}$$

The complex values of the clutter returns are $\alpha_{i,k}$, and the clutter space-time snapshot is

$$\boldsymbol{\chi}_{\mathrm{C}} = \sum_{\mathrm{i}=1}^{\mathrm{N_r}} \sum_{\mathrm{k}=1}^{\mathrm{N_c}} \alpha_{\mathrm{i,k}} \mathbf{v_{i,k}} \tag{12.20}$$

The clutter-to-noise ratio corresponding to the i,kth clutter bin is $\xi_{i,k}$. Because echoes from different clutter bins are decorrelated,

$$E(\alpha_{i,k}\alpha_{j,l}^*) = \sigma^2 \xi_{i,k} \delta_{i,j} \delta_{k,l} \tag{12.21}$$

where the δs are Kronecker deltas (Section 8.2.2). The clutter covariance matrix is then

$$\mathbf{R_C} = E(\boldsymbol{\chi}_\mathbf{C}\boldsymbol{\chi}_\mathbf{C}^\mathbf{H}) = \sigma^2 \sum_{i,k} \xi_{i,k} \mathbf{v_{i,k}}(\Theta_{i,k}, \omega_{i,k}) \mathbf{v_{i,k}^H}(\Theta_{i,k}, \omega_{i,k}) \quad (12.22)$$

We now want more explicit expressions for $\xi_{i,k}$ and $\omega_{i,k}$. We estimate $\xi_{i,k}$ from the radar equation. (To avoid confusion with the standard deviation of the received noise energy, we temporarily denote RCS by ϵ instead of σ.) We have

$$RCS_i = \epsilon_i = (\gamma \sin \psi_i) R_i \Delta\phi \cdot \Delta R \sec \psi_i \quad (12.23)$$

For flat-earth geometry, $\psi_i = \theta_i$; for spherical earth, $\psi_i(\theta_i)$ can be calculated from Section 3.2.2. Then

$$\xi_{i,k} = \frac{P_{\text{peak}} G(\theta_i, \phi_k) g(\theta_i, \phi_k) \lambda_c^2 \epsilon_i \tau}{(4\pi)^3 R_i^4 LkT_{\text{sys}}} \quad (12.24)$$

where G and g represent the gains of the full aperture and one subaperture, respectively, and the other symbols are as defined as in Section 1.11.

To examine the behavior of $\omega_{i,k}$, we assume that the crab angle is zero (see Problem 12.2). Thus, the doppler frequency shift of a clutter patch is

$$\begin{aligned} f_{di,k} &= \frac{2V_a}{\lambda_c} \cos\theta_i \sin\phi_k \\ \omega_{i,k} &= \left\{\frac{f_{di,k}}{f_R}\right\} = \left\{\frac{2V_a/\lambda_c}{f_R} \cos\theta_i \sin\phi_k\right\} = \left\{\frac{2V_a t_R}{d} \Theta_{i,k}\right\} \end{aligned} \quad (12.25)$$

When range is large compared to platform altitude and $\theta_i \sim 0$, then

$$\omega_{i,k} \approx \left\{\frac{2V_a \sin\phi_k / \lambda_c}{f_R}\right\} \quad (12.26)$$

We define

$$\beta \equiv \frac{2V_a t_R}{d}, \ \omega_{i,k} = \{\beta \Theta_{i,k}\} \quad (12.27)$$

β is twice the number of subaperture lengths traversed during a PRI. (If the antenna elements are spaced at a half-wavelength, $d = \lambda/2$, then β is the number of times the clutter doppler spectrum aliases into the unambiguous doppler space). A plot of $\omega_{i,k}$ versus $\Theta_{i,k}$ results in a straight line, called a clutter ridge. Figure 12.3 (from Ward [1]) illustrates clutter ridges for several values of β. When $\beta = 0$, the ridge is horizontal; when $\beta = 1$, the ridge extends over the full spatial frequency interval; when $\beta > 1$, the clutter is said to be doppler ambiguous.

12.2.5 Summary of Target and Interference

Table 12.1 summarizes the equations that represent noise, target, jamming, and clutter.

12.3 STAP Algorithm

The fully adaptive STAP procedure is the following:

1. Choose a training strategy and use it to select training data, which is then used to estimate the interference covariance matrix $\mathbf{R_u}$.
2. Compute an adaptive weight vector $\mathbf{w}$ ($MN \times 1$), to test for a target at a specified azimuth and velocity (the procedure for computing $\mathbf{w}$ will be discussed).
3. Using the test data $\boldsymbol{\chi}$, compute a test statistic $= z = \mathbf{w}^{\mathbf{H}}\chi$ (z is a complex scalar).
4. Compare $|z|$ with a threshold to decide the presence or absence of target.

The number of elements in the training data "support" usually must be a few times the number of elements in the test data, and usually must be obtained during the same CPI, possibly from other range bins. Meeting those conditions is a particularly challenging aspect of the design of STAP radars.

The adaptive weight vector $\mathbf{w}$ is calculated using the adaptive matched filter, discussed in Section 9.2 in the context of target detection. The result is

$$\mathbf{w} = \mathbf{R_u^{-1}} \mathbf{v_T} \tag{12.28}$$

where, from (12.3),

$$\mathbf{R_u} = \mathbf{E}\{\boldsymbol{\chi}_\mathbf{u} \boldsymbol{\chi}_\mathbf{u}^\mathbf{H}\} \tag{12.29}$$

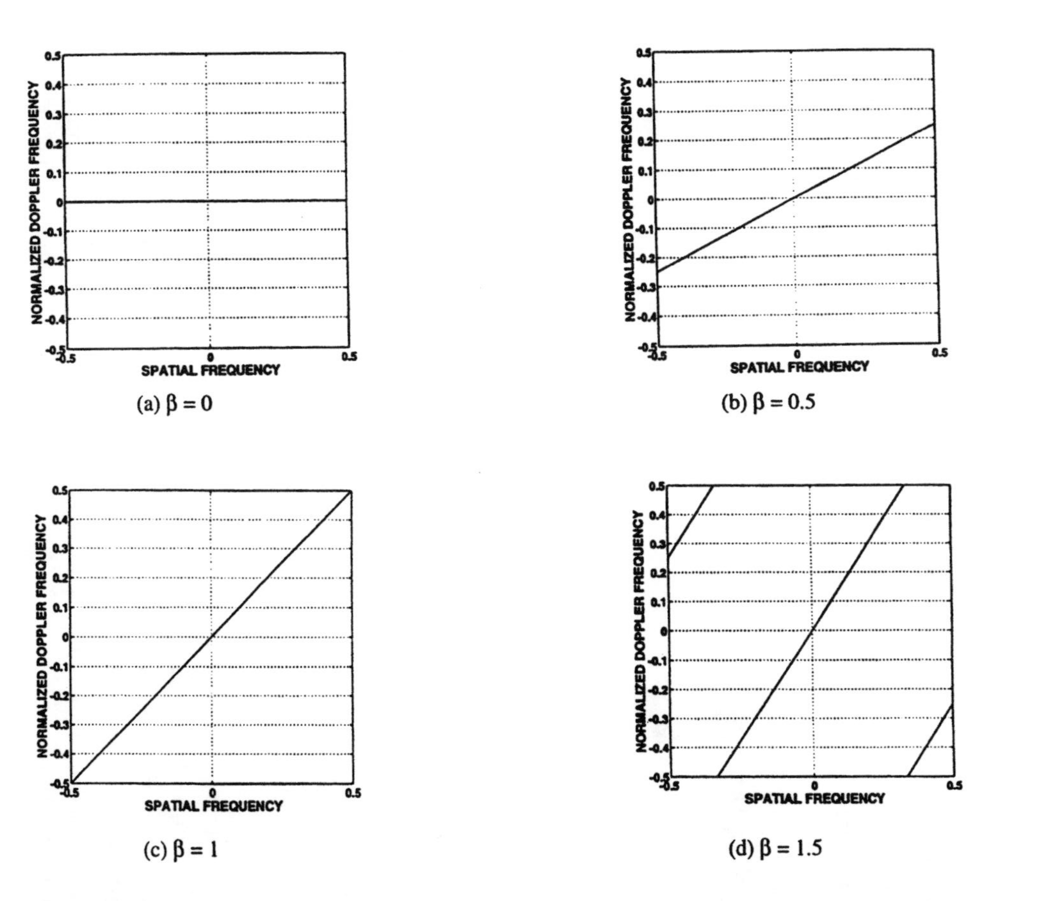

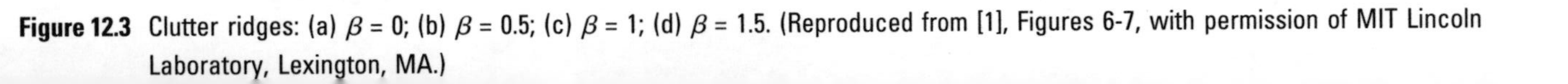
Figure 12.3 Clutter ridges: (a) $\beta = 0$; (b) $\beta = 0.5$; (c) $\beta = 1$; (d) $\beta = 1.5$. (Reproduced from [1], Figures 6-7, with permission of MIT Lincoln Laboratory, Lexington, MA.)

Table 12.1
STAP Expressions for Noise, Target, Jamming, and Clutter

Noise	Target	Jamming (Single Jammer)	Clutter (Smooth Earth)
$\mathbf{R}_n = E(\chi_\mathbf{n}\chi_\mathbf{n}^\mathbf{H}) = \sigma^2\mathbf{I}$	$\Theta_T \equiv \frac{d}{\lambda_c}\cos\theta_T \sin\theta_T$	$\Theta_J \equiv \frac{d}{\lambda_c}\cos\theta_J \sin\phi_J$	$i = 1, \ldots, N_r;\ k = 1, \ldots, N_C;\ N_C = \frac{2\pi}{\Delta\phi}$
	$\{\omega_T \equiv f_T/f_R\}$	$\mathbf{a}_J = [1;\ e^{j2\pi\Theta_J}; \ldots;\ e^{j2\pi(N-1)\Theta_J}]$	$\Theta_{ik} = \frac{d}{\lambda_c}\cos\theta_i \sin\phi_k$
	$\mathbf{a}_T = [1;\ e^{j2\pi\Theta_T}; \ldots;\ e^{j2\pi(N-1)\Theta_T}]$	$\mathbf{b}_J = [\alpha_0;\ \alpha_1; \ldots;\ \alpha_{M-1}]$	$\mathbf{a}_{ik} = [1;\ e^{j2\pi\Theta_{ik}}; \ldots;\ e^{j2\pi(N-1)\Theta_{ik}}]$
	$\mathbf{b}_T = [1;\ e^{j2\pi\omega_T}; \ldots;\ e^{j2\pi(M-1)\omega_T}]$	$\xi_J = JNR$	$\mathbf{b}_{ik} = [1;\ e^{j2\pi\omega_{ik}}; \ldots;\ e^{j2\pi(M-1)\omega_{ik}}]$
	$\mathbf{v}_T = \mathbf{a}_T \otimes \mathbf{b}_T$	$E(\mathbf{b}_J\mathbf{b}_J^H) = \sigma^2\xi_J\mathbf{I}_M$	$\mathbf{v}_{ik} = \mathbf{a}_{ik} \otimes \mathbf{b}_{ik}$
		$\chi_J = \mathbf{a}_J \otimes \mathbf{b}_J$	α_{ik} = complex return from clutter patch i, k
		$\mathbf{R}_J = E(\chi_J\chi_J^\mathbf{H}) = \sigma^2\xi_J\mathbf{I_M} \otimes \alpha_\mathbf{J}\alpha_\mathbf{J}^\mathbf{H}$	$\chi_\mathrm{C} = \sum_{i=1}^{N_r}\sum_{k=1}^{N_c}\alpha_{ik}\mathbf{v}_{ik}$
			$\xi_{ik} = CNR$
			$E(\alpha_{ik}\alpha_{jl}^*) = \sigma^2\xi_{ik}\delta_{ij}\delta_{kl}$
			$\mathbf{R}_C = E(\chi_\mathbf{C}\chi_\mathbf{C}^\mathbf{H}) = \sigma^2\sum_i\sum_k \xi_{ik}\mathbf{v}_{ik}\mathbf{v}_\mathbf{ik}^\mathbf{H}$
			$RCS_i = \epsilon_i = (\gamma\sin\psi_i)R_i\Delta\phi\Delta R\sec\psi_i$
			$\xi_{ik} = \frac{P_\mathrm{peak}G(\theta_i,\ \phi_k)g(\theta_i,\ \phi_k)\lambda_c^2\epsilon_i\tau}{(4\pi)^3R_i^4 LkT_0F}$
			No crab: $f_{dik} = \frac{2V_a}{\lambda_c}\cos\theta_i\sin\phi_k,\ \omega_{ik} = \frac{f_{dik}}{f_R} = \frac{2V_a t_R}{d}\Theta_{ik}$
			$\beta \equiv \frac{2V_a t_R}{d},\ \omega_{ik} = \beta\Theta_{ik}$

Interference is here defined as clutter and jamming, but not noise. With respect to the target for which the filter is designed, **w** maximizes the SINR and maximizes P_D for given P_{FA} [3]. However, it has high sidelobes in both angle and doppler, resulting in relatively high probability of incorrect declaration that a target in an angle or doppler sidelobe is at the mainlobe location.

To reduce sidelobes, we typically apply a tapering function (Section 2.3.4) to the target steering vector before computing z. In that case,

$$\mathbf{w} = \mathbf{R_u^{-1}} \mathbf{g_T} \tag{12.30}$$

where $\mathbf{g_t}$ is the target steering vector modified via tapering to produce low sidelobes. We use

- $\mathbf{t_a}$ = angle tapering vector ($N \times 1$)
- $\mathbf{t_b}$ = doppler tapering vector ($M \times 1$)

Those typically correspond to a standard tapering function, such as Taylor or Hann. Then

- $\mathbf{t} = \mathbf{t_b} \otimes \mathbf{t_a}$ = overall tapering vector ($MN \times 1$)
- $\mathbf{g_T} = \mathbf{t} \cdot \mathbf{v_T}$ = tapered target steering vector ($MN \times 1$)

12.4 Summary of Fully Adaptive STAP

The fully adaptive STAP procedure can be summarized as follows.

1. Transmit M pulses from full aperture.
2. Receive M pulses on N subapertures; sample in range.
3. Form $\boldsymbol{\chi} = MN \times 1$ space-time snapshot. $\boldsymbol{\chi} = \alpha_T \boldsymbol{v}_\mathbf{T} + \boldsymbol{\chi}_\mathbf{u}$ = target return plus interference return.
4. Form interference covariance matrix from training data: $\mathbf{R_u} = E(\boldsymbol{\chi}_\mathbf{u} \boldsymbol{\chi}_\mathbf{u}^\mathbf{H}) = \mathbf{R_N} + \mathbf{R_J} + \mathbf{R_C}$.
5. Form adaptive weight vector:
   a. Unweighted: $\mathbf{w} = \mathbf{R_u^{-1}} \mathbf{v_T}$.
   b. Weighted: $\mathbf{w} = \mathbf{R_u^{-1}} \mathbf{g_T}$, $\mathbf{g_T} = \mathbf{t} \cdot \mathbf{v_T}$, $\mathbf{t}$ = tapering vector.
6. Compute test statistic and compare it with threshold

$$|z| = |\mathbf{w}^{\mathbf{H}}\boldsymbol{\chi}| \underset{H_0}{\overset{H_1}{\gtrless}} T, \text{ with corresponding } P_D \text{ and } P_{FA} \tag{12.31}$$

7. Iterate over Θ, ω of potential targets.

The fully adaptive STAP procedure is summarized in Figure 12.4. As the figure indicates, if the interference magnitude is varying in a measurable way, the threshold T can be varied also, so as to hold P_{FA} constant; that is the CFAR procedure (Section 9.1).

Figure 12.5 is an overview of the principle of STAP. The set of possible radar echoes from a point target is represented by a two-dimensional surface (we assume $\theta \sim 0$, $H << R$). One axis is labeled sin(azimuth) (= sin ϕ), representing the fact that the radar can receive echoes from $-\pi/2 < \phi < \pi/2$, or $-1 < \sin\phi < 1$. The other axis is $\omega_T = \{f_T/f_R\}$, proportional to LOS target velocity in the absence of velocity ambiguity. Its values are -0.5 to $+0.5$, representing the doppler frequency interval from $-f_R/2$ to $+f_R/2$. The example reflects the simplifying assumptions that $\beta = 1$ and that $d = \lambda/2$.

In this case, the clutter is along a straight clutter ridge from the lower left to the upper right. Its amplitude follows the antenna pattern. (In this particular case, it reflects the assumption that the antenna is pointed broadside to the flight path; $\phi = 0$). A noise jammer is located at a fixed ϕ and radiates signals at all ω. A moving target at zero azimuth ($\phi = 0$) is indicated, with $\omega_T \neq 0$. A single-aperture radar pointing directly at the target cannot detect it without doppler processing, because it is buried in the mainlobe clutter. Even if exoclutter doppler processing is performed (Section 11.4), the target still is not detected, because it is buried in the jammer signal. However, STAP can produce a two-dimensional plot of signal strength versus azimuth and doppler frequency, revealing the target.

12.5 Example of Fully Adaptive STAP

Still following Ward [1], this section presents an example of fully adaptive STAP. Table 12.2 summarizes the assumed parameters.

12.6 STAP Performance Metrics

Several performance metrics are used to evaluate a STAP algorithm, including the following. Each is illustrated using the example.

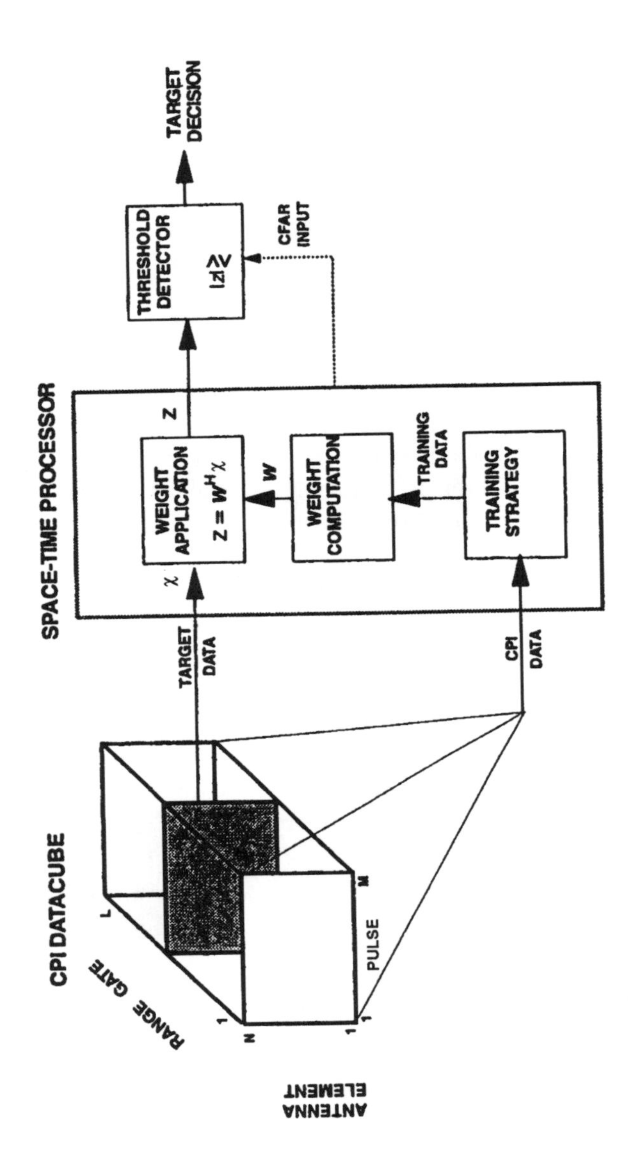

Figure 12.4 STAP summary. (After [1], Figure 21, with permission of MIT Lincoln Laboratory, Lexington, MA.)

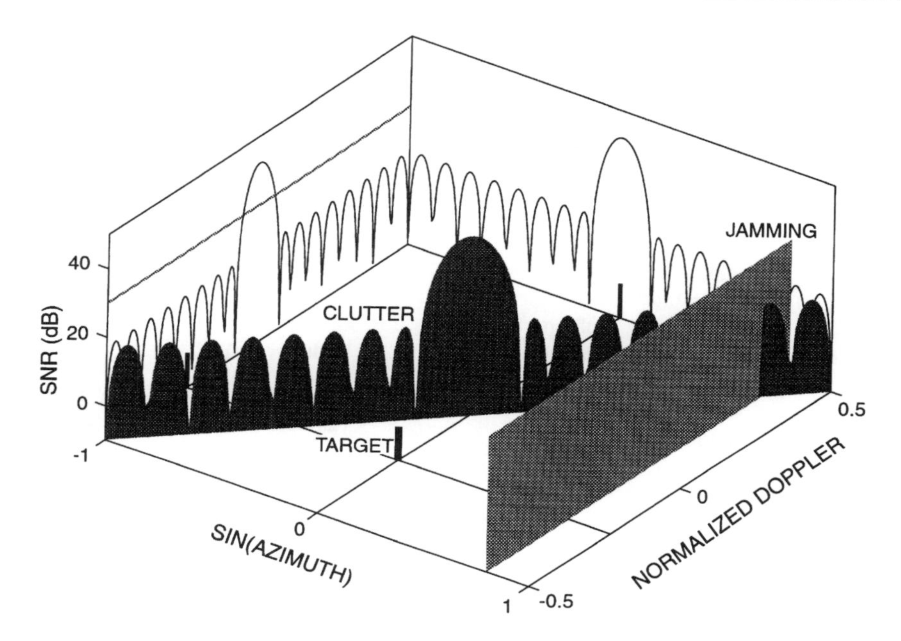

Figure 12.5 STAP principle. (Reproduced from [1], Figure 1, with permission of MIT Lincoln Laboratory, Lexington, MA.)

- Adapted pattern;
- SINR;
- SINR loss;
- SINR improvement factor;
- MDV and usable doppler space fraction (USDF).

12.6.1 Adapted Pattern

We compute **w** using the given target, clutter, and jammer parameters and test for the presence of a target at the full range of values of Θ and ω. The adapted pattern is defined as

$$P_w(\Theta, \omega) \equiv |\mathbf{w}^{\mathbf{H}}(\Theta_T, \omega_T)\mathbf{v}(\Theta, \omega)|^2 \tag{12.32}$$

Ward [1, Figs. 23 and 24] illustrates the adapted pattern for both optimum fully adaptive STAP and tapered fully adaptive STAP. The pattern creates nulls at the (Θ, ω) of the clutter and jammers and produces a high value at Θ_T and ω_T. The "optimum" STAP is optimum insofar as it produces the maximum

Table 12.2
Assumed Parameters for STAP Example

Radar System Parameters	Values
Operating frequency	450 MHz
Wavelength	0.667m
Peak power	200 kW
Duty factor	6%
Transmit gain (full aperture)	22 dB
Receive gain (one subaperture)	10 dB
Instantaneous bandwidth	4 MHz
Noise figure	3 dB
System losses	4 dB
PRF	300 Hz
Number of pulses/CPI	18
Pulse width	200 μs
Antenna Array Parameters	
Number of elements	18
Element spacing	0.33m = $\lambda/2$
Element pattern	Cosine
Transmit taper	Uniform
Platform Parameters	
Platform altitude	9,000 m
Platform velocity	50 m/s
Number of clutter foldovers	$\beta = 1$
Velocity misalignment angle	0
Jamming Parameters	
Number of jammers	2
Azimuth angles (degrees)	–40, 25
Elevation angles (degrees)	0, 0
Effective radiated power density (ERPD)	1,000 W/MHz
Range	370 km
Clutter Parameters	
Number of patches	360
Range	130 km
Reflectivity γ	–3 dB
Intrinsic velocity standard deviation σ_v	0 m/s

Table 12.2 (continued)
Assumed Parameters for STAP Example

Target Parameters	**Values**
Azimuth, elevation: Θ_t	0
Doppler frequency: ω_T (= $2v/\lambda$; at 450 MHz, v = 33.3 m/s = 65 kn)	100 Hz
Tapering Parameters	
Space (antenna element)	–30 dB Chebyshev
Time (pulse)	–40 dB Chebyshev

Source: Ward [1], Tables 2 and 4, with permission of MIT Lincoln Laboratory, Lexington, MA.

SINR at the correct target location in azimuth-doppler space; however, it also produces relatively high SINR for targets at other locations—the sidelobes. The tapered STAP produces slightly less SINR at the correct location but much lower SINR in the sidelobes; thus, it probably is more useful for practical applications.

12.6.2 SINR

We have

$$\begin{aligned} z(\Theta, \omega) &= \mathbf{w}^{\mathbf{H}}(\Theta_T, \omega_T)\boldsymbol{\chi}(\Theta, \omega) \\ &= z_T + z_u = \alpha_T \mathbf{w}^{\mathbf{H}}\mathbf{v_T} + \mathbf{w}^{\mathbf{H}}\boldsymbol{\chi_u} \\ p_T(\Theta, \omega) &= |z_T|^2 = \text{output (power) corresponding to target} \\ p_u(\Theta, \omega) &= E\{|z_u|^2\} = \text{output (power) corresponding} \\ &\quad \text{to interference plus noise} \end{aligned} \tag{12.33}$$

The SINR is then

$$\text{SINR}(\Theta, \omega) = \frac{p_T(\Theta, \omega)}{p_u(\Theta, \omega)} \tag{12.34}$$

We define $\xi_T \equiv$ target signal-to-noise ratio (SNR) for 1 pulse and 1 array element, and set $\xi_T = 1$. In the absence of clutter and jamming, a perfect matched filter would coherently integrate the results of the M pulses and N subapertures and produce an output of SINR = SNR = MN = 18 × 18 = 324, or 10 log(MN) = 25.1 dB. Figure 12.6 illustrates the SINR versus target doppler frequency for zero azimuth. Indeed, the peak response (which is obtained over most of the doppler interval) is 25 dB.

In Figure 12.6, a very large number of doppler filters was assumed. For the case of M = 18 doppler filters, the SINR is the same as before for the center of each filter and falls off near the filter edges, a phenomenon known as doppler straddling loss (illustrated in Figure 12.7). (Range straddling loss also can occur.)

12.6.3 SINR Loss

For noise only (interference = 0), the optimum SINR is $\text{SINR}_0 = MN\xi_T = MN$. For nonzero interference, SINR loss is defined as

$$\text{SINR loss} = \frac{\text{SINR}}{\text{SINR}_0} \tag{12.35}$$

SINR loss includes unsuppressed interference, taper loss, and straddling loss.

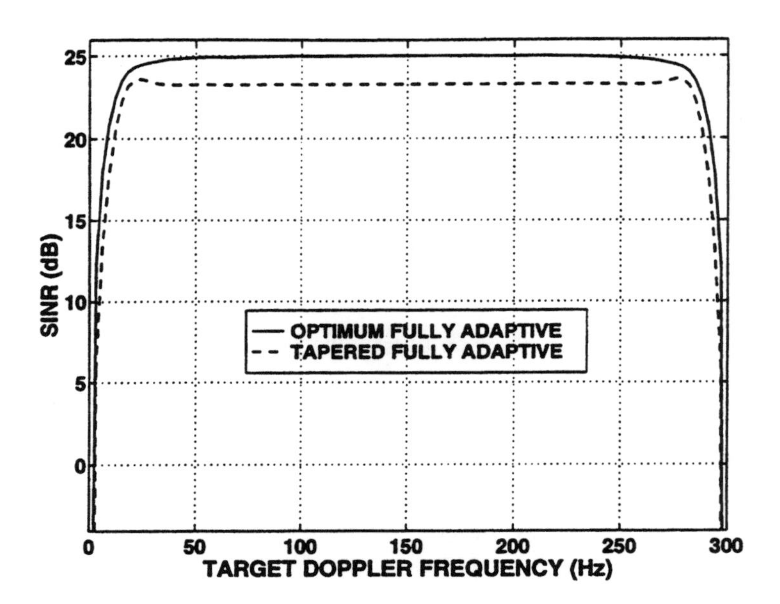

Figure 12.6 SINR for fully adaptive STAP. (Reproduced from [1], Figure 25, with permission of MIT Lincoln Laboratory, Lexington, MA.)

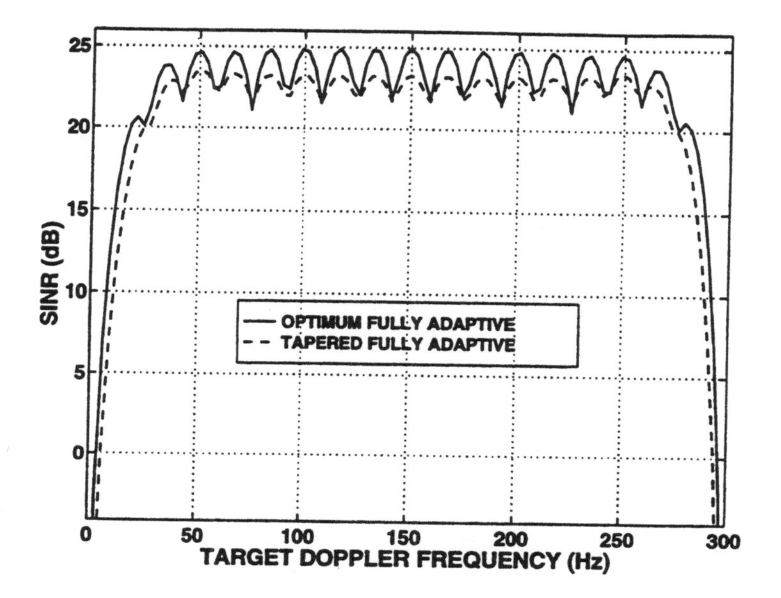

Figure 12.7 SINR for fully adaptive STAP, including doppler-straddling losses. (Reproduced from [1], Figure 26, with permission of MIT Lincoln Laboratory, Lexington, MA.)

Figure 12.8 expresses the STAP result as SINR loss versus target doppler frequency. Evidently, it is the same as Figure 12.6 except for a change of scale.

12.6.4 SINR Improvement Factor

We define the following:

- ξ_C = ratio of received clutter power to noise (preprocessing);
- ξ_J = ratio of received jammer power to noise (preprocessing).

and we relax the assumption that $\xi_T = 1$. Then, before processing,

$$\text{SINR}_{\text{in}} = \frac{\xi_T}{1 + \xi_C + \xi_J} \equiv \frac{\xi_T}{1 + \xi_I} << 1 \text{ (typically)} \tag{12.36}$$

The SINR improvement factor, I_{SINR}, is the ratio of SINR (postprocessing) to SINR (preprocessing):

$$I_{\text{SINR}} = \frac{\text{SINR}}{\text{SINR}_{\text{in}}} \tag{12.37}$$

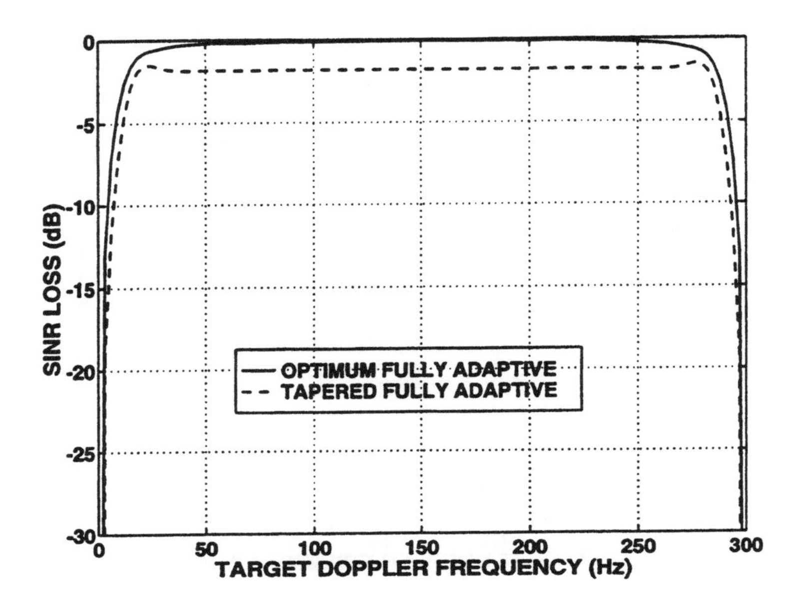

Figure 12.8 SINR loss for fully adaptive STAP. (Reproduced from [1], Figure 27, with permission of MIT Lincoln Laboratory, Lexington, MA.)

Included in SINR is processing gain from both coherent integration, MN, and from interference suppression, in principle as much as $\xi_I + 1$.

Figure 12.9 illustrates I_{SINR} versus target doppler frequency. Its peak is indeed essentially the ideal value, $10 \log[MN(\xi_I + 1)] = 73$ dB, and it has the same shape as Figures 12.8 and 12.6.

12.6.5 Minimum Detectable Velocity and Usable Doppler Space Fraction

Using some appropriate criterion, we define "acceptable SINR." For example, the criterion could be based on chosen values of P_D, P_{FA}, a target model (e.g., Swerling-n), and distributions of interference (e.g., Gaussian, log-normal, etc.), as discussed in Section 4.1.2.

The MDV is the velocity closest to the mainlobe clutter at which acceptable SINR is achieved. Using multiple subapertures and STAP processing, MDV can be reduced so as to detect targets that would be buried in clutter if conventional exoclutter MTI (Section 11.4.2) were used. Thus, the STAP technique is often called endoclutter processing.

The processor can place nulls at additional doppler frequencies besides that of the mainlobe clutter. The UDSF is the fraction of doppler space such that SINR is acceptable.

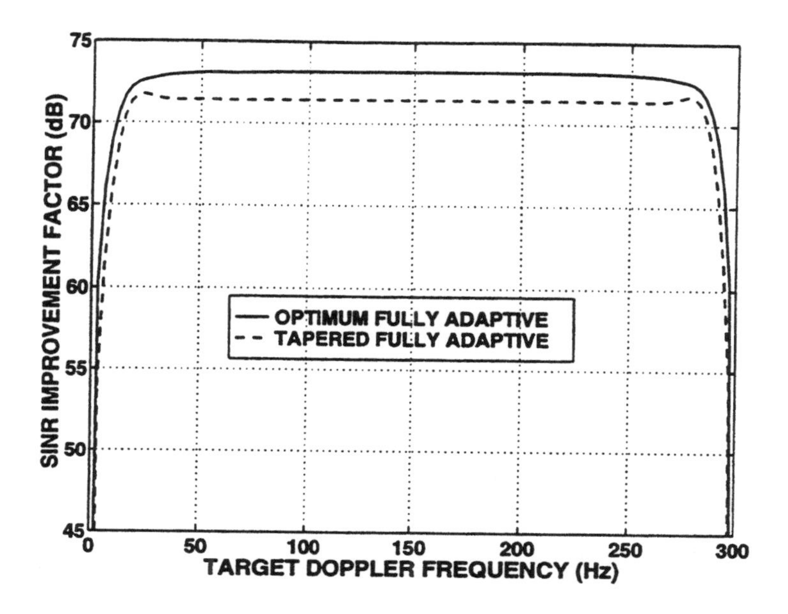

Figure 12.9 SINR improvement factor for fully adaptive STAP. (Reproduced from [1], Figure 28, with permission of MIT Lincoln Laboratory, Lexington, MA.)

Table 12.3 summarizes MDV and UDSF for the example being considered. The values in parentheses indicate the ratio of acceptable SINR to maximum SINR. As acceptable SINR decreases, MDV decreases and UDSF increases.

12.7 Partially Adaptive STAP

Fully adaptive STAP requires obtaining and then inverting an MN-dimensional covariance matrix. Whereas the number of subapertures N is typically 2, 3, or

Table 12.3
MDV and UDSF for Fully Adaptive STAP

	Optimum	Tapered Fully Adaptive
MDV(–12 dB)	1.2 m/s	2.0 m/s
UDSF(–12 dB)	96.0%	94.9%
MDV(–5 dB)	2.7 m/s	3.8 m/s
UDSF(–5 dB)	94.1%	92.1%

Source: Reproduced from [1], Table 5, with permission of MIT Lincoln Laboratory, Lexington, MA.

perhaps up to 10, the number of pulses M is typically tens, hundreds, or thousands. Obtaining such detailed support for covariance estimation can be prohibitively difficult. Furthermore, the computational complexity may be beyond current (2000) digital processor technology. Therefore, much research focuses on ways to reduce the effective dimensions of the problem. Partially adaptive STAP is a procedure that, using some criteria, reduces the dimension of the $MN \times 1$ space-time snapshot and therefore of the interference covariance matrix.

We have previously defined

$$\mathbf{R_u} = \mathbf{R_N} + \mathbf{R_J} + \mathbf{R_C} \tag{12.38}$$

and we attempt to reduce the rank (number of independent rows) of $\mathbf{R_u}$. We can utilize the following assumptions.

- Noise is random zero-mean Gaussian, independent of subaperture number (n) or pulse number (m).
- Jamming signals arrive from a small number of fixed angular directions; their magnitude is independent of m or n, and they are mutually decorrelated in time.
- There may be clutter-free range bins (for low PRF, beyond the horizon [Figure 12.10]), or clutter-free doppler bins (for high PRF, at velocities greater in absolute value than the platform velocity [Figure 12.11]).

Ward [1] defines a taxonomy of partially adaptive STAP algorithms, as shown in Figure 12.12. The raw data can be described as "element-space, pre-doppler," that is, a function of subaperture (element) number and pulse number. The STAP final result can be described as "beam-space, post-doppler," that is, a function of Θ and ω. By performing a series of detailed calculations described in [1], Ward computes the performance of STAP algorithms described as "element-space, pre-doppler," "element-space, post-doppler," "beam-space, pre-doppler," and "beam-space, post-doppler," expressed as SINR loss versus target doppler frequency. The results are summarized in Table 12.4. The partially adaptive algorithms all produce somewhat inferior results relative to the fully adaptive algorithm but require much less computational support.

12.8 Other STAP Results

12.8.1 STAP-SAR

The preceding discussion assumed that, if the clutter intensity varied with location, the data were collected over a short enough time that the change in

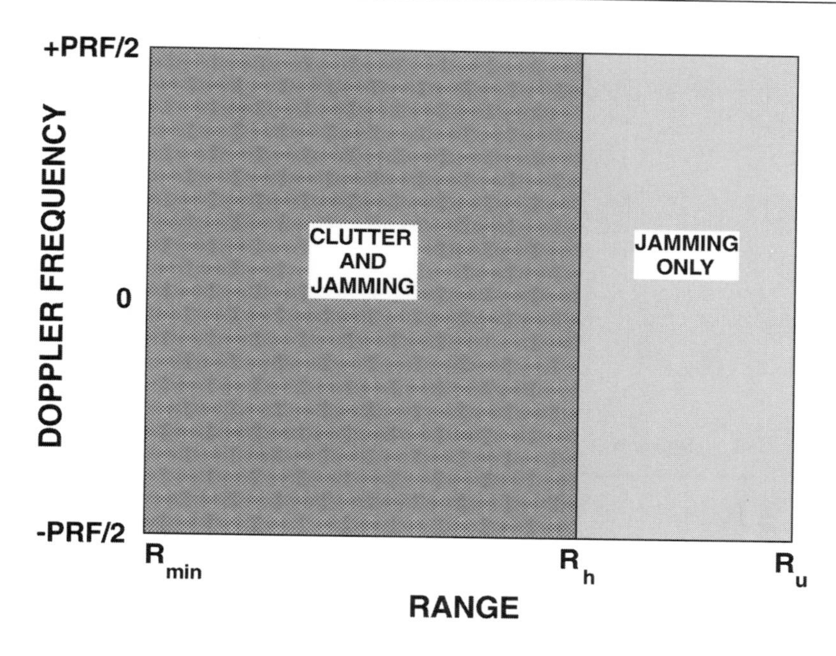

Figure 12.10 Range-doppler view of interference: LPRF. (Reproduced from [1], Figure 18, with permission of MIT Lincoln Laboratory, Lexington, MA.)

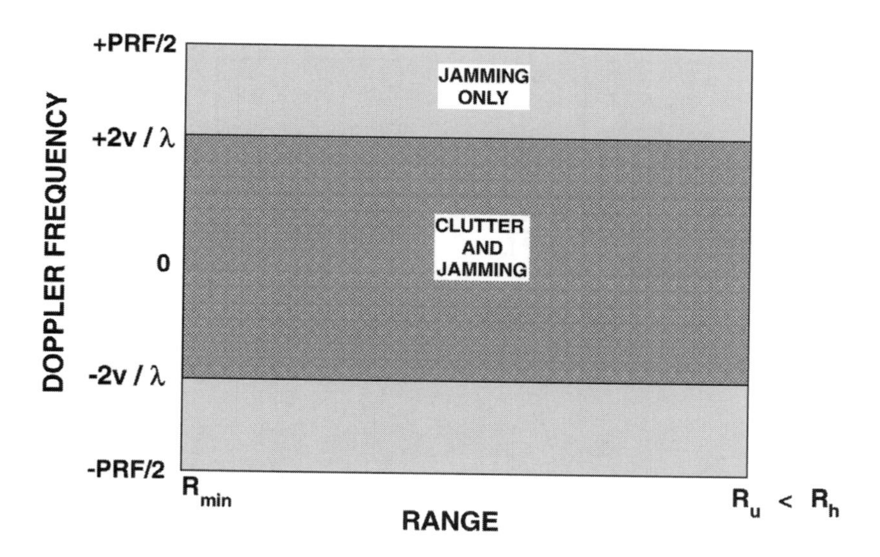

Figure 12.11 Range-doppler view of interference: HPRF. (Reproduced from [1], Figure 19, with permission of MIT Lincoln Laboratory, Lexington, MA.)

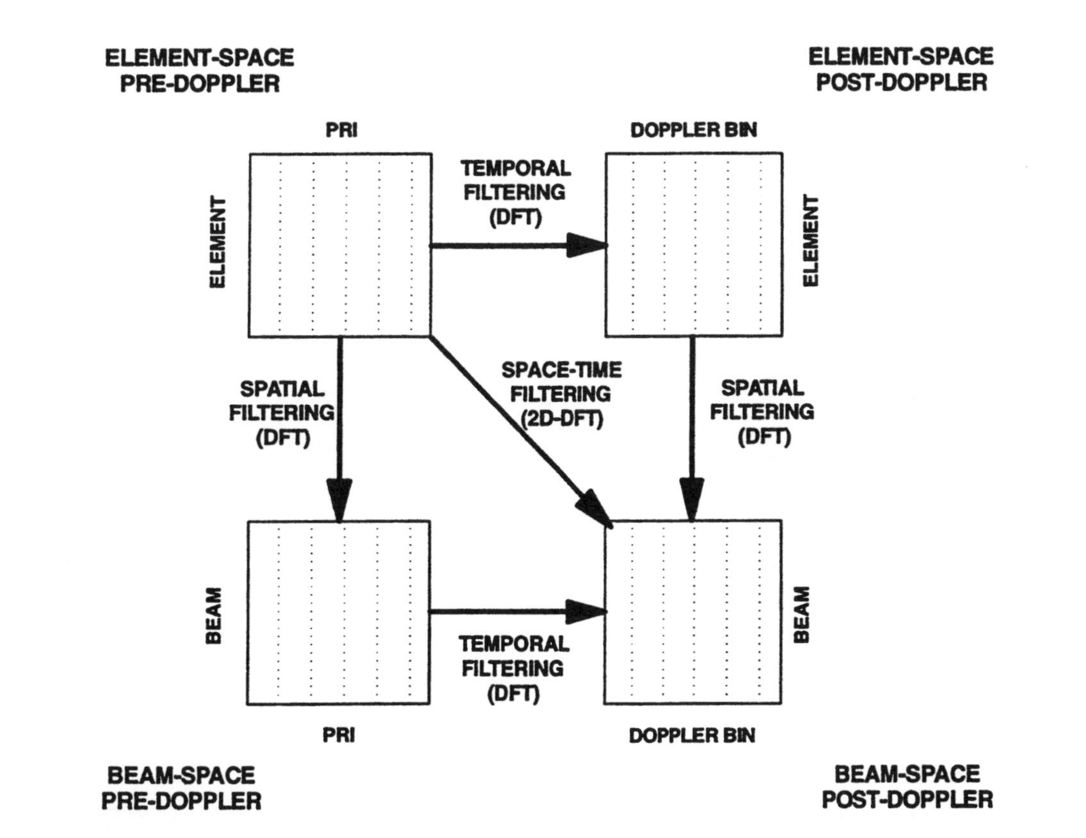

Figure 12.12 Taxonomy of partially adaptive STAP algorithms. (Reproduced from [1], Figure 33, with permission of MIT Lincoln Laboratory, Lexington, MA.)

platform location could be neglected during the collection interval. Barbarossa and Farina [4] discuss the theory of relaxing that assumption to use SAR techniques (Chapter 7) to produce a focused image of the ground, while using multiple subapertures to detect and focus moving targets and possibly suppress jammers. Figure 12.13(a) shows the result for a simulated moving point target, using STAP only; Figure 12.13(b) shows the result for STAP and SAR combined.

12.8.2 Endoclutter Airborne MTI and SAR

The AN/APG-76 Ku-band radar [5], mounted on an F-4E aircraft, can perform SAR with 1m or 0.3m resolution and also perform endoclutter GMTI with three subapertures using a STAP technique. The radar can collect SAR and GMTI data simultaneously and overlay moving target locations on the SAR image. By collecting data from more than one location, and using intersecting layover circles (Section 7.5.3.2), azimuth/elevation monopulse (Section 5.2), and INS/GPS data (Section 7.1.3), precise location of both fixed and moving targets may be obtained.

12.8.3 Knowledge-Based STAP

Melvin et al. [6] describe knowledge-based STAP (KBSTAP). A priori knowledge about the environment (target, jamming, clutter) is used to preadapt the STAP algorithm and to help decide what type of STAP processing to use (e.g., element space versus beam-space, pre-doppler versus post-doppler). Melvin et al. use an example of data collected from the multichannel airborne radar measurement (MCARM) testbed STAP radar installed in a British Aerospace BAC1-11 aircraft operated by the U.S. Air Force (USAF) Rome Laboratory and Northrop-Grumman Corporation [7]. They use their a priori knowledge that highways with many moving vehicles are at known locations in the scene and adapt the filter to account for that. Moving targets detected near the highway are given higher probability of being actual vehicles than targets detected in regions known to be vegetation. Similarly, clutter covariance matrices are computed using clutter known to be relatively homogeneous. Targets are shown to be more readily detectable when a priori knowledge is utilized.

12.8.4 Sigma-Delta STAP

Section 11.4.6 discussed the DPCA concept [8]. Brown et al. [9] and Zhang et al. [10] describe a generalization of DPCA: a two-aperture STAP technique for which the returns from the two subapertures are summed and differenced.

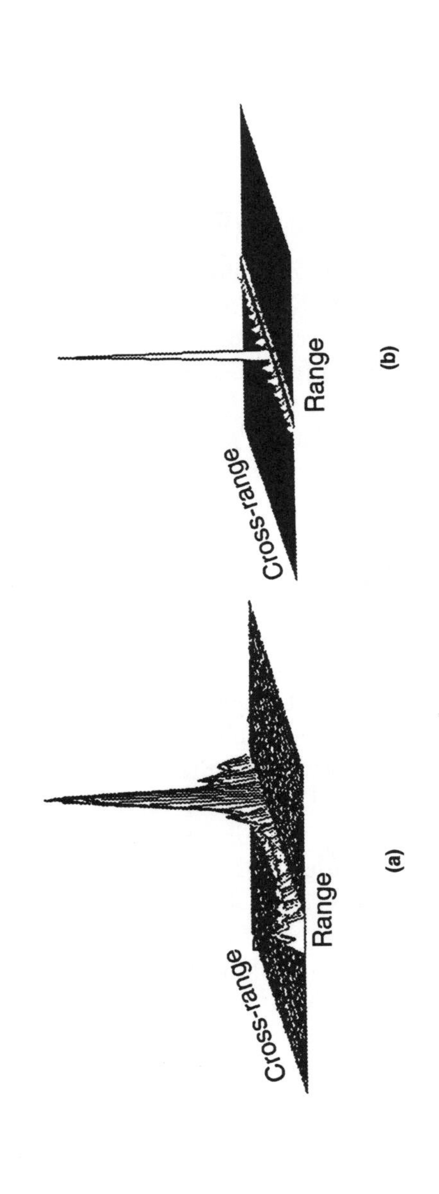

Figure 12.13 STAP and SAR: (a) image of moving target, after clutter cancelation; (b) image of moving target, after range migration and phase compensation. (After [4], Figures 14 and 15, copyright 1999, IEEE.)

The technique sigma-delta STAP ($\Sigma\Delta$-STAP) has many advantages over conventional STAP, including the following.

- The technique is relatively simple and affordable.
- Covariance matrix estimation requires considerably fewer training data.
- Channel calibration errors are less significant.
- Some existing systems can be upgraded using software modifications only or with minimal hardware modifications.
- Computational requirements are considerably reduced.

A disadvantage is poor jammer cancellation.

12.8.5 Partially Adaptive CFAR STAP

Goldstein et al. [11] have described a partially adaptive CFAR STAP technique. From Section 9.11 the adaptive-matched-filter CFAR test is

$$\Lambda = \frac{|\mathbf{v}_\mathbf{T}^\mathbf{H}\mathbf{R}_\mathbf{u}^{-1}\boldsymbol{\chi}|^2}{\mathbf{v}_\mathbf{T}^\mathbf{H}\mathbf{R}_\mathbf{u}^{-1}\mathbf{v}_\mathbf{T}} \underset{H_0}{\overset{H_1}{\gtrless}} T \tag{12.39}$$

The denominator represents the (variable) ratio of target signal vector to expected interference; the numerator represents the output of the adaptive matched filter operating on the measured data. The procedure is CFAR if each voltage component of the interference is independent zero-mean multivariate Gaussian. Goldstein et al. illustrate the procedure using results from the mountaintop radar surveillance technology experimental radar (RSTER) and MCARM radar (see Section 12.8.3).

12.8.6 Three-Dimensional STAP and Bistatic STAP

Techau et al. [12] have developed a three-dimensional STAP methodology for determining performance bounds for the joint mitigation of cold clutter (ordinary radar clutter resulting from scattering of transmitted radar energy from terrain) and hot clutter (energy emitted by a jammer and scattered from terrain). Faute [13] discusses the possibility of bistatic spaceborne STAP.

References

[1] Ward, J., *Space-Time Adaptive Processing for Airborne Radar*, Technical Report 1015, Lexington, MA: Lincoln Laboratory, Massachusetts Institute of Technology, 1994.

[2] Klemm, R., *Space-Time Adaptive Processing Principles and Applications*, Stevenage, Hertfordshire, United Kingdom: Institute of Electrical Engineers, 1998.

[3] Brennan, L. E., and I. S. Reed, "Theory of Adaptive Radar," *IEEE Trans. Aerospace and Electronic Systems*, Vol. AES-9, 1973, pp. 237–252.

[4] Barbarossa, S., and Farina, A., "Space-Time-Frequency Processing of SAR Signals," *IEEE Trans. Aerospace and Electronic Systems,* Vol. 30, No. 2, April 1994, pp. 341–358.

[5] Tobin, M., and M. Greenspan, "Smuggling Interdiction Using an Adaptation of the AN/APG-76 Multimode Radar," *IEEE-AES Systems Magazine*, Nov. 1996, pp. 19–24.

[6] Melvin, W., et al., "Knowledge-Based STAP for Airborne Early Warning Radar," *IEEE-AES Systems Magazine*, April 1998, pp. 37–42.

[7] Little, M. O., and W. P. Berry, "Real-Time Multichannel Airborne Radar Measurements," *Proc. IEEE 1997 National Radar Conf.*, pp. 138–142.

[8] Staudaher, F. M., "Airborne MTI," Chap. 16 of M. Skolnik (ed.), *Radar Handbook*, 2nd ed., New York: McGraw-Hill, 1990.

[9] Brown, R. D., et al., "A STAP Approach for Improved Performance and Affordability," *Proc. 1996 IEEE National Radar Conf.*, pp. 321–326.

[10] Zhang, Y., and Wang, H., "Further Results of ?D-STAP Approach to Airborne Surveillance Radars," *Proc. 1997 IEEE National Radar Conf.*, pp. 337–342.

[11] Goldstein, J. S., I. S. Reed, and P. A. Zulch, "Multistage Partially-Adaptive STAP CFAR Algorithm," *IEEE Trans. Aerospace and Electronic Systems*, Vol. 35, No. 2, April 1999, pp. 645–661.

[12] Techau, P. M., et al., "Performance Bounds for Hot and Cold Clutter Mitigation," *IEEE Trans. Aerospace and Electronic Systems*, Vol. 35, No. 4, Oct. 1999, pp. 1253–1265.

[13] Fante, R. L., "Ground and Airborne Target Detection with Bistatic Adaptive Space-Based Radar," *IEEE-AES Systems Magazine,* Oct. 1999, pp. 39–44.

Problems

Problem 12.1

Assume a rectangular array of subapertures with spacing d_x horizontally and d_z vertically. Find the expression for the target steering vector.

Problem 12.2

Find the clutter ridge for nonzero aircraft crab. (Hint: See Ward [1, Section 2.6.4].)

13

Bistatic Radar and Low-Probability-of-Intercept Radar

This chapter examines two special radar topics: bistatic radar and low-probability-of-intercept (LPI) radar. We assume that $T_{ant} = T_{radar} = T_0$ (see Section 2.2.11).

13.1 Bistatic Radar

Until now we have assumed that the radar transmitter and receiver are at the same location. We now relax that assumption and consider the possibility that they are at different positions. The former configuration is monostatic radar, the latter is bistatic radar. If more than one receiver is used, the arrangement is termed multistatic radar. (The etymology is mixed: *bis, multus*, and *status* are Latin for "twice," "many," and "position," respectively. *Monos* ($\mu o \nu o \varsigma$) is Greek for "alone.")

13.1.1 Fundamentals of Bistatic Radar

This discussion of the fundamentals of bistatic radar follows that of Willis [1, 2], who provides a history of many bistatic radar proposals and programs and a summary of many technical analyses. Figure 13.1 introduces some basic terminology. Instead of two locations of interest (the radar and the target), we now have three (transmitter, Tx; receiver, Rx; and target) forming the vertices of the bistatic triangle, which defines the bistatic plane. All three vertices may vary with time, as in an air warfare scenario. The line between Tx and Rx is

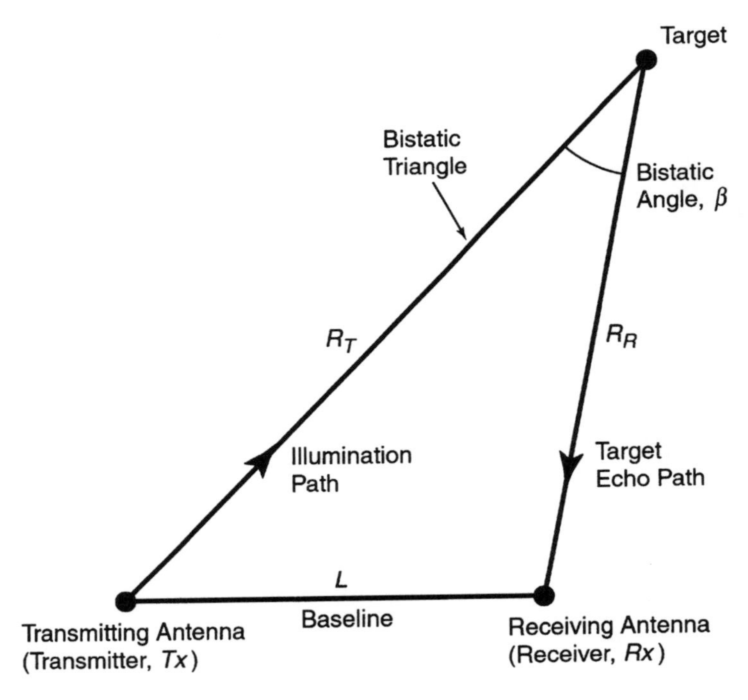

Figure 13.1 Bistatic radar geometry. (After [1], Figure 1.1, with permission of Artech House.)

the baseline. From the target location, the angle between the vector to the Tx and the vector to the Rx is the bistatic angle.

From the initial position shown in Figure 13.1, we now consider the locus of points along an arc in the bistatic plane so that the sum of the Tx-target range (R_T) and the Rx-target range (R_R) is constant, as illustrated in Figure 13.2. Clearly the locus of target locations is an ellipse with the Tx and Rx at the foci. More generally, in three-dimensional space, the locus of target locations is an ellipsoid. Figure 13.3 illustrates a set of confocal iso-range ellipses, along with corresponding orthogonal confocal hyperbolas.

The general equation for an ellipse (with major axis parallel to the x axis) is

$$\frac{x^2}{a^2} + \frac{y^2}{b^2} = 1 \tag{13.1}$$

where a is the semimajor axis and b is the semiminor axis. A set of confocal ellipses can be represented as (Problem 13.1)

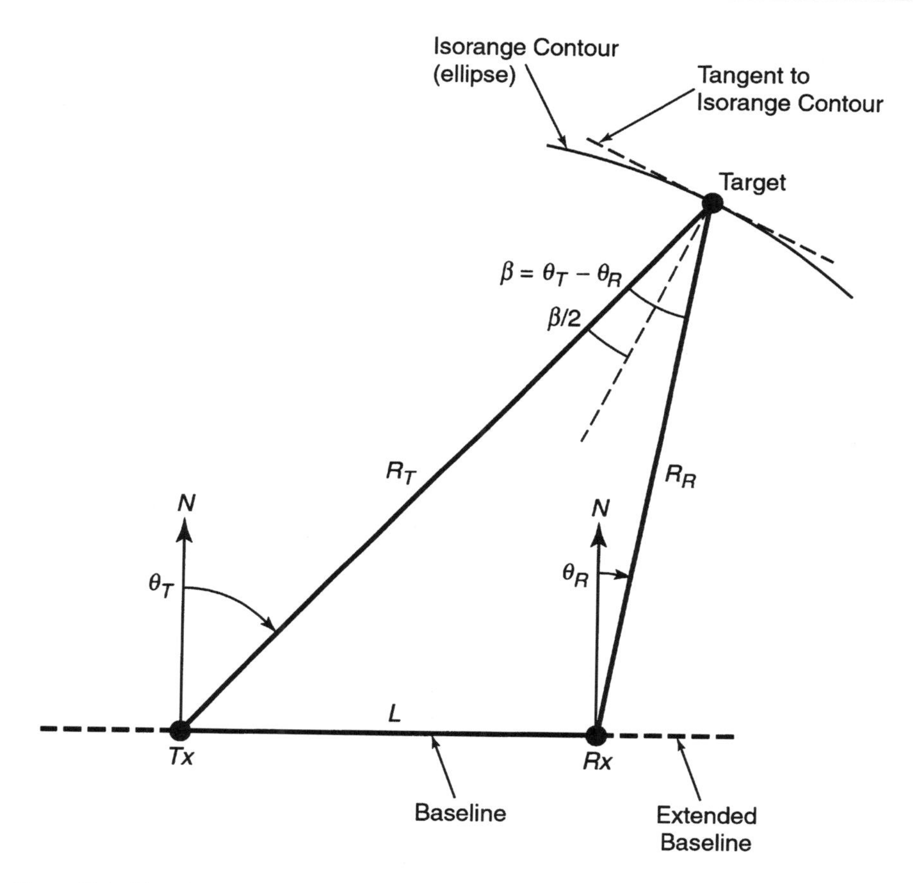

Figure 13.2 Bistatic radar terminology. (After [1], Figure 3.1, with permission of Artech House.)

$$a_i^2 - b_i^2 = \frac{L^2}{4} \tag{13.2}$$

where L is the distance between foci. The eccentricity of an ellipse is (Problem 13.2)

$$e \equiv \frac{L}{2a} = \frac{L}{R_T + R_R} = \frac{(a^2 - b^2)^{1/2}}{a} \tag{13.3}$$

and $0 \leq e \leq 1$. Clearly, if $e = 0$, the ellipse is a circle and the two foci are coincident. As $e \rightarrow 1$, $a \rightarrow \infty$, and the ellipse becomes a parabola (Problem 13.3). In polar coordinates, the ellipse is (Problem 13.4)

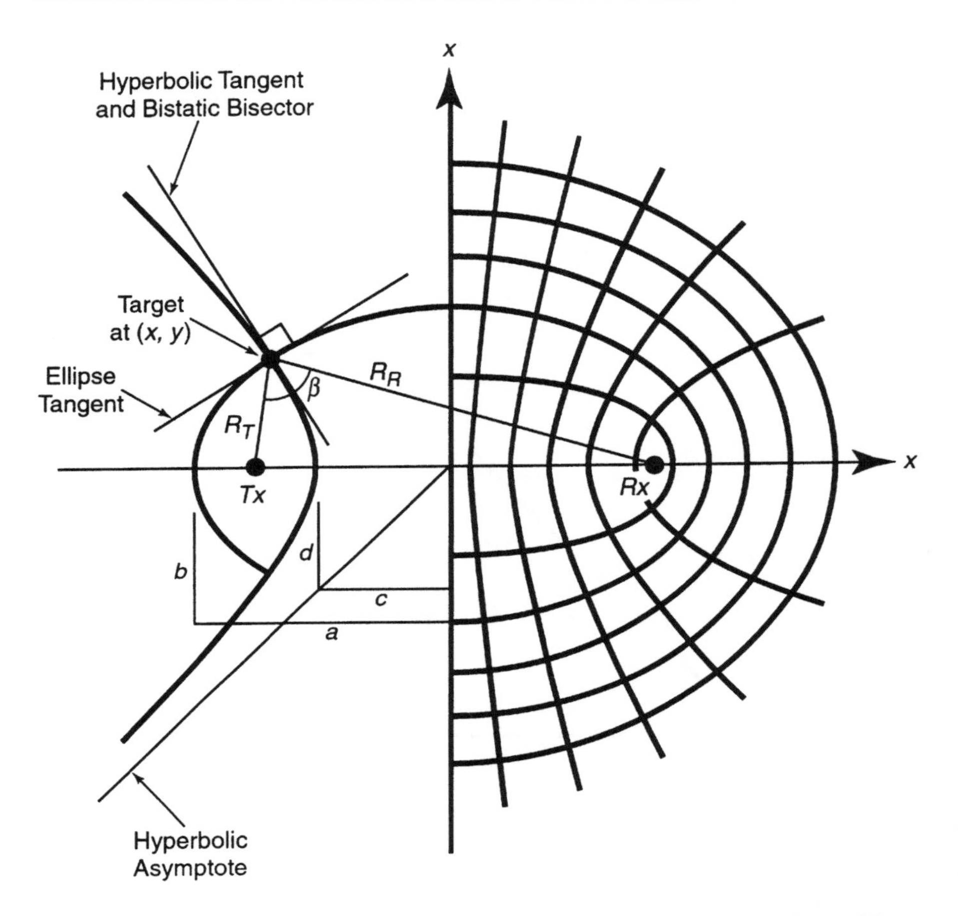

Figure 13.3 Confocal iso-range ellipses and orthogonal confocal hyperbolas. (After [1], Figure 3.2, with permission of Artech House.)

$$r = \frac{a(1 - e^2)}{1 - e\cos\phi} \tag{13.4}$$

with the origin at one focus.

Referring again to Figure 13.1, the maximum bistatic angle is (Problem 13.5)

$$\beta_{\max} = 2 \sin^{-1} e \tag{13.5}$$

Contours of constant bistatic angle β, as shown in Figure 13.4, are circles (Problem 13.6) of radius

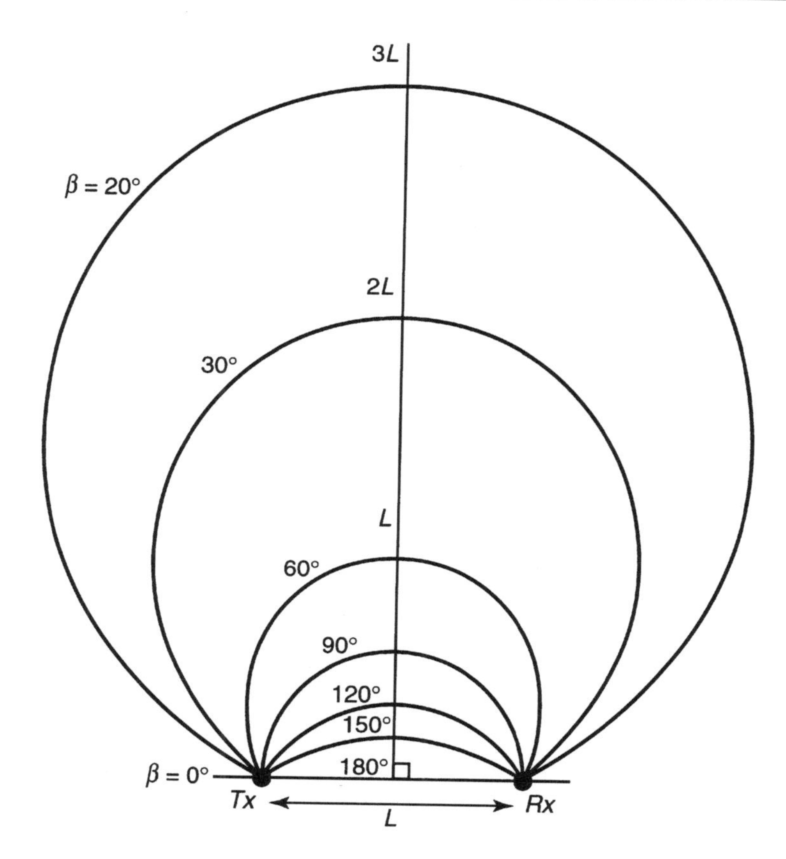

Figure 13.4 Contours of constant bistatic angle. (After [1], Figure 3.3, with permission of Artech House.)

$$r_\beta = \frac{L}{2\sin\beta} \tag{13.6}$$

with center on the bisector of L at a distance from L of

$$d_\beta = \frac{L}{2\tan\beta} \tag{13.7}$$

Other bistatic radar relationships are summarized in Table 13.1.

We generalize the radar equation (1.22) and its notation as follows:

$$SNR = \frac{PG_{Tx}G_{Rx}\lambda^2\sigma\Delta t}{(4\pi)^3 R_T^2 R_R^2 k T_0 F C_B L_{Tx} L_{Rx}} \tag{13.8}$$

Table 13.1
Bistatic Radar Relationships

$$R_R = \frac{(R_T + R_R)^2 - L^2}{2(R_T + R_R + L\sin\theta_R)}$$

$$= \frac{L(1 - e^2)}{2e(1 + e\sin\theta_R)}$$

$$R_T = (R_R^2 + L^2 + 2R_R L\sin\theta_R)^{1/2}$$

$$= \frac{L(e^2 + 1 + 2e\sin\theta_R)}{2e(1 + e\sin\theta_R)}$$

$$R_T = \frac{(R_T + R_R)^2 - L^2}{2(R_T + R_R - L\sin\theta_T)}$$

$$= \frac{L(1 - e^2)}{2e(1 - e\sin\theta_T)}$$

$$R_R = (R_T^2 + L^2 - 2R_T L\sin\theta_T)^{1/2}$$

$$= \frac{L(e^2 + 1 - 2e\sin\theta_T)}{2e(1 - e\sin\theta_T)}$$

$$\frac{R_R}{R_T} = \frac{\cos\theta_T}{\cos\theta_R} = \frac{d\theta_T}{d\theta_R} = \frac{1 - e\sin\theta_T}{1 + e\sin\theta_R}$$

$$= \frac{1 - e^2}{1 + e^2 + 2e\sin\theta_R} = \frac{1 + e^2 - 2e\sin\theta_T}{1 - e^2}$$

$$R_T = L\cos\theta_R/\sin\beta$$

$$R_R = L\cos\theta_T/\sin\beta$$

$$L = (R_T^2 + R_R^2 - 2R_T R_R\cos\beta)^{1/2}$$

Source: Reproduced from [1], Table 3.1, with permission of Artech House.

Here P may be peak or average power, as long as Δt is interpreted correctly. The gains are in the direction of the target, which is not necessarily the boresight direction.

If the Tx and Rx antenna gains and the target RCS are all isotropic (an unrealistic but perhaps interesting assumption), then the locus of target locations in a plane such that the received power is constant is given by (Problem 13.7)

$$R_T^2 R_R^2 = constant = \left(r^2 + \frac{L^2}{4}\right)^2 - r^2 L^2 \cos^2\phi \qquad (13.9)$$

That curve is known as an oval of Cassini. Figure 13.5 illustrates a family of ovals of Cassini. Unlike ellipses, those closer to the foci break into two separate parts, each of which is called a leminiscate.

13.1.2 Bistatic Radar Cross Section

Crispin and Siegel [3, Sec. 5.3] prove the following, known as the bistatic RCS theorem [1]:

> For vanishingly small wavelengths, the bistatic RCS of a sufficiently smooth, perfectly-conducting target is equal to the monostatic RCS measured on the bisector of the bistatic angle.

The definition of "sufficiently smooth" is given in [3]; essentially it states that the wavelength is much less than the characteristic size of any protuberance on the target.

Figure 13.6 illustrates measurements of the bistatic RCS of a large jet aircraft, as a function of aspect angle (between aircraft nose and bistatic angle

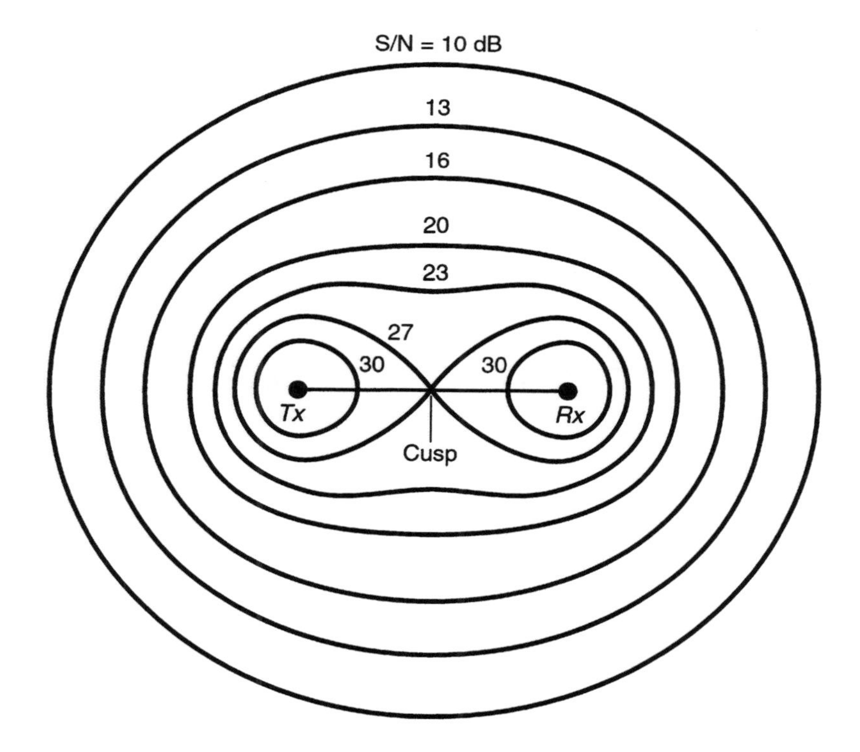

Figure 13.5 Ovals of Cassini. (After [1], Figure 4.2, with permission of Artech House.)

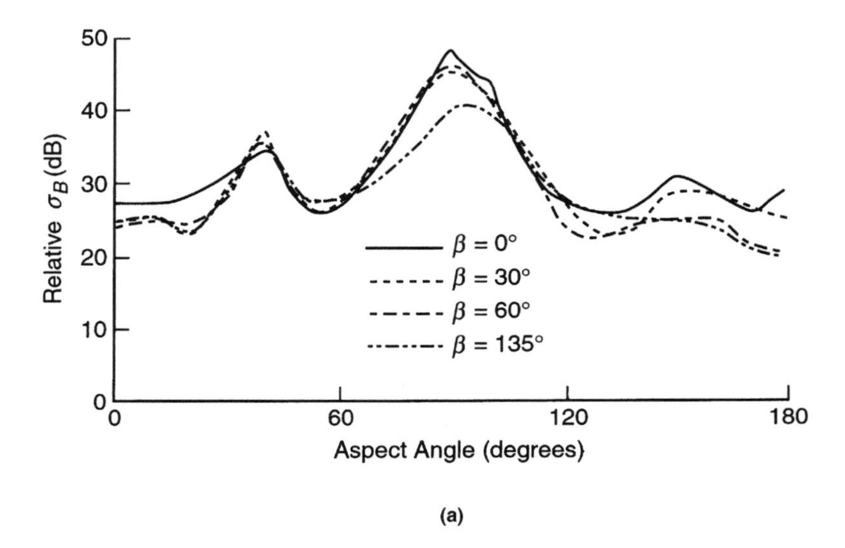

(a)

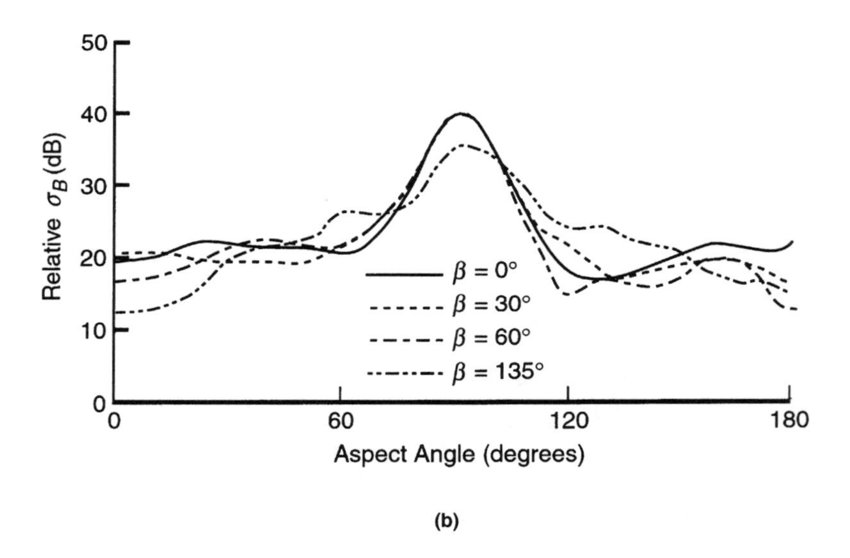

(b)

Figure 13.6 Bistatic RCS measurement: (a) 10-degree maximum values; (b) 10-degree median values. (Figure 81 from METHODS OF RADAR CROSS-SECTION ANALYSIS [3] by J. W. Crispin and K. M. Siegel, copyright © 1991 by Academic Press, reproduced by permission of the publisher.)

bisector) and bistatic angle. The aircraft is stationary on the ground, and the elevation angle is zero. Measurements were performed over 10-degree aspect bins, and both the maximum RCS in the bin, shown in Figure 13.6(a), and its median, shown in Figure 13.6(b), are plotted. The abscissa represents the aspect angle of the bistatic angle bisector, the ordinate depicts the relative RCS, and curves are plotted for several values of the bistatic angle. If the bistatic RCS theorem applied, the curves would be identical. Clearly a large aircraft does not satisfy the conditions of the bistatic RCS theorem. Nevertheless, the curves for different bistatic angles are similar over a wide range of bistatic angles (0 degrees, i.e., monostatic, to 135 degrees).

13.1.3 Other Bistatic Radar Considerations

A particularly challenging option is bistatic SAR. Successful results were achieved by the Environmental Research Institute of Michigan (ERIM) in 1984 [1, 4]. Figure 13.7 shows the configurations. Two aircraft were used, and SAR images were obtained at bistatic angles of 2, 40, and 80 degrees. The experiment illustrated that the many technical challenges can be overcome and that bistatic SAR images can be produced.

A number of bistatic radars have been tested; Willis [1, Chap. 2] gives a detailed history. However, apparently because of the complexity of coordinating pulse transmit and receive times, other technical challenges, and cost, bistatic radars have yet to be developed for routine operational use by military or civilian agencies.

Willis [1, 2] also summarizes other considerations concerning bistatic/multistatic radar, including area coverage and clutter cell size, doppler relationships, clutter characteristics, timing considerations, resolution (range, velocity, and angle), and RCS effects. He discusses the procedure of pulse-chasing, in which an agile receive beam rapidly scans the volume covered by the transmit beam, essentially chasing the pulse as it propagates.

Barton [5, Sec. 3.5] points out that, at bistatic angles approaching 180 degrees, the wave front, after passing a target, contains a hole or shadow of the target plus a forward-scatter lobe of relatively high intensity. The resulting increased RCS may potentially be used for target detection. That is challenging, however, because (1) it is difficult to arrange such geometry in a combat situation, (2) it is difficult to separate the scattered echo from the radar transmission, and (3) if the target is moving, the forward-scattered signal will not exhibit a significant doppler shift compared with a stationary target (since the total path length is not significantly varying with time), making it difficult to separate the target from clutter.

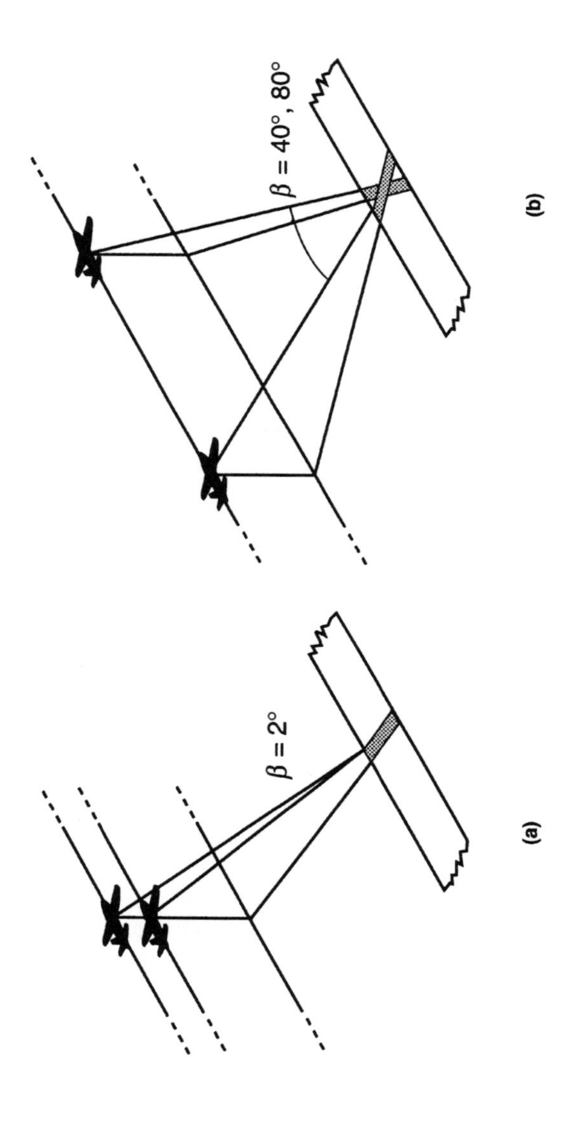

Figure 13.7 Bistatic SAR geometry: (a) vertical separation; (b) horizontal separation. (After [1], Figure 2.11, and [4], copyright 1984 IEEE.)

Kingsley and Quegan [6, Chap. 13] discuss the fact that the first radars were bistatic and that the transmitter need not be cooperative. During World War II, the Germans received reflections from the British "chain home" radar transmissions on British bombers and obtained warning of bombing raids without revealing the location of their receivers. Swerling [7] discusses a "silent sentry" concept that "utilizes terrestrial [television] and radio broadcast transmitters as its emitters of opportunity for purposes of tracking air targets."

Steinberg [8] constructed an X-band monostatic-bistatic multiaperture "radio camera," with which he obtained radar images of (noncooperative) commercial aircraft flying into Philadelphia International Airport. He discusses three modes.

- "Spatial radio camera" [8, Sect. III and Fig. 1]: transmit from a 1.2m-diameter paraboloid, receive on an array of 32 19-cm × 14-cm horns, essentially at one instant. This can be described as bistatic SAR.
- "Temporal radio camera" [8, Sec. IV and Fig. 2]: transmit and receive on the 1.2m dish over a time interval. The synthetic aperture is formed by the aircraft rotation relative to the radar. This is ISAR with a noncooperative target (see Section 6.12).
- "Spatial-temporal radio camera" [8, Sec. V and Fig. 3]: transmit on one antenna, receive on two antennas 25m apart, over a period of time, with the synthetic aperture formed by the aircraft rotation relative to the radar. This can be described as multistatic ISAR. In principle, more than two receive antennas could have been used.

13.2 Low-Probability-of-Intercept Radar

It is sometimes suggested that a radar may employ special techniques so that an enemy will have difficulty detecting it, a technique known as low-probability-of-intercept (LPI) radar [9–12]. Suggested LPI techniques include "power management" (reducing the transmitted power to a level just high enough to allow successful radar operation but still to prevent detection of the radar by an enemy interceptor), ultralow sidelobe antennas (to prevent detection by an enemy interceptor in the sidelobes), and spread-spectrum or coded waveforms (to prevent an enemy from recognizing the waveform as a radar). One suggested technique is to use bistatic radar; the radar receiver need not emit radiation at all, and the enemy need not know that the transmitter is working with a separate receiver. However, unless otherwise specified, the term LPI radar implies a monostatic radar, which is always subject to the problem that received power at the enemy interceptor falls off as $1/R^2$, whereas received power at the radar falls off as $1/R^4$.

We assume the existence of a monostatic LPI radar and examine some consequences. We assume that the radar emits average power P_{avg} for a dwell time t_d and detects a target and that an enemy interceptor receives the emitted radiation but cannot detect the radar.

By assumption, the radar can detect the target; therefore, from the radar equation

$$SNR_R = \frac{P_{avg} G_R^2(\theta_T, \phi_T) \lambda^2 \sigma t_d}{(4\pi)^3 R_T^4 k T_0 F_R L_R L_{RA} L_{RP}} \geq SNR_{tR} \tag{13.10}$$

Here the subscripts R and T indicate radar and target, respectively, and

- $G_R(\theta_T, \phi_T)$ = radar gain in target direction;
- σ = target RCS;
- t_d = dwell time;
- R_T = distance from radar to target;
- F_R = radar noise figure;
- L_R = radar hardware losses;
- L_{RA} = atmospheric loss between radar and target (two-way);
- L_{RP} = radar processing loss relative to the matched filter (= C_B; see Section 1.10);
- SNR_{tR} = detection threshold for radar, that is, if $SNR_R \geq SNR_{tR}$, the target is detected.

Also by assumption, the enemy receiver cannot detect the radar signal. Therefore

$$SNR_I = \frac{P_{avg} G_R(\theta_I, \phi_I) A_{effI}}{4\pi R_I^2 k T_0 F_I L_I L_{IA} L_{IP}} \cdot t_d \cdot f_{overlap} \leq SNR_{tI}$$
$$A_{effI} = \frac{G_I(\theta_R', \phi_R') \lambda^2}{4\pi} \tag{13.11}$$

where

- $G_R(\theta_I, \phi_I)$ = radar gain in interceptor direction (assumed to be constant over period of interest; interceptor is presumably in radar sidelobe);

- A_{effI} = effective area of interceptor antenna in direction of radar;
- f_{overlap} = fraction of time interceptor is directed at or near radar, for example, if interceptor is scanning or searching;
- $G_I(\theta_R', \phi_R')$ = interceptor gain in radar direction when interceptor is directed at or near radar; the primes indicate spherical coordinates with the interceptor, rather than the radar, at the origin;
- R_I = distance from radar to interceptor;
- F_I = interceptor noise figure;
- L_I = interceptor hardware losses;
- L_{IA} = atmospheric loss between radar and interceptor (one-way);
- L_{IP} = interceptor processing loss relative to a filter matched to the radar signal (the latter would require precise knowledge of the radar signal);
- SNR_{tI} = detection threshold for interceptor, that is, if $SNR_I \geq SNR_{tI}$, the radar is detected.

From (13.10) and (13.11) we have

$$\frac{4\pi(SNR_{tR})R_T^4F_RL_RL_{RA}L_{RP}}{G_R^2(\theta_T, \phi_T)\sigma} \leq \frac{P_{\text{avg}}\lambda^2 t_d}{(4\pi)^2kT_0}$$

$$\frac{(SNR_{tI})R_I^2F_IL_IL_{IA}L_{IP}}{G_R(\theta_I, \phi_I)G_I(\theta_R', \phi_R')f_{\text{overlap}}} \geq \frac{P_{\text{avg}}\lambda^2 t_d}{(4\pi)^2kT_0} \tag{13.12}$$

$$\frac{(SNR_{tI})R_I^2F_IL_IL_{IA}L_{IP}}{G_R(\theta_I, \phi_I)G_I(\theta_R', \phi_R')f_{\text{overlap}}} \geq \frac{4\pi(SNR_{tR})R_T^4F_RL_RL_{RA}L_{RP}}{G_R^2(\theta_T, \phi_T)\sigma}$$

so that, for successful LPI operation, we require

$$\frac{SNR_{tR}}{SNR_{tI}}\frac{F_RL_RL_{RA}L_{RP}}{F_IL_IL_{IA}L_{IP}}\frac{G_I(\theta_R', \phi_R')}{G_R(\theta_T, \phi_T)}\frac{G_R(\theta_I, \phi_I)}{G_R(\theta_T, \phi_T)}f_{\text{overlap}} \leq \frac{R_I^2\sigma}{4\pi R_T^4} \tag{13.13}$$

We assume that $F_R \sim F_I$, $L_R \sim L_I$, and $SNR_{tR} \sim SNR_{tI}$. Then we have

$$\frac{L_{RA}L_{RP}}{L_{IA}L_{IP}} \cdot \frac{G_I(\theta_R', \phi_R')}{G_R(\theta_T, \phi_T)} \cdot \frac{G_R(\theta_I, \phi_I)}{G_R(\theta_T, \phi_T)}f_{\text{overlap}} \leq \frac{R_I^2\sigma}{4\pi R_T^4} \tag{13.14}$$

For example, if we assume $R_T = R_I = 10^5$m, and $\sigma = 10$ m^2, then the left side must be less than $\sim 10^{-10}$. Clearly, higher values of σ and R_I/R_T make LPI operation easier.

From (13.14), we can conclude that, for LPI operation to be possible, one or more of the following must be true.

- $R_T << R_I$. From the radar, the range to the target is much less than the range to the interceptor.
- $L_{RA} << L_{IA}$. The two-way radar-to-target atmospheric loss is much less than the one-way radar-to-interceptor atmospheric loss. That may possibly be arranged if the radar is observing a nearby target with a frequency that is strongly absorbed by the atmosphere, such as 22 GHz [12, p. 514].
- $L_{IP} >> L_{RP}$. The interceptor processor is not well matched to the radar signal. That may possibly be arranged by using power management to preclude peak-power detection by the interceptor and using spread-spectrum and/or coding techniques to prevent the interceptor from correctly interpreting the radar waveform.
- $G_I(\theta_R', \phi_R') << G_R(\theta_T, \phi_T)$. The interceptor mainlobe gain is much less than that of the radar.
- $G_R(\theta_I, \phi_I) << G_R(\theta_T, \phi_T)$. The interceptor is in a (low) radar sidelobe.
- $f_{\text{overlap}} << 1$. The interceptor is usually not looking in the direction of the radar.

As indicated in Figure 13.8, design of an LPI radar would typically take advantage of one or more of those principles. It would probably be based on the following assumptions:

- A low-technology interceptor in the sidelobes of the radar, perhaps at relatively long range, perhaps scanning with a low overlap function;
- A high-technology radar with very low or ultralow antenna sidelobes; perhaps a strongly absorbed carrier frequency; power management plus a sophisticated, coded waveform, the interceptor being unaware of the details.

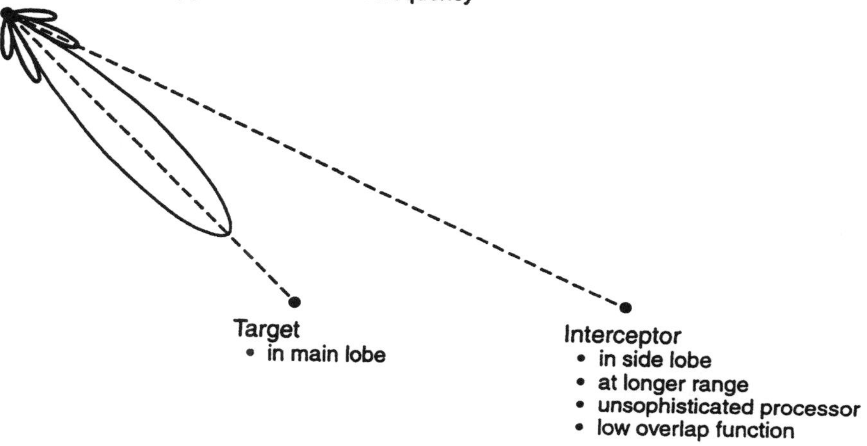

Figure 13.8 Successful LPI operation must utilize one or more of the principles illustrated here.

References

[1] Willis, N. J., *Bistatic Radar*, Norwood, MA: Artech House, 1991.

[2] Willis, N. J., "Bistatic Radar," Chap. 25 in M. Skolnik (ed.), *Radar Handbook*, 2nd ed., New York, McGraw-Hill, 1990.

[3] Crispin, J. W., and Siegel, K. M. (eds.), *Methods of Radar Cross-Section Analysis*, New York: Academic, 1968.

[4] Auterman, J. L., "Phase Stability Requirements for a Bistatic SAR," *Proc. 1984 National Radar Conf.*, Atlanta, pp. 48–52.

[5] Barton, D. K., *Modern Radar System Analysis*, Norwood, MA: Artech House, 1988.

[6] Kingsley, S., and S. Quegan, *Understanding Radar Systems*, Mendham, NJ: SciTech, 1999.

[7] Swerling, P., "Radar Into the Next Millennium," *IEEE AES Systems Magazine*, Aug. 1999, pp. 7–11.

[8] Steinberg, B., "Microwave Imaging of Aircraft," *Proc. IEEE*, Vol. 76, No. 12, Dec. 1988, pp. 1578–1592.

[9] Toomay, J. C., *Radar Principles for the Non-Specialist*, 2nd ed., New York: Van Nostrand Reinhold, 1989.

[10] Edde, B., *Radar: Principles, Technology, Applications*, Upper Saddle River, NJ: Prentice Hall, 1993.

[11] Stimson, G. W., *Introduction to Airborne Radar*, 2nd ed., Mendham, NJ: SciTech, 1998.

[12] Wehner, D. R., *High-Resolution Radar*, 2nd ed., Norwood, MA: Artech House, 1995.

Problems

Problem 13.1

Show that a set of confocal ellipses can be represented as

$$a_i^2 - b_i^2 = \frac{L^2}{4}$$

where a is the semimajor axis, b is the semiminor axis, and L is the distance between the foci.

Problem 13.2

Show that the eccentricity of an ellipse is given by

$$e \equiv \frac{L}{2a} = \frac{L}{R_T + R_R} = \frac{(a^2 - b^2)^{1/2}}{a}$$

and $0 \leq e \leq 1$.

Problem 13.3

Show that, as $e \rightarrow 1$, $a \rightarrow \infty$, and the ellipse becomes a parabola.

Problem 13.4

Show that, in polar coordinates, the ellipse is described by

$$r = \frac{a(1 - e^2)}{1 - e \cos\phi}$$

with the origin at one focus.

Problem 13.5

Show that the maximum bistatic angle is

$$\beta_{\max} = 2 \sin^{-1} e$$

Problem 13.6

Show that contours of constant bistatic angle β, as shown in Figure 13.4, are circles of radius

$$r_\beta = \frac{L}{2\sin\beta}$$

with center on the bisector of L at a distance from L of

$$d_\beta = \frac{L}{2\tan\beta}$$

Problem 13.7

Show that the locus of target locations in a plane such that the received power is constant is given by an oval of Cassini.

$$R_T^2 R_R^2 = constant = \left(r^2 + \frac{L^2}{4}\right)^2 - r^2 L^2 \cos^2\phi$$

Problem 13.8

Assume that an LPI radar occasionally points in the direction of an interceptor, with an overlap function of f_{overlapR}, and modify (13.14) accordingly. How does one now interpret $G_R(\theta_I, \phi_I)$?

14

Weather Radar and Ground-Penetrating Radar

This chapter examines two topics of modern radar: meteorological radar and ground-penetrating radar.

14.1 Weather Radar

Of all modern applications of radar, perhaps none is so universally practical as the use of radar to observe weather patterns.

14.1.1 Radar Beam Geometry

Consider a real-beam radar observing a rainstorm [1], with azimuth and elevation beamwidths ϕ and θ, respectively. Because the area of an ellipse is πab, where a and b are the semimajor and semiminor axes, at a distance R the beam area is

$$A_{\text{beam}} = \frac{\pi\theta\phi R^2}{4} \tag{14.1}$$

A given time sample of the return signal will include information about a range gate of extent $c\tau/2$ (Section 1.18; we assume an unmodulated pulse) and therefore a volume of

$$V = \frac{\pi\theta\phi R^2 c\tau}{8} \tag{14.2}$$

From Section 3.2.8, the RCS of the rain in that volume is

$$\sigma = \eta V \tag{14.3}$$

where the units of η are m^2/m^3. η can be expressed as

$$\begin{aligned} \eta &= \rho_n \overline{\sigma} \\ \rho_n &= \text{raindrop number density} \\ \overline{\sigma} &= \text{average raindrop RCS} \end{aligned} \tag{14.4}$$

14.1.2 Scattering by an Individual Raindrop

Ishimaru [2, Chap. 2] summarizes Rayleigh's derivation of the scattering of plane EM waves by a small dielectric sphere (raindrop), where the sphere diameter D is much less than the wavelength λ. First, he makes the following dimensional-analysis argument. The scattered electric field $|E_s|$ must be proportional to the volume of the drop and thus be related to the incident electric field $|E_i|$ by

$$|E_s| = K_1 |E_i| \frac{V}{R} = K_2 |E_i| \frac{D^3}{R} \tag{14.5}$$

where K_1 and K_2 are factors dependent on the wavelength and scattering angle, V and D are the volume and diameter of the drop, and R is the distance from the drop. K_2 must have the dimensions of length^{-2}. Because the dependence on D has been considered and only one other length is involved, K_2 must be proportional to λ^{-2}. The scattered power must thus be proportional to D^6/λ^4.

More precisely, Ishimaru derives the differential scattering cross section σ_d—the fraction of total scattering cross-section that results in scattering in a particular direction; its units are meter^2/steradian. The integral of scattering cross-section over 4π steradians is the total scattering cross-section, such that if one multiplies the area by the incident flux density (W/meter^2), the result is the power scattered away from the original plane-wave beam. The RCS (1.14) is 4π times the differential scattering cross-section. For Rayleigh scattering,

$$\begin{aligned} \sigma_d(\hat{E}, \hat{R}) &= \frac{9k^4}{(4\pi)^2} \left| \frac{m^2 - 1}{m^2 + 2} \right|^2 V^2 \sin\chi \\ \sin^2\chi &= 1 - (\hat{E} \cdot \hat{R})^2 \end{aligned} \tag{14.6}$$

where $\hat{E}$ and $\hat{R}$ are the unit vectors representing the directions of the incident electric field polarization and the scattering direction, $k = 2\pi/\lambda$ is the wave number, m is the complex index of refraction, and χ is the scattering angle relative to the initial polarization (the pattern is that of dipole radiation). Because

$$V = \frac{4\pi a^3}{3} = \frac{\pi D^3}{6} \tag{14.7}$$

(a is the drop radius), then for backscattered radiation ($\sin\chi = 1$)

$$\sigma_d = \frac{16\pi^4}{\lambda^4 \cdot 16\pi^2} \cdot 9|K|^2\left(\frac{\pi D^3}{6}\right)^2 = \frac{\pi^4 D^6}{4\lambda^4}|K|^2 \tag{14.8}$$

where

$$|K|^2 = \left|\frac{m^2 - 1}{m^2 + 2}\right|^2 \tag{14.9}$$

The RCS is 4π times σ_d, or[1]

$$\sigma = \frac{\pi^5 |K|^2 D^6}{\lambda^4} = \frac{64\pi^5 a^6}{\lambda^4}|K|^2 \tag{14.10}$$

At temperatures between 0° and 20°C, at microwave frequencies, $|K|^2$ is approximately 0.93 for liquid water (rain) and 0.20 for ice (hail or sleet) [1].

14.1.3 Volume Reflectivity of Rain

Figure 3.16 (Figure 3.6.6 in Barton [3]) summarizes volume reflectivity of rain. To provide a rough comparison with (14.10), we make the oversimplified approximation that all raindrops have the same volume V. Then the rainfall rate γ (mm/hr) is

$$\gamma = \rho_n V v \tag{14.11}$$

1. (14.10) does not reduce to the value for a small, perfectly conducting sphere (Section 3.1.1.1) as $m \to \infty$, but rather to 4/9 that value; see [14, Vol. 1, footnote on p. 163].

where v is the drop velocity (note that γ and v have the same dimensions). The RCS per unit volume (m^2/m^3) is

$$\eta = \rho_n \sigma = \frac{\sigma}{Vv} \cdot \gamma \tag{14.12}$$

From [2, p. 46], we assume that a = 3 mm, V = 113 mm^3 = 1.13 · 10^{-7} m^3, v = 10 m/s, λ = 0.032m, and from (14.10) σ = 1.26 · 10^{-5} m^2. Then σ/Vv = 11.2 m^{-1}/(m/s) = 3.1 · 10^{-6} m^{-1}/(mm/hr). For γ = 10 mm/hr, that implies η = 3.1 · 10^{-5} m^2/m^3. That fortuitously agrees with Figure 3.16, which is nonlinear in γ and was determined using more sophisticated methods than the simple approximation of (14.11).

14.1.4 dBZ

Meteorologists use a quantity Z [1, 4–6] to describe the density of rain as well as the size distribution of D^6. From (14.4) and (14.10),

$$\eta = \frac{\rho_n \pi^5 |K|^2 \langle D^6 \rangle}{\lambda^4} \equiv \frac{\pi^5 |K|^2}{\lambda^4} Z \tag{14.13}$$

where the angle brackets indicate a mean. Z is defined as the expected value of the sixth power of the drop diameter, per unit volume; its dimensions are $(\text{length})^3$:

$$Z \equiv \int_0^\infty N(D) D^6 dD \tag{14.14}$$

where $N(D)$ is the number density of drops between diameter D and $D + dD$.

We now have

$$\begin{aligned} \sigma = \eta V &= \frac{\pi^5 |K|^2 Z}{\lambda^4} \cdot \frac{\pi \theta \phi R^2 c\tau}{8} \\ &= \frac{\pi^6 \theta \phi c\tau |K|^2}{8\lambda^4} Z R^2 \end{aligned} \tag{14.15}$$

and, from (1.22),

$$P_{\text{recd}} = \frac{P_{\text{trans}} G^2 \lambda^2 \sigma}{(4\pi)^3 R^4 L}$$

$$= \frac{P_{\text{trans}} G^2 \lambda^2}{64\pi^3 R^4 L} \cdot \frac{\pi^6 \theta\phi c\tau |K|^2}{8\lambda^4} Z R^2 \tag{14.16}$$

$$= \frac{\pi^3 P_{\text{trans}} G^2 \theta\phi c\tau |K|^2}{512\lambda^2 L} \cdot \frac{Z}{R^2}$$

The received power (peak or average) is inversely proportional to R^2 and directly proportional to Z (see Problem 14.1).

The gain is not uniform over the beamwidth, and we should alter (14.16) slightly to take that into account. The actual expression depends on the beam shape. Typically, this shape is approximated by a Gaussian, in which case P_{recd} is reduced by a factor of 2 ln(2) [1]:

$$P_{\text{recd}} = \frac{\pi^3 P_{\text{trans}} G_{\text{peak}}^2 \theta\phi c\tau |K|^2}{(2\ \ln 2) 512\lambda^2 L} \cdot \frac{Z}{R^2} \tag{14.17}$$

Z is typically expressed in mm^6/mm^3, and the most commonly used quantity is dBZ, or 10 log(Z) in those units. The U.S. National Weather Service (NWS) terminology for different levels of dBZ is given in Table 14.1 [6]. Nonprecipitating clouds produce a reflectivity of about −40 dBZ [1].

14.1.5 The NEXRAD Weather Radar Network

In the United States, weather is monitored by the Next Generation Weather Radar (NEXRAD) system [7–9], completed in 1998, which measures both dBZ and LOS velocity using doppler processing. The network consists of

Table 14.1
National Weather Service (NWS) Rain Levels [6]

Level	dBZ	Description
1	18–30	Light rain
2	30–41	Moderate rain
3	41–46	Heavy rain
4	46–50	Very heavy rain
5	50–57	Intense rain
6	>57	Extreme rain

138 S-band WSR-88D radars, each with an effective range of ~230 km. The radars perform three-dimensional volume scans. Table 14.2 summarizes the radar parameters. The radar information is processed by the Advanced Weather Interactive Processing System (AWIPS).

The NEXRAD network has greatly improved weather forecasting, compared with the previous radar network. For example, the number of tornadoes that hit without warning has been reduced from about 33% to about 13%, and the NWS is issuing fewer severe thunderstorm and tornado warnings that turn out to be false alarms [10].

14.2 Ground-Penetrating Radar (GPR)

Radar can be useful for penetrating the ground and detecting and/or identifying underground objects. The primary difficulties concern strong attenuation of the radar signal and confusion of the target signature with (1) the very large number of false-alarm signals arising from other underground objects (natural and human-made) and (2) scattering from the nonhomogeneous ground itself.

14.2.1 Location of an Underground Object

Consider the simple case of a small underground object (target) that can be easily seen by the radar; we want to determine the target location. Assume that the (small) antenna of a GPR moves along a straight line on the (planar) surface of the (homogeneous) ground, using a waveform that yields target range versus radar position, as shown in Figure 14.1(a). If the antenna passes directly above the target, then for a target depth d and an antenna position x, the measured range r is given by

$$r^2 = d^2 + x^2 \tag{14.18a}$$

$$\frac{r^2}{d^2} - \frac{x^2}{d^2} = 1 \tag{14.18b}$$

Equation (14.18b) is in the standard form for a hyperbola, showing that this is the shape of a plot of r versus x, as illustrated in Figure 14.1(b). This type of plot is often used to display the results of GPR scans; examples are given by Daniels [11, pp. 230–231]. If the antenna does not pass directly above the target, the shape is still hyperbolic, with d corresponding to the minimum antenna-target distance.

Table 14.2
Parameters of the WSR-88D Weather Radar

Antenna Subsystem		
Radome		
Type	Fiber glass skin foam sandwich	
Diameter	11.89m	
rf loss—two way	0.3 dB at 2995 MHz	
Pedestal		
Type	Elevation Over Azimuth	
	Azimuth	Elevation
Scanning rate—maximum	30° s^{-1}	30° s^{-1}
Acceleration	15° s^{-2}	15° s^{-2}
Mechanical limits	–1° to +60°	
Reflector[a]		
Type	Paraboloid of revolution	
Polarization	Linear[b]	
Reflector diameter	8.54m	
Gain[c]	44.5 dB	
Beam width	1°	
First sidelobe level	–26 dB (with radome)	
Transmitter and Receiver Subsystem		
Transmitter		
Type	Master oscillator power amplifier	
Frequency	2700 MHz to 3000 MHz	
Pulse power[c]	475 kW	
Pulse width	1.57 μs and 4.57 μs	
rf duty cycle	0.002 maximum	
PRFs		
short pulse (eight selectable)	320 Hz to 1300 Hz	
long pulse:	320 Hz and 450 Hz	
Receiver		
Type	Linear	
Dynamic range	93 dB	
Intermediate frequency	57.6 MHz	
System noise power[c]	–113 dBm	
Filter		
Short pulse:	Analog filter; bandwidth (3 dB): 0.63 MHz bandwidth (6 dB): 0.80 MHz Additional digital filtering; 3 samples (spaced 0.25 km) of *I* and *Q* are averaged. Output samples are spaced at 0.5 km intervals.	

Table 14.2 ***(continued)***
Parameters of the WSR-88D Weather Radar

System performance
A reflectivity factor of –7.5 dBZ at 50 km must produce an SNR > 0 dB.

[a]Antenna specifications include the effects of the radome.
[b]Initially the first radars will have circular polarization which will be changed.
[c]Transmitted power, antenna gain, and receiver noise power are measured at the antenna port.
Source: Table 3.1 from DOPPLER RADAR AND WEATHER OBSERVATIONS [4], Second Edition, by Richard J. Doviak and Dusan S. Zrnic, copyright © 1993 by Academic Press, reproduced by permission of the publisher.

We again consider the case for which the antenna passes directly above the target. As discussed in [11, Sec. 2.2.10], for any two points along the scan, taken as $x = 0$ and $x = x_A$, we can use the two measured ranges r_0 and r_A to solve for the target position (x_T, z_T), as follows (Figure 14.2):

$$\begin{aligned} r_0^2 &= x_T^2 + z_T^2 \\ r_A^2 &= (x_T - x_A)^2 + z_T^2 \end{aligned} \tag{14.19}$$

We have two equations in two unknowns, x_T and z_T, which are solved as follows:

$$\begin{aligned} z_T^2 &= r_0^2 - x_T^2 \\ r_A^2 &= (x_T^2 - 2x_T x_A + x_A^2) + r_0^2 - x_T^2 = -2x_T x_A + x_A^2 + r_0^2 \\ x_T &= \frac{r_0^2 + x_A^2 - r_A^2}{2x_A} \\ z_T &= (r_0^2 - x_T^2)^{1/2} \end{aligned} \tag{14.20}$$

That solution is easily extended to three dimensions (Figure 14.3). If we use three noncollinear antenna positions on the surface, denoted as $(x, y) = (0, 0)$, $(x_A, 0)$, and (x_B, y_B), to measure ranges r_0, r_A, and r_B, then

$$\begin{aligned} r_0^2 &= x_T^2 + y_T^2 + z_T^2 \\ r_A^2 &= (x_T - x_A)^2 + y_T^2 + z_T^2 \\ r_B^2 &= (x_T - x_B)^2 + (y_T - y_B)^2 + z_T^2 \end{aligned} \tag{14.21}$$

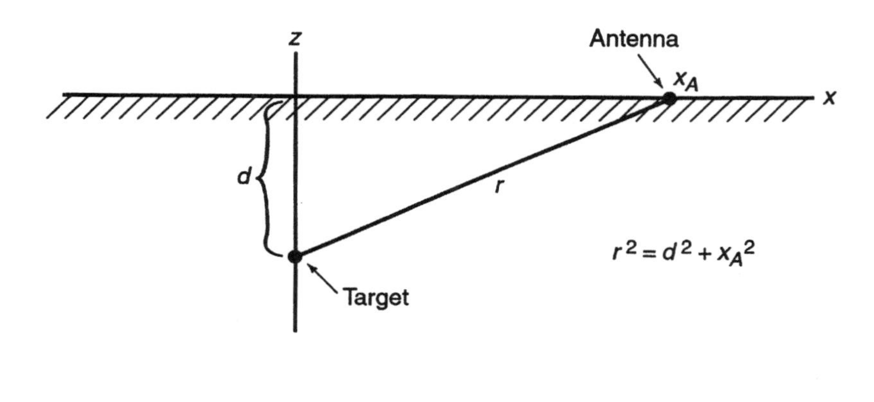

(a)

(b)

Figure 14.1 Range versus antenna position for underground target: (a) principle; (b) example (curve is hyperbola).

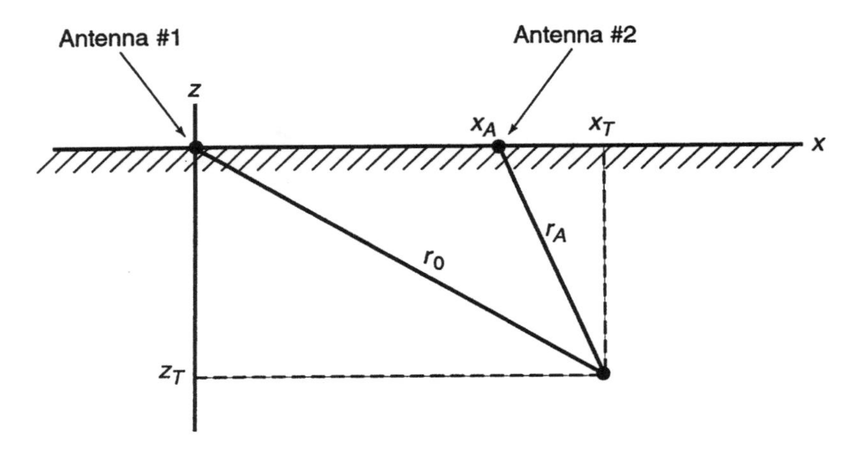

Figure 14.2 Two-dimensional determination of target position from two antenna positions.

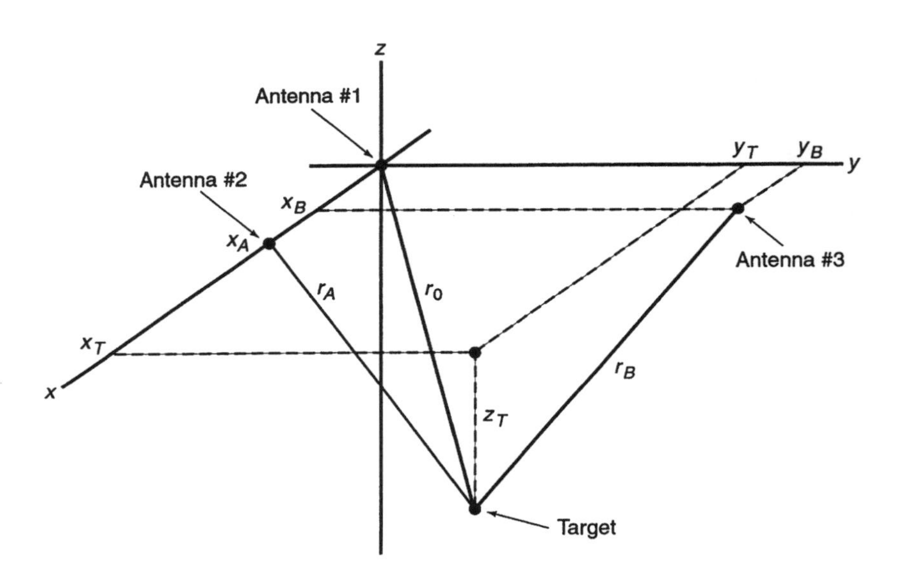

Figure 14.3 Three-dimensional determination of target position from three antenna positions.

The solution is

$$
\begin{aligned}
z_T^2 &= r_0^2 - x_T^2 - y_T^2 \\
r_A^2 &= x_T^2 - 2x_T x_A + x_A^2 + y_T^2 + r_0^2 - x_T^2 - y_T^2 = -2x_T x_A + x_A^2 + r_0^2 \\
x_T &= \frac{r_0^2 + x_A^2 - r_A^2}{2x_A} \\
r_B^2 &= x_T^2 - 2x_T x_B + x_B^2 + y_T^2 - 2y_T y_B + y_B^2 + r_0^2 - x_T^2 - y_T^2 \\
&= -2x_T x_B + x_B^2 - 2y_T y_B + y_B^2 + r_0^2 \\
y_T &= \frac{r_0^2 + x_B^2 + y_B^2 - r_B^2 - 2x_T x_B}{2y_B} \\
z_T &= (r_0^2 - x_T^2 - y_T^2)^{1/2}
\end{aligned}
\tag{14.22}
$$

Equations (14.20) and (14.22) were obtained by assuming that the antenna is in direct contact with the ground and that the speed of microwaves in the soil is known. If the antenna is elevated above the ground, reflection and refraction effects must be considered, and the solution becomes much more complicated. Furthermore, the electrical properties of the soil must be known to determine the wave speed [11, Chaps. 2 and 3].

14.2.2 Attenuation of Radar Signal in Ground

As discussed by Daniels [11, Chap. 3], microwaves are strongly attenuated by soil, especially if the water content is significant. Table 14.3 summarizes the

Table 14.3
Attenuation of Some Common Materials (After Daniels [11] Table 2.3, with permission of IEE.)

Material	Loss at 100 MHz, dB/m	Loss at 1 GHz, dB/m
Clay (moist)	5–300	50–3,000
Loamy soil (moist)	1–60	10–600
Sand (dry)	0.01–2	0.1–20
Ice	0.1–5	1–50
Fresh water	0.1	1
Sea water	1,000	10,000
Concrete (dry)	0.5–2.5	5–25
Brick	0.3–2.0	3–20

attenuation (decibels per meter) of some common soils and other materials. Evidently the attenuation coefficient at 1 GHz is about 10 times that at 100 MHz.

14.2.3 GPR Applications

Daniels [11, Chap. 7] presents a useful summary of GPR applications. The following applications all involve radars placed on, or very near, the ground or other obscuring medium:

- Detection of buried mines (metallic and nonmetallic)
- Detection of buried pipes and cables
- Internal inspection of pipes
- Nondestructive testing of concrete and masonry
- Geophysical applications:
  - Peatland
  - Geological structures, for example, groundwater
  - Soil erosion behind/under dikes
  - Subterranean coal, salt, oil shale, gypsum, and rock structures
  - Snow, permafrost, and glaciers
  - Archaeological structures
  - Integrity of tunnel linings
  - Thickness of layers on roads

Daniels also summarizes the following remote-sensing applications of GPR, in which the radar is far removed from the ground.

- The Apollo lunar sounder experiment, flown on the manned lunar mission Apollo 17 (1971), operating at 5 MHz, detected interfaces at depths of 1,000m and 1,400m.
- NASA shuttle imaging radar (SIR-A), an L-band SAR, in 1984 revealed previously unknown stream valleys and erosional surfaces in Egypt and Sudan that are currently covered by 1m to 2m of sand.

Jao et al. [12] describe analysis and results of GPR SAR measurements performed in 1995 in Yuma, Arizona. Two GPR SARs were used: the CARABAS, a 20–90 MHz (VHF) H-pol SAR developed by the Swedish National Defence Research Establishment, and FOLPEN, a 200–400 MHz (UHF) HH- and VV-pol impulse radar (i.e., short pulse, wide bandwidth) developed by SRI International, Menlo Park, California. The authors state that "most buried and surface targets, including mines, vehicles, wires, pipes, ammunition boxes, oil barrels, etc., and targets of opportunity, such as bunkers, were observed. . . . Metallic M-20 mines laid on the ground surface or shallow-buried up to a depth of six inches were observed. . . . Detection of relatively deep underground targets may also be feasible in soils of very low loss such as dry soil with high sand but low salt content."

Also used at the Yuma experiment was the U.S. Army Research Laboratory (ARL) ultra-wideband polarimetric BoomSAR, described by Carin et al. [13] (that paper also provides a good bibliography of GPR literature). The BoomSAR antenna is mounted at the top of a vehicle with a vertically telescoping boom lift, capable of moving at ~1 km/hr while the basket is elevated to a height of 45m. The short pulse has frequency components from 50 to 1,200 MHz. The authors describe several techniques used to process the SAR data, including a matched-filter-like detector that directly incorporates the target signatures themselves. A theoretical model is also used to describe wave phenomenology in various soil environments. A ROC curve (see Section 4.1.2) is given for M-20 metallic mines, both on the surface and buried 6 inches deep; single-polarization data were used. The curve indicates a P_d of ~0.7 with ~200 false alarms per square kilometer. The authors also state [13, p. 30]: "Buried *plastic* mines were virtually invisible to the radar and were *not* detected [emphasis in original]. Since many mines have only trace metal content, this failure is particularly troubling. . . . We have examined the utility of applying water to a suspected mine field, thereby increasing the electrical contrast between the target and soil. . . . This technique potentially has significant utility."

References

[1] Serafin, R. J., "Meteorological Radar," Chap. 23 of M. Skolnik (ed.), *Radar Handbook*, 2nd ed., New York: McGraw-Hill, 1990.

[2] Ishimaru, A., *Wave Propagation and Scattering in Random Media*, Piscataway, NJ: IEEE Press, 1997.

[3] Barton, D. K., *Modern Radar System Analysis*, Norwood, MA: Artech House, 1988.

[4] Doviak, R. J., and D. S. Zrnic, *Doppler Radar and Weather Observations*, 2nd ed., New York: Academic, 1993.

[5] Battan, L. J., *Radar Observation of the Atmosphere*, Chicago: University of Chicago Press, 1973.

[6] Mahapatra, P., *Aviation Weather Surveillance Systems*, New York: American Institute of Aeronautics and Astronautics, 1999.

[7] Crum, T. D., and R. L. Alberty, "The WSR-88D and the WSR-88D Operational Support Facility," *Bulletin of the American Meteorological Society*, Vol. 74, No. 9, Sept. 1993, pp. 1669–1687.

[8] Klazura, G. E., and D. A. Imy, "A Description of the Initial Set of Analysis Products Available From the NEXRAD WSR-88D System," *Bulletin of the American Meteorological Society*, Vol. 74, No. 7, July 1993, pp. 1293–1311.

[9] Friday, E. W., "The Modernization and Associated Restructuring of the National Weather Service: An Overview," *Bulletin of the American Meteorological Society*, Vol. 75, No. 1, Jan. 1994, pp. 43–52.

[10] Williams, J., "Doppler Effects," *Weatherwise*, Aug.-Sept. 1994, pp. 43–46.

[11] Daniels, D. J., *Surface-Penetrating Radar*, London: Institution of Electrical Engineers, 1996. See also Conyers, L. B., and D. Goodman, *Ground-Penetrating Radar: An Introduction for Archaeologists*, Walnut Creek, CA: Altamira Press, 1997.

[12] Jao, J. K., et al., "Analysis and Results of the 1995 Yuma Ground Penetration SAR Measurements," *IEEE AES Systems Magazine*, June 1999, pp. 5–9.

[13] Carin, L., et al., "Ultra-Wide-Band Synthetic Aperture Radar for Mine-Field Detection," *IEEE Antennas and Propagation Magazine*, Vol. 41, No. 1, Feb. 1999, pp. 18–33.

[14] Ruck, G. T., et al., *Radar Cross Section Handbook*, 2 vols., New York: Plenum, 1970.

Problems

Problem 14.1

Is there an unambiguous correspondence between dBZ and rainfall in mm/hr? Assume that all raindrops are 3 mm in diameter and fall at 10 m/s and calculate this relationship.

Glossary[1]

A/D whitening — ensuring that the noise level is high enough to "toggle" the *least significant bit* of the A/D, thereby permitting measurement of any bias (Section 2.2.14)

absorptivity — for a surface, the fraction of incident radiation absorbed (see *Kirchhoff's law*) (Section 3.3)

active array — phased array such that each *array element* contains its own *transmit-receive (T/R) module,* which produces the microwave radiation locally within the element (Section 2.4)

adaptive matched filter (AMF) — theoretically ideal algorithm for testing for the presence of a target in interference, where the target and interference are vectors instead of scalars (Section 9.2)

adaptive matched filter (AMF) — *AMF* test (used in *STAP*) that allows for variable interference: the denominator represents the (variable) ratio of target signal vector to expected interference; the numerator represents the output of the adaptive matched filter operating on the measured data. (Section 12.8.5)

adaptive sidelobe reduction — (*ASR*) procedure developed by S. DeGraaf of ERIM (now at Northrop-Grumman Corporation) for forming a fine-resolution image, based on simplifying the *minimum-variance method (MVM)* (Section 8.2.6)

---

1. Definitions followed by a [1] are reprinted with permission from IEEE Std. 686-1997 "IEEE Standard Radar Definitions" Copyright 1997, by IEEE. The IEEE disclaims any responsibility or liability resulting from the placement and use in the described manner. All rights reserved.

additive noise — thermal noise (used to describe a SAR image; see *multiplicative noise*) (Section 7.3.4)

agile-beam technique — use of an *electronically-scanned array* to vary the direction of the transmitted radiation rapidly from one direction to another, perhaps on a pulse-to-pulse basis (Section 2.4.2, 7.5.7)

airborne moving-target indication (AMTI) radar — "an MTI radar flown in an aircraft or other moving platform with corrections applied for the effects of platform motion, which include the changing clutter Doppler frequency and the spread of the clutter Doppler spectrum" [1]; see next definition (Section 11.0)

airborne-moving-target indication (AMTI) — (note the different arrangement of the hyphens) detection of moving targets that are airborne; see previous definition (Section 11.0)

aliasing — term referring to the fact that a target at some true position and velocity may appear at some other apparent position or velocity; the target is "aliased" from its correct position (or velocity) to the apparent position (or velocity) (Section 10.3)

altitude return — the radar echo from the ground directly underneath the *ownship* (Section 11.2.3)

ambiguity function — "the squared magnitude $|\chi(\tau, f_d)|^2$ of the function which describes the response of a radar receiver to targets displaced in range delay τ and Doppler frequency f_d from a reference position, where $\chi(0,0)$ is normalized to unity" [1] (Section 4.2.4)

Ament's formula — expression describing the coefficient of forward scattering of microwaves (Section 3.2.5)

analog-to-digital (A/D) converter — component that converts a final analog DC signal voltage into digital form, for recording or transmitting to signal processor (Section 2.2.14)

angular frequency (ω) — frequency measured in radians/sec ($\omega = 2\pi f$) (Section 1.3)

antenna — "that part of a transmitting or receiving system which is designed to radiate or to receive electromagnetic waves" [2] (Section 1.6, 2.3, 2.4)

antenna temperature (T_{ant}) — effective temperature of scene; for an isothermal blackbody scene, $T_{ant} = T_{scene}\epsilon$, where ϵ is the scene *emissivity* (Section 2.2.11) aperture illumination efficiency (η_i) — the ratio of the actual directivity to the theoretically maximum directivity; see *aperture efficiency* (Section 1.6)

aperture illumination function of an aperture antenna — pattern of radiation intensity (W/m^2) immediately beyond the plane of the aperture (Section 1.6, 2.3.1)

aperture antenna — an antenna such that the radiation essentially is emitted from a planar aperture (Section 1.6, 2.3.1)

aperture efficiency (η) — of an aperture antenna ratio of *effective area* (A_e) to physical area (A) ($A_e/A \leq 1$), where A_e is measured normal to the aperture; see *aperture illumination efficiency; radiation efficiency* (Section 1.6)

aperture time (t_A) — for spotlight SAR, time required to collect the data to produce an image; $t_A = \lambda R/(2V\delta_{cr}\cos\theta_{sq})$ (Section 7.1.1)

apparent range (R_{app}) — for a waveform with a single PRF, the actual range of a target modulo the maximum unambiguous range R_u. In measuring target range, the radar cannot distinguish between R_{app} and $R_{app} + NR_u$, where N is an integer (except perhaps by echo strength, or extrinsic knowledge about the environment, such as the existence of a horizon). (Section 10.3)

array elements — small antennas that operate with coordinated phases to comprise a larger *phased-array* antenna (Section 2.4)

array factor — voltage radiation pattern that would result from a set of isotropic point emitters located at the configuration of the *array element* positions (Section 2.4)

articulated vehicle — vehicle with a portion moved relative to its "normal state," such as an open door or hatch (Section 9.0)

A-display — "a display in which targets appear as vertical deflections from a horizontal line representing a time base." [1] (Section 1.19)

autofocus — the use of information in the (complex) SAR image itself to estimate and correct phase errors, then reprocess and sharpen the image (Section 7.2.2)

automatic target recognition (*ATR*) — procedure by which a computer algorithm detects, discriminates, and classifies (or recognizes) cultural targets according to target type (Section 9.3)

average power (P_{avg})	transmitted power averaged over an extended series of pulses (Section 1.4)
azimuth-elevation (AZ-EL) coordinates	coordinate system generally preferred by experimentalists in describing a radiation pattern. The polar axis (z) of the coordinate system is *parallel* to the aperture plane (in fact, in the aperture plane). See *theta-phi coordinates.* (Section 2.3.2)
back lobe	a side lobe more than 90 degrees away from the main lobe. (Section 2.3.2)
band-limited function	a function of frequency that is zero except over a finite frequency interval (or intervals) (Section 2.1)
bandpass signal	signal with narrow bandwidth $\Delta\omega << \omega_c$, centered about a carrier $\pm\omega_c$ (Section 4.2.1)
bandwidth correction factor (C_B)	ratio of *SNR* achievable with an ideal *matched filter* to that achieved with an actual filter (processor) (Section 1.10)
Barker code	"a binary phase code used for pulse compression, in which a long pulse is divided into n subpulses with the phase of each subpulse being 0 or π radians." [1] (Section 7.2.1)
baseband	demodulated signal with center frequency equal to zero (Section 2.2.4)
baseline	in bistatic radar, the line segment between the transmitting and receiving locations (Section 13.1.1)
Bayes criterion	criterion used in a detection procedure stating that, in determining the threshold voltage, the "risk" should be minimized (Section 4.1.2)
beamshape loss	"a loss factor included in the radar equation to account for the use of the peak antenna gain in the radar equation instead of the effective gain that results when the received train of pulses is modulated by the two-way pattern of a scanning antenna." [1] (Section 4.1.4.1)
beam-steering computer (*BSC*)	computer that calculates the phase vs. time of each *array element* in an *ESA* (Section 2.4)
binary integration	"integration in which each pulse is applied to a threshold, and the number M of threshold crossings is used as the criterion for target detection" ([3], p. 69) (Section 4.1.2)

biphase modulation	phase modulation such that the phase is switched between 0° and 180° in a coded fashion, e.g., *Barker code* (Section 7.2.1)
bistatic angle	in bistatic radar, angle with vertex at target and rays in the direction of transmitter and receiver (Section 13.1.1)
bistatic plane	plane of the *bistatic triangle* (Section 13.1.1)
bistatic radar	"a radar using antennas for transmission and reception at sufficiently different locations that the angles or ranges from those locations to the target are significantly different." [1] (Section 1.1, 13.1)
bistatic radar cross section	RCS of a target as measured by a bistatic radar (Section 1.7, 13.1.2)
bistatic RCS theorem	theorem that states: "for vanishingly small wavelengths, the bistatic RCS of a sufficiently smooth, perfectly-conducting target is equal to the monostatic RCS measured on the bisector of the bistatic angle" ([4], p. 145) (Section 13.1.2)
bistatic triangle	in bistatic radar, the triangle whose vertices are the transmitting platform, receiving platform, and target (Section 13.1.1)
blind speed	"radial velocity of a target with respect to the radar for which the MTI response is approximately zero." [1] (Section 1.16, 10.3)
blind zones in range	"a range corresponding to the time delay of an integral multiple of the interpulse period plus a time less than or equal to the transmitted pulse length." [1] (Section 10.3)
Boltzmann's constant (k)	fundamental constant; $k = 1.38 \times 10^{-23}$ joule/Kelvin (Section 1.9)
boresight direction	of an antenna radiation pattern; direction of highest-intensity region of the main lobe (Section 1.6)
bounce gain	on an *outdoor range,* increase in received power due to multipath from the flat ground; used to advantage, to increase *SNR* (Section 3.2.6)
brightness distribution $B(\theta,\phi)$	emitted radiation from a scene, measured in W m^{-2} Hz^{-1} sr^{-1} (Section 3.3)
broadside direction	direction perpendicular to a particular plane; depending on the context, the plane could be that

of an antenna aperture or the face of an array, or alternatively the side of the fuselage of an aircraft (Section 2.4)

B-display — "a rectangular display in which each target appears as an intensity-modulated blip, with azimuth indicated by the horizontal coordinate and range by the vertical coordinate." [1] (Section 1.19)

calibration of RCS measurement — measurement of the RCS of a target (σ_t (meas)) by comparing its apparent RCS (σ_{ta}) with the apparent RCS of a well-understood target (σ_{ca}), the correct RCS of which (σ_c) is known quite accurately (Section 1.7)

carrier frequency — frequency of the unmodulated radar wave. Modulation—amplitude, frequency, and/or phase—is typically superimposed on the carrier, resulting in a signal with a carrier plus modulation. (Section 1.3)

cell-averaging CFAR — *CFAR* procedure where the threshold for testing for a target at a particular cell (location in some space; e.g., pixel in a SAR image) is determined by evaluating the mean value of nearby cells (Section 9.1)

change detection — *ATR* step that compares the image with previous images taken of the same region and attempts to identify regions that have changed (Section 9.4)

Chinese Remainder Theorem — theorem (from number theory) useful for determining actual target range using several PRFs, given the maximum unambiguous range and the apparent range (and neglecting eclipsing time and measurement errors); the analogous procedure may be used to disambiguate target velocity (Section 10.6)

chip — (A) portion of a pulse, e.g., with a particular frequency, (B) portion of an image, e.g., to be scrutinized for the presence of a target (Section 4.2.7.3)

chirp — "a form of pulse compression that uses frequency modulation (usually linear) during the pulse" [1], originally coined because of the analogy to the chirp of a bird (Section 4.2.7.3)

chirp-diverse waveform — waveform where the *LFM* chirp slope γ varies from one pulse to the next (Section 4.2.7.3)

chirp-scaling algorithm (CSA)	simpler, faster version of *range-migration algorithm* (Section 7.2.2)
circularly-polarized EM wave	EM wave with an electric field vector that describes a circular helix in space (Section 1.3)
circulator	a type of duplexer (Section 2.2.7)
classification	last step in typical *ATR* algorithm, which attempts to specify the particular type of target, e.g. tank, truck, etc. Terminology varies; "recognition" and "identification" are also used, often to indicate still greater precision. (Section 9.3.3)
classification tree	a series of yes-no questions about a *ROI* such that their answers specify its specific region of *feature space* and therefore, with high probability, the target type (Section 9.3.3)
closing	in airborne radar, term describing target that is moving toward *ownship*; see *opening* (Section 11.1)
clutter	"unwanted echoes, typically from the ground, sea, rain or other precipitation, chaff, birds, insects, meteors, and aurora." [1] Evidently, one person's clutter may be another person's target. Echoes from the earth's surface (land or sea) are often called "clutter" even though they may be the desired signal. (Section 3.2.7)
clutter-to-noise ratio (CNR)	in a SAR image, the ratio of the received power from the earth's surface (land or sea) to the thermal noise (Section 7.3.2)
coherent detector	detector that obtains both the amplitude and phase of the voltage signal (see *envelope detector*) (Section 4.1.1)
coherent integration	"integration of radio frequency (RF), intermediate frequency (IF), or bipolar envelope signals over an interval in which phase or polarity is preserved." [1] (Section 4.1.2)
coherent oscillator (COHO)	oscillator driven by the *reference oscillator*; the COHO oscillates at the *intermediate frequency* (Section 2.2.1)
coherent processing interval (CPI)	(or *coherent dwell*) "the time during which the radar signal is received and processed coherently." [1] (Section 4.2.8)
coherent waveform	waveform consisting of a sinusoidal signal that is generated by an oscillator within the radar that maintains

a constant frequency over many pulses; the phases of returned pulses may then be compared with the oscillator phase to determine relative phases between pulses (Section 1.13)

cold clutter — ordinary radar clutter resulting from scattering of transmitted radar energy from terrain; see *hot clutter* (Section 12.8.6)

complex image — SAR/ISAR image formed by Fourier-transform processing, characterized by a (usually 2-D) set of pixels, each possessing an amplitude V and a phase ϕ (or alternatively I and Q components, which may be thought of as real and imaginary parts of a complex pixel) (see *voltage image, power image*) (Section 8.1.1)

complex-image analysis (CIA) — a procedure for identifying fixed or moving targets in SAR or ISAR images, emphasizing the fact that, in radar images as opposed to optical images, the phase, as well as the magnitude, is available for each pixel (Section 9.5)

computer-aided tomography (CAT) — medical imaging technique using an array of (noncoherent) x-ray emitters and a parallel array of x-ray detectors on the other side of a target (e.g. part of a human body), producing projections of the density function of the target at a series of angles; these projections are processed using the *convolution backprojection (CBP) algorithm* to form an image of a cross-section of the target. (Section 6.9)

confusion matrix — matrix summarizing results of a *classification* algorithm. The row labels correspond to the true target, the column labels correspond to possible declarations by the algorithm suite. (Section 9.3.3)

conical scan — "a form of angular tracking in which the antenna beam is offset from the tracking axis of the antenna." [1] (Section 5.1)

constant false-alarm ratio (CFAR) receiver — "a radar receiver that maintains the output false-alarm rate constant in spite of the varying nature of the receiver noise level, echoes from the clutter environment, or from electronic countermeasures." [1] (Section 9.1)

constant-range-contours — for airborne radar, concentric spheres with the radar at the center. On the ground, the contours are the

	intersection of the spheres with the ground—a set of concentric circles with the subradar point at the center. (Section 7.1.2, 11.1)
constant-velocity cone	for airborne radar (*ownship* velocity = V), the locus of all possible positions of a stationary target with apparent velocity $v_{app} < V$: a right-circular cone with generating angle $\theta = \cos^{-1}(-v_{app}/V)$ and axis parallel to the ownship velocity vector (see *isodop*) (Section 7.1.2, 11.1)
continuous-wave (CW) EM wave	an EM wave that does not employ pulses; transmitted power is approximately constant (Section 1.4)
contrast ratio (CR)	for a SAR image of terrain, ratio of the brightest terrain to the faintest (e.g., a shadow). Typically thermal noise would be included in the numerator and denominator of the ratio. If care is taken to insure that numerator and denominator are each considerably brighter than the thermal noise, the ratio becomes the *multiplicative noise ratio.* (Section 7.3.4)
convolution	a mathematical procedure for computing the integral of a function multiplied by an inverted, shifted version of itself (Section 2.1)
convolution backprojection (CBP) algorithm	algorithm used in *computer-aided tomography (CAT)* to produce an image of the target cross section. "The CBP algorithm is the one employed in virtually all modern medical CAT scanners" ([5], p. 59). (Section 6.9)
cosecant-squared antenna pattern	"a vertical-plane antenna pattern in which the transmitting and receiving power gains vary as the square of the cosecant of the elevation angle. *Note:* The unique property of this pattern is that it results in the received echo signal being independent of range if —The target is of constant radar cross section —The target moves at constant altitude —The earth's surface can be considered flat" [1] (Section 2.3.5)
Costas pulse	pulse consisting of N contiguous *chips,* each of duration T, and each at a different frequency, selected from N frequencies spaced $1/T$ apart. The order of the frequencies determines the ambiguity function (AF). If the correct "Costas sequence" is chosen, the AF in

	one dimension will approximate a *thumbtack,* with sidelobes no greater than $1/N$. *Staggered Costas signals,* i.e. trains of M pulses, each with a different Costas sequence, can have an AF that closely approximates a 2D thumbtack AF with sidelobe level $\sim 1/MN$ [6]. (Section 4.2.7.3)
Cramer-Rao bound	a theoretical lower bound on the variance in the measurement error of a parameter (Section 4.3.2)
creeping wave	EM wave that propagates along the surface of a target, often from one side to the other, causing a delayed radar echo that can appear to indicate non-physical scatterer(s) behind the physical target (Section 6.2)
crossed-field amplifier (CFA)	transmitter based on the interaction between a DC electric field and a DC magnetic field at right angles to each other (Section 2.2.5)
cross-range	direction normal to the radar *line-of-sight* (LOS) to the target, as in "cross-range resolution" (see *down-range*). The term is confined to 2D imagery; for 3D images, it is necessary to specify whether the normal to the LOS is along the azimuth or elevation (or other) direction. (Section 6.5)
C-display	a display of target azimuth vs. elevation (Section 1.19)
cumulative detection	"detection using *binary integration,* where $M = 1$ is the target-detection criterion" ([3], p. 69) (Section 4.1.2)
cutoff frequency	minimum frequency that will propagate in a waveguide (Section 2.2.6)
cylindrical EM wave	(Section 2.3.1) EM wave such that the electric vector is the same at all points on the surface of a cylinder (this property remains as the wave propagates) (Section 2.2.6)
datacube	*STAP* term describing a 3D array of complex input data, with dimensions L (range bin), M (pulse number), and N (antenna element) (Section 12.1)
dB isotropic (dBi)	dB relative to the radiation intensity of an isotropic antenna (Section 1.6)
dBZ	quantity describing density of rain. $\text{dBZ} = 10\log(Z)$; the units of Z are mm^6/mm^3. (Section 14.1.4)

dechirp (also called deramp)	SAR procedure where the received echo is mixed with a delayed replica of the transmitted pulse. The region of nonzero information in (f, t) space is still a parallelogram, but now the constant-range contours are parallel to the time axis (see Figure 7.9). (Section 7.2.1.2)
decibels (dB)	measure of the ratio of two power levels; $dB = 10\log_{10}(P_1/P_2)$ (Section 1.5)
delta function (Dirac delta function)	a special "function" that is zero everywhere except at one point, but the integral of which is unity; also called "impulse function" (Section 2.1)
demodulation	process by which the modulation on a signal at frequency f_{RF} is extracted to produce another signal at a different frequency (possibly baseband) with the modulation still included; see *detection* (Section 2.2.4)
depression angle	for airborne radar, the angle between the horizontal and a ray emanating from the radar. For flat earth, referring to a particular ray, depression angle equals *grazing angle.* (Section 2.3.5)
deramp	*dechirp* (Section 7.2.1.2)
deskew	SAR operation that transforms the received "data function" to the desired rectangular form (see Figure 7.10). After deskew, a time sample includes the results of all ranges across the swath, at a single transmitted frequency. (Section 7.2.1.3)
detectability factor (D)	"in pulsed radar, the ratio of single-pulse signal energy to noise power per unit bandwidth that provides stated probability of detection for a given false-alarm probability." [1] (Section 4.1.2) See *dechirp.*
detection	"(A) determination of the presence of a signal; (B) demodulation: the process by which a wave corresponding to the modulating wave is obtained in response to a modulated wave" [2] (Section 4.1.1)
detector	"a device for the indication of the presence of EM fields" [1]; "that portion of the radar receiver from the output of the IF amplifier to the input of the indicator or data processor" ([7], p. 382) (Section 4.1.1)
difference channel (monopulse radar)	"a receiving channel in which the response, as a function of a given radar coordinate, approximates the first

derivative of the response of the main (sum) channel, to indicate the displacement of the target from the center of the main channel." [1] (Section 5.2)

differential scattering cross-section (σ_d) (meter^2/steradian) — the fraction of *total scattering cross-section* that results in scattering in a particular direction; the *radar cross-section* (RCS) is 4π times the differential scattering cross-section. (Section 14.1.2)

diffraction — term referring to the fact that waves do not strictly travel in straight lines, but, to some extent, bend around corners (Section 2.3.1)

diffuse scattering — scattering from a surface such that the incident radiation is scattered uniformly into the solid-angle not containing the surface (a hemisphere if the surface is flat) (see *specular reflection, Lambert's law*) (Section 3.2.5)

dihedral corner reflector — reflector, sometimes used for polarimetric radar calibration, consisting of two metal sheets fastened together in a right dihedral angle. For the case of a radar observing a dihedral along a line of sight (LOS) normal to the dihedral edge and bisecting the dihedral right angle, the *polarimetric scattering matrix* varies in a simple way with the angle of rotation of the dihedral about the LOS. (Section 3.1.4.2)

direct digital synthesis (DDS) — procedure for producing detailed analog waveform information directly from a digital input, e.g., from a computer (Section 2.2.2)

directivity [$D(\theta,\phi)$] of an antenna — ratio of radiation intensity in a particular direction (usually boresight) to the average radiation intensity (Section 1.6)

discrete Fourier transform (DFT) — digital FT procedure for transforming a finite set of time-domain data into a finite set of frequency-domain data (*inverse* DFT performs the transform in the other direction) (Section 8.1.2)

discrimination — second step in typical SAR ATR algorithm, involving a decision as to whether a *region of interest* selected by *prescreening* is similar enough to a real target to be investigated further (Section 9.3.2)

displaced-phase-center antenna (DPCA) — antenna with two subapertures capable of performing *endoclutter GMTI*; see *physical DPCA* and *electronic DPCA* (Section 11.4.6)

doppler beam sharpening (DBS)	"a form of squint-mode synthetic-aperture radar (SAR) employed in a sector-scanning air-to-ground radar. *Note:* It usually has less resolution than a conventional SAR, since it employs a shorter processing (integration) time and varies this time as a function of beam squint angle so as to keep the resolution constant. Often displayed in near-real time on a plan-position indicator (PPI)." [1] (Section 7.1)
doppler effect	effect describing the fact that the echo from a target moving at line-of-sight velocity v will be shifted in frequency (*doppler shift*) from the carrier frequency by an amount $f_d = -2vf/c = -2v/\lambda$ (positive v corresponds to movement away from the radar); if the target is spontaneously emitting the radiation, the factor of two is omitted. (Section 1.14)
doppler radar	"a radar that uses the Doppler effect to determine the radial component of relative radar target velocity or to select targets having particular radial velocities." [1] (Chapters 10, 11)
downconversion	mixing that results in a signal with lower carrier frequency (see *upconversion*) (Section 2.2.4)
down-range (or simply range)	direction along the radar *line-of-sight* to the target, as in "down-range resolution" (see *cross-range*) (Section 6.5)
duplexer	a single-pole, double-throw switch to the antenna. During transmission, the duplexer connects the antenna to the transmitter. During reception, the duplexer connects the antenna to the components leading to the receiver. (Section 2.2.7)
duty cycle (f_D) or duty factor	"the ratio of the active or ON time within a specified period to the duration of the specified period. *Note:* For a pulsed radar, the ratio of transmitted pulse width to pulse repetition interval." [1] (Section 1.4)
dwell time (or dwell)	time during which a radar is receiving echoes from a particular direction, which echoes are to be processed to determine the presence of (or properties of) a target (see *coherent processing interval*) (Section 4.1.4.1)
dynamic range	the ratio of the maximum measurable power of a signal to its minimum measurable power (Section 2.2.14)

eccentricity of an ellipse (e)	parameter describing the extent to which an ellipse deviates from a circle; $e = L/(2a)$, where a = semi-major axis and L = distance between foci. $0 \leq e < 1$. A circle has $e = 0$; as $e \rightarrow 1$, the curve becomes a parabola. (Section 13.1.1)
echo	"the portion of energy of the radar signal that is reflected to a receiver." [1] (Section 1.8)
eclipsing	"the loss of information on radar echoes at ranges when the receiver is blanked because of the occurrence of a transmitter pulse. Numerous such blankings can occur in radars having high pulse-repetition frequencies." (Section 10.3) [1]
effective area of antenna [$A_e(\theta,\phi)$]	area that is multiplied by returned intensity (W/m^2) to produce received power (except for correction due to losses); typically less than physical aperture area (Section 1.6)
electric field (E)	field representing amount of force applied to an electric charge (Section 1.3)
electromagnetic (EM) waves	waves consisting of electric and magnetic fields, which propagate through free space or other matter; predicted by *Maxwell's Equations* (Section 1.3)
electronic DPCA	*DPCA* procedure where the PRF need not be matched exactly to the antenna spacing and platform velocity, and phase corrections are applied. Transmission is typically on the sum channel and reception is on the sum and difference channels—the *monopulse* technique. See *physical DPCA.* (Section 11.4.6)
electronically-scanned array (ESA)	phased array, the elements of which are electronically-controlled to steer the beam into many angular directions in rapid succession (or even simultaneously, especially with receive beams) (Section 2.4)
element factor	voltage radiation pattern from a single *array element* (Section 2.4)
element function	illumination function (electric field) of an individual *array element* (Section 2.4)
elliptically-polarized EM wave	EM wave with an electric field vector that describes an elliptical helix in space (Section 1.3)
emissivity (ϵ)	for a blackbody surface, the ratio of the emitted radiation to that corresponding to *Planck's law* (see also *Kirchhoff's law*) (Section 2.2.11)

endoclutter	*GMTI* term describing a moving target or a processing procedure, when the moving target may *not* be detected in the presence of clutter by a single-aperture radar using Doppler processing; see *exoclutter.* Endoclutter moving targets may sometimes be detected by a radar with multiple subapertures and *DPCA* or *STAP processing* (Section 11.4.2)
energy ratio $2E/N_0$	"the ratio of signal energy to [double sided] noise power spectral density in the receiver, at a point where the noise factor has been established and prior to filtering that would exclude components of the input signal." [1] (Section 5.2)
envelope detector	detector that obtains the envelope (magnitude), but not the phase, of the voltage signal (see *coherent detector*) (Section 4.1.1)
exciter	waveform generator (Section 2.2.2)
exoclutter	*GMTI* term describing a moving target or a processing procedure, when the moving target may be detected in the presence of clutter by a single-aperture radar using Doppler processing; see *endoclutter* (Section 11.4.2)
fall time	of a pulse, ideally, the time between the time the power begins to decline from its full value and the end of transmission; may be defined more precisely (e.g., time between 90% and 10% of full transmitted power; see *rise time*) (Section 1.4)
false alarm	"an erroneous radar target detection decision caused by noise or other interfering signals exceeding the detection threshold." [1] (Section 4.1.2)
far field	region of the radiated field pattern far enough from the antenna so that the EM fields essentially correspond to a plane wave; also called *Fraunhofer region.* The usual criterion is $R > 2L^2/\lambda$, where L is the larger of the aperture and the characteristic target dimension. See *near field.* (Section 2.3.1)
fast Fourier transform	DFT algorithm that requires much less computation time than a direct calculation. If N is a power of 2, then direct calculation requires N^2 complex multiplications, whereas an FFT of the same N requires only $N(\log_2 N)/2$ complex multiplications. (Section 8.1.2)

fast time	term describing time increments within one pulse repetition interval (PRI) (see *slow time*) (Section 7.2)
feature of a ROI	quantitative characteristic, such as length, area, brightness, certain polarimetric characteristics, etc., that may be used to compare SAR *ROIs* with similar "chips" of known targets, to decide whether further investigation is warranted (Section 9.3.2)
feature space	multidimensional space, whose axes are *features,* used to compare *feature vectors* of ROIs with feature vectors corresponding to known targets, e.g., by measuring the multidimensional distance between them, or by dividing the feature space into regions corresponding to the typical feature-space locations of known targets (Section 9.3.2)
feature vector	vector whose components are different *features* of a *ROI* (Section 9.3.2)
feed	a small antenna located at the focal point (or focal line) of a parabolic reflector antenna (Section 2.3.2)
fill time	time required for the first pulse in a waveform to reach the target and return, interpreted as the time required to "fill the space with pulses." Because the transmitter and receiver typically are controlled by the same timing procedure, the fill time (1 ms for R = 150 km) may be lost to the detection procedure. (Section 10.6)
fire-control radar	"a radar whose prime function is to provide information for the manual or automatic control of artillery or other weapons." [1] (Section 1.19)
fluctuating target	"a radar target whose echo amplitude varies as a function of time." [1] (Section 4.1.4)
fluctuation loss	"the change in radar detectability or measurement accuracy for a target of given average echo return power due to target fluctuation. *Note:* It may be measured as the change in required average echo return power of a fluctuating target as compared to a target of constant echo return, to achieve the same detection probability or measurement accuracy." [1] (Section 4.1.4.2)
flux density	watts per square meter; see *intensity* ([8], p. 5) (Section 1.6)

foliage-penetration (FOPEN) SAR	SAR that uses relatively low frequencies (typically UHF) to penetrate foliage; see *ultra-wideband SAR* (Section 7.5.5)
forward-look SAR	technique of forming a SAR image of an object (such as a vehicle) in the forward-look direction at θ_{sq} = 90°. The apparent rotation axis is now parallel to the ground. If a target vehicle is oriented parallel to the platform flight path, the SAR-image view will be a "side view," rather than the "top view" that would be obtained in a broadside image. (Section 7.5.4)
Fourier transform (FT)	a mathematical procedure that converts a signal representation from time domain to frequency domain, or vice versa (latter is *inverse Fourier transform*) (Section 2.1)
fractal dimension	feature of a *ROI* that represents its similarity to a point, line, area, etc. Fractal dimensions of some simple shapes are: dim(point) = 0, dim(line) = 1, dim(area) = 2. If the region has gaps, "peninsulas," complicated "wiggles" in its boundaries, etc., the fractal dimension may increase beyond 2, or assume a non-integer value. (Section 9.3.2)
fractional bandwidth	ratio of the signal bandwidth to the carrier frequency (Section 6.4)
Frank code	phase code such that individual pulse *chips* may have a phase of 0°, 90°, 180°, or 270° (Section 7.2.1)
Fraunhofer criterion	criterion for the far field: that, over the distance L (equal to the larger of the aperture and the characteristic target dimension), the wave front deviates from a plane by no more than $\lambda/16$ (Section 2.3.1)
Fraunhofer region	the far field (Section 2.3.1)
frequency (f) of an EM wave	number of wave crests per unit time (second) passing a given point; measured in cycles per second (hertz) (Section 1.3)
frequency space	coordinate system, analogous to *wavenumber space,* where the axes are the components of the wave number **k** multiplied by $c/2\pi$ (see Figure 6.5) (Section 6.8)
frequency-jump burst (FJB)	waveform such that each pulse is *LFM,* and center frequencies vary linearly; individual pulse bandwidths may be overlapped, tangent, or gapped (Section 7.2.1)

Fresnel region	region of the *near field* where the EM fields essentially correspond to spherical waves (Section 2.3.1)
gain of antenna [$G(\theta,\phi)$]	factor multiplied by $P_{trans}/(4\pi R^2)$ to produce radiation intensity at θ,ϕ; $G(\theta,\phi) = \eta D(\theta,\phi)$ (Section 1.6, 2.3.3)
gain of amplifier	ratio of output signal to input signal levels (voltage or power, as specified) (Section 2.2.10)
geometric theory of diffraction (GTD)	a hybrid system of optics based on a combination of *geometrical optics* and the concept of "diffracted rays" (see *physical optics*) (Section 3.1.3)
geometrical optics (GO)	principle of optics that assumes that radiation travels in straight lines (assuming homogeneous media); uses classical ray-tracing (Section 3.1.3)
glint	"the inherent component of error in measurement of position and/or Doppler frequency of a complex target due to interference of the reflections from different elements of the target." [1] (Section 5.4)
grating lobe(s)	in an array pattern, additional major lobe(s) that appear in addition to the desired major lobe; named after the analogous effect that occurs for a diffraction grating (Section 2.4.1)
grazing angle (ψ)	angle between the surface of the ground (locally flat) and the line of sight to the radar (the complement is the "incidence angle") (Section 2.3.5)
ground-moving-target indication (GMTI)	detection of ground-moving targets from an airborne or spaceborne radar (Section 11.0)
ground-penetrating radar (GPR)	radar designed to detect underground objects such as mines, pipes, or objects of archaeological interest (Section 14.2)
ground-plane image	SAR image formed from slant-plane image by appropriate interpolation and resampling. Ground-plane imagery with minimal distortion is necessary if comparison is to be made with maps or with images obtained from other sensors, such as optical sensors or other SARs. (Section 7.1.4)
group velocity	effective velocity of energy transfer in a wave (Section 2.2.6)

guard channel	"one or more auxiliary parallel processing channels to control the main processing channel in order to reject interference that is partly in, but not centered on, the main channel." [1] (Section 11.3.2)
Hadamard (vector) product (a · b)	element-by-element product of two $M \times 1$ vectors to produce a third $M \times 1$ vector (Section 12.2)
Hermitian adjoint	conjugate transpose (of a matrix) (Table 9.1)
Hermitian matrix	square matrix such that the conjugate transpose equals the matrix itself (Table 9.1)
heterodyne process (from the Greek for "different strength")	process by which a signal is mixed to a non-zero IF (Section 2.2.3)
high-definition vector imaging (HDVI)	procedure developed by G. Benitz of MIT Lincoln Laboratory for forming a fine-resolution image using a covariance matrix that has reduced *rank* relative to that used in the *minimum-variance method (MVM).* The "vector aspect" of HDVI utilizes a series of matched filters applied to the phase history data, each tuned for a different elementary target type, including the point scatterer, flat plate, dihedral, trihedral, and cylinder on a ground plane. (Section 8.2.5)
high-pulse-repetition-frequency (HPRF) waveform	"a waveform whose pulse-repetition frequency (PRF) is high enough to have no Doppler ambiguities for a given maximum-speed target." [1] (see *low PRF, medium PRF*) (Section 1.17, 11.1–3)
high-range-resolution (HRR)	target-recognition procedure based on determining one-dimensional downrange target profile, typically by using pulse-compression (Section 9.5)
homodyne	heterodyne process using a zero IF frequency; resulting signal is at baseband (Section 2.2.4)
horizontally polarized EM wave	EM wave with electric field vector essentially perpendicular to local gravitational field vector (Section 1.3)
hot clutter	energy emitted by a jammer and scattered from terrain; see *cold clutter* (Section 12.8.6)
Huygens' principle	the principle stipulating that the future behavior of a wave front may be predicted by considering each instantaneous wave front as a set of sources for new waves (Section 2.3.1)

IF amplifier	amplifier following the mixer that demodulates the signal to IF (Section 2.2.12)
IF bandwidth	in SAR, the bandwidth of the received signal after dechirp (Section 7.2.1.2)
illumination efficiency of an antenna	ratio of actual directivity to theoretically maximum directivity (Section 1.6)
image processing	processing performed on an image that has been produced via *signal processing,* e.g., to improve the contrast, resolution, etc. (Section 8.0)
image-quality metric (IQM)	parameter that characterizes the quality of a SAR image, e.g., impulse-response 3-dB width or multiplicative noise ratio (Section 7.3)
imaging radar	"a high-resolution radar whose output is a representation of the radar cross section within the resolution cell . . . from the object or scene resolved in two or three spatial dimensions. *Note:* The radar may use real aperture (such as a sidelooking airborne radar), synthetic-aperture radar (SAR), inverse synthetic aperture radar (ISAR), interferometric SAR, or tomographic techniques." [1] (Section 6.0)
impulse-response (IPR) function	in a SAR/ISAR image, the *point-spread function* (Section 7.3.1)
index of refraction (n)	property of a medium: the ratio of the speed of an EM wave in vacuum to its speed in the medium (Section 3.2.1)
indoor range	indoor facility for studying target RCS. It includes an instrumentation radar and a facility for mounting a target. See *outdoor range.* (Section 6.2)
inertial measurement unit (IMU)	instrument on board aircraft smaller than conventional *INS,* relying on the same general principles, typically "strapped down" very near the antenna (Section 7.1.3)
inertial navigation system (INS)	instrument on board aircraft that uses accelerometers and gyroscopes to measure the small-scale deviations of the flight path from a constant-velocity straight-line path (Section 7.1.3)
in-phase (I) signal	signal produced by mixing the IF echo signal with the IF from the COHO (Section 2.2.13, 4.2.1)

instrumentation radar	radar used to measure properties of targets that are under the control of the user (Section 1.19)
integrated sidelobe ratio (ISLR)	for a *point-spread function* (power), the integral over the side lobes divided by the integral over the main lobe (Section 7.3.3)
integration (of radar signals)	"the combination by addition (or the logical equivalent) of echo pulses or signal samples obtained by a radar as it illuminates a target so as to increase the output signal-to-noise ratio beyond that available from a single pulse or sample." [1] (Section 4.1.2)
integration loss	"the loss incurred by integrating a signal noncoherently (postdetection) instead of coherently." [1] (Section 4.1.3.4)
intensity	watts per steradian; see *flux density* ([8], p. 5) (Section 1.6)
interferometer	"an antenna and receiving system that determines the angle of arrival of a wave by phase comparison of the signals received at widely separated antennas. *Note:* In radar, the angle measurement made by an interferometer is generally ambiguous, and means must be used to resolve the ambiguities." [1] (Section 5.3)
interferometric SAR (IFSAR)	technique utilizing two complex SAR images taken from antennas at slightly different locations and compared coherently to obtain fine-resolution information regarding the height of terrain or targets in the image. IFSAR may be performed using a single platform with two antennas (*single-pass IFSAR*) or by the same platform making two passes over the same terrain (*two-pass IFSAR*). (Section 7.5.3.4)
interleaved modes	radar technique for performing two or more modes simultaneously, by transmitting a pulse for mode 1, then a pulse for mode 2, etc. Pulse characteristics and (using *agile-beam techniques*) beam directions can be quite different. Example modes that could be interleaved from an airborne platform are SAR, ISAR, and GMTI. (Section 7.5.7)
intermediate frequency (IF)	center frequency of echo signal after it is demodulated from the carrier; frequency of *coherent oscillator (COHO)* (Section 2.2.1)

inverse-synthetic-aperture radar (ISAR)	"an imaging radar in which cross-range resolution (angular resolution) of a target (such as a ship, aircraft, or other reflecting object) is obtained by resolving in the Doppler domain the different Doppler frequencies produced by echoes from the individual parts of the object, when these different Doppler frequencies are caused by the object's own angular rotation relative to the radar." [1] (Section 6.2)
IQ-circularity	the degree to which the *IQ-curve* is circular (Section 2.2.13)
IQ-curve	the locus of I, Q values obtained as target range varies over $\lambda/2$, ideally a circle (Section 2.2.13)
isodop	in airborne radar, the hyperbola formed by the intersection of a *constant-velocity cone* with the (assumed flat) ground (Section 7.1.2, 11.1)
isotropic antenna	hypothetical antenna required for a radar that radiates power uniformly over all 4π steradians of solid angle (Section 1.6)
jammer steering vector (χ_J)	in *STAP,* the $MN \times 1$ vector representing the complex jammer signal received with the M pulses at the N subapertures (Section 12.2)
Kirchhoff's law	principle that the *emissivity* of a surface is equal to its *absorptivity* (Section 3.3)
knowledge-based STAP	*STAP* procedure in which *a priori* knowledge about the environment (target, jamming, clutter) is used to pre-adapt the STAP algorithm, and to help decide what specific type of STAP processing to use (Section 12.8.3)
Kronecker (vector) product of matrices	product of an $M \times 1$ vector and an $N \times 1$ vector yielding an $MN \times 1$ vector (Section 12.2)
Kronecker delta	quantity equal to either zero or unity: $\delta_{ab} = 1$ if $a = b$, $\delta_{ab} = 0$ if $a \neq b$, where a and b are integers (Section 8.2.2)
Lambert's law (the adjective is Lambertian)	relationship describing diffuse scattering, viz. the intensity versus θ, the angle from the perpendicular to the surface, is $\sim \cos\theta$, due to the projection of the surface as seen from the direction of scattered radiation. (Section 3.2.5)

layover	for a SAR image, the phenomenon that an elevated object (e.g., the top of a tower) will appear to be "laid over" onto a pixel location different from the location of the ground underneath it (the tower base) (Section 7.5.3.2)
layover contour	the locus of points in 3-D space such that an object at any of the points will be assigned to the same location in a SAR image. Disregarding ambiguities, this contour is the intersection of a constant-range sphere and a constant velocity cone—a *layover circle.* (Section 7.5.3.2)
leading edge	of a pulse, the first part of a pulse to be radiated (Section 1.12)
least significant bit (LSB)	the bit in an A/D representing the units place in the binary number output of the A/D; the roundoff error corresponding to the value of this bit results in *quantization noise* (Section 2.2.14)
LFM ridge	aspect of the ambiguity function for a LFM waveform: that ambiguities exists along diagonal lines with slope $d\nu/d\tau = B/t_p$, where t_p is pulse width. The ambiguity is of no practical consequence in SAR. (Section 7.2.1.4)
likelihood ratio $\Lambda(V)$	ratio between $p_1(V)$, the *PDF* for the target-present case, and $p_0(V)$, the PDF for the target-absent case. The likelihood ratio is a useful *test statistic.* (Section 4.1.2)
limiter	a non-linear device that blocks passage of radiation if its power is greater than a level that might damage the receiver components (Section 2.2.9)
line of sight (LOS)	line segment between radar and target (Section 6.1, 7.1.2)
linear detector	envelope detector with output voltage proportional to the signal input voltage (Section 4.1.1)
linear-FM (LFM) pulse	a pulse whose frequency changes linearly with time over a bandwidth B; by means of *pulse-compression,* a PSF peak-to-first-null value of $c/2B$ may be obtained. Pulses of this type are widely used in modern radar. (Section 4.2.6)

linearly-polarized EM wave	EM wave such that the direction of electric field vector is (except for a ± sign) constant (Section 1.3)
linear-odd illumination	term referring to weighting function applied to a *monopulse difference beam*; weighting function is linear and equal to zero at the center of the full aperture, thus an odd function with absolute-value maxima at the edges of the full aperture (see *uniform-odd*) (Section 5.2.1)
lobe switching	"a means of direction finding in which a directive radiation pattern is periodically shifted in position so as to produce a variation of the signal at the target." [1] (Section 5.1)
log-normal distribution	*PDF* described by a Gaussian (normal) distribution; the variable that is Gaussian is $\ln(V/\mu)$, where V is the received signal voltage magnitude and μ is its mean. The log-normal PDF is sometimes used to describe clutter. See *Weibull distribution.* (Section 4.1.7)
long side	the edge of a triangular trihedral between two dihedral vertices (Section 3.1.1.3)
loss (L) in a passive electronic component	factor representing reduction in power when radiation passes through any passive electronic component (including passive antennas); L is usually expressed as a number greater than one by which the input power is divided to obtain the output power. *Loss* is also often used to describe other mechanisms that result in reduced received power. (Section 1.6)
low-pulse-repetition-frequency (LPRF) waveform	"a pulsed-radar waveform whose pulse-repetition frequency is such that targets of interest are unambiguously resolved with respect to range." [1] (see *high PRF, medium PRF*) (Section 1.17, 11.1)
low-noise amplifier (LNA)	first amplifier to receive echo signal; it is imperative to keep its noise level low to maintain high final SNR (Section 2.2.10)
low-probability-of-intercept (LPI) radar	radar employing special techniques so that an enemy will have difficulty detecting it (Section 7.2.1, 13.2)
Luneburg lens	device used to calibrate radars, consisting of a sphere having an index of refraction that increases from the surface to the center (Section 3.1.1.4)

magnetic field (magnetic intensity) (H)	field representing amount of force applied to a moving electric charge (Section 1.3)
magnetron	a type of noncoherent transmitter using a DC magnetic field (Section 2.2.5)
main lobe	of an antenna radiation pattern the angular region of relatively high radiation intensity (Section 2.2.5)
major lobe	in an array pattern, an angular region of relatively high radiation intensity (Section 2.4.1)
majors	several PRFs spaced fairly widely apart, sometimes chosen to mitigate velocity ambiguity; see *minors* (Section 10.6)
master-oscillator power-amplifier (MOPA)	term describing a wide variety of radars including a reference oscillator to maintain coherence and a high-power amplifier (transmitter) to generate the full transmitted signal; an example of a non-MOPA radar is one including an active *electronically-scanned array,* where each antenna element has its own *T/R module.* (Section 2.2)
matched filter	"a filter that maximizes the output ratio of peak signal power to mean noise power" [1]. Its output ratio of *mean* sinusoidal signal power to mean noise power is the collected energy divided by the noise power density, kT_s: $SNR = E/N_0$. The matched filter is the optimum processor; no other processor (filter) can produce a higher SNR. (Section 4.2.2)
Maxwell's equations	fundamental equations of electricity and magnetism, which predict the existence of EM waves (Section 1.3)
medium-pulse-repetition-frequency (MPRF) waveform	"a pulsed-radar waveform whose pulse-repetition frequency (PRF) is such that targets of interest are ambiguous with respect to both range and Doppler shift." [1] (see *high PRF, low PRF*) (Section 1.17, 11.1)
microwaves	"a term used rather loosely to signify radio waves in the frequency range from about [1 GHz] upwards" [2]; in this book, "microwaves" means radio waves of frequency approximately 0.1 to 100 GHz (VHF through W band) (Section 1.3)
Mie scattering	scattering from a sphere, the radius a of which is comparable to a wavelength; the RCS fluctuates around πa^2 (Section 3.1.1.1)

minimum detectable velocity (MDV) — in *GMTI,* the velocity such that, for lower velocities, σ_{target} (dBsm) < $\sigma_{clutter}$ (dBsm) + $SINR_{min}$ (dB), and the target is "buried" in the clutter (Section 11.4.2)

minimum-variance method (MVM) — (also known as the *maximum-likelihood method,* or *Capon's method*) for a point target at a particular pixel location, a theoretically ideal procedure for maximizing the energy in the correct pixel (mainlobe) and minimizing the energy outside it (sidelobes); based on the *adaptive matched filter* (Section 8.2.4)

minor lobe — in an array pattern, an angular region of relatively low radiation intensity (Section 2.4.1).

minors — several PRFs spaced fairly close together, sometimes chosen to mitigate range ambiguity; see *majors* (Section 10.6)

missed detection — error such that target is present but detector says target is absent (Type II error) (Section 4.1.2)

mixer — component that produces output signals at both the sum and the difference of two input signals. The output signal is typically passed through a filter to remove the unwanted signal (either the sum or the difference). The overall mixer output signal has a frequency that is either the sum or difference of the frequencies of the two input signals, depending on the application. (Section 2.2.3)

mode — a sequence of *waveforms* (or a single waveform) processed to achieve a particular objective (e.g., PRF switching); PRF and beam position may vary. (Section 11.3.3)

modulation — process by which a sinusoidal signal at frequency f_{RF} is modified (information is added via amplitude-modulation, frequency-modulation, or phase-modulation) to produce a signal at f_{RF} with the modulation information included. (Section 2.2.4)

monochromatic pulse — as used in this book (not necessarily in others), a pulse of width τ formed by multiplying a pure continuous tone by a rectangular function; the peak-to-first-null bandwidth is $1/\tau$. (Section 1.18, 4.3.1)

monopulse — "a radar technique in which information concerning the angular location of a target is obtained by compari-

	son of signals received in two or more simultaneous antenna beams." [1] (Section 5.2)
monopulse slope (k_m)	quantity related to the slope of the *difference beam* at the center of the full aperture (Section 5.2)
monostatic radar	"a radar system that transmits and receives through either a common antenna or through collocated antennas." [1] (Section 1.1)
monostatic radar cross section (RCS)	RCS of a target as measured by a monostatic radar (Section 1.7)
motion compensation (mocomp)	in SAR, procedure for measuring, and correcting for, deviations of the platform from the nominal straight-line constant-velocity path. After appropriate filtering of the data, the true flight path must be estimated to within a fraction of a wavelength. (Section 7.1.3)
motion compensation point (MCP)	central reference point within a SAR scene (Section 7.2.1.1)
moving-target indication (MTI)	"a technique that enhances the detection and display of moving radar targets by suppressing fixed targets. *Note:* Doppler processing is one method of implementation." [1] (Section 11.0)
multipath	"the propagation of a wave from one point to another by more than one path. When multipath occurs in radar, it usually consists of a direct path and one or more indirect paths by reflection from the surface of the earth or sea or from large man-made structures." [1] (Section 3.2.6)
multiplicative noise	noise, in a SAR image, that is proportional to the average scene intensity, due to clutter sidelobes, ambiguous returns, artifacts, etc. (Section 7.3.4)
multiplicative noise ratio (MNR)	for a SAR image, the ratio of the image intensity (power) in *no-return area (NRA)* divided by the average image intensity in a relatively bright surrounding area. In principle both these power levels should be strong enough to be much greater than thermal noise (see *contrast ratio*). (Section 7.3.4)
multistatic radar	"a radar system having two or more transmitting or receiving antennas with all antennas separated by large

	distances when compared to the antenna sizes." [1] (Section 1.1, 13.1)
near field	region closer to the radar than the beginning of the far field; sometimes divided into the *Fresnel region* and the very near field (see *far field*) (Section 2.3.1)
negative frequency	frequency such that signal vector moves clockwise in *I, Q* diagram (see *positive frequency*). (Section 4.2.1)
Next Generation Weather Radar (NEXRAD)	network of weather-observation radars across the United States, which measures both dBZ and line-of-sight velocity using Doppler processing. The network consists of 138 S-band WSR-88D radars. (Section 14.1.5)
noise	power generated by thermal motion of electrons, or other sources, that interferes with the received signal (Section 1.9)
noise figure (or noise factor) (F)	defined in terms of the standard temperature $T_0 = 290\text{K}$: $T_{\text{rcvr}} = (F - 1)T_0$, $F > 1$ (Section 2.2.11)
noise-equivalent clutter ($\text{NE}\sigma^0$)	that clutter (echo from earth's surface) level (σ^0) which produces a received power equal to the thermal noise power, i.e., a *CNR* of unity (Section 7.3.2)
noncoherent integration	"integration in which each pulse is envelope detected, and the resulting video pulses are added together prior to application of thresholding" (same as *video integration*) ([3], p. 69) (Section 4.1.2)
noncoherent detector	envelope detector (Section 4.1.1)
nonfluctuating target	target from which the radar echoes are constant in time (Section 4.1.3)
nonlinear FM modulation	FM modulation more complicated than simple linear, e.g. piecewise linear (Section 7.2.1)
no-return area (NRA)	for a SAR image, an area with essentially zero return—usually a very smooth area such as a calm lake or a specially constructed large sheet of aluminum (see *multiplicative noise ratio*) (Section 7.3.4)
null	with respect to an antenna radiation pattern or a PSF, a portion of the pattern where the intensity is approaching zero, and is considerably less than the intensities at the peaks of the lobes (Section 1.6)

Nyquist frequency	the sampling frequency necessary to recover a full signal (Section 2.1)
opening	in airborne radar, term describing target that is moving away from *ownship*; see closing (Section 11.1)
optical scattering	scattering from a sphere, the radius a of which is significantly greater than a wavelength; the RCS is essentially πa^2 (Section 3.1.1.1)
optimum fully-adaptive STAP	ideal *STAP* algorithm producing theoretically optimum *SINR*; requires much computational complexity and typically produces high sidelobes (Section 12.6.1)
orthogonal matrix	square matrix such that the transpose equals the inverse (Table 9.1)
oscillator group	the RO, COHO, and STALO considered together (Section 2.2.1)
outdoor range	outdoor facility for studying target RCS. It includes an instrumentation radar and a facility for mounting a target; the target is typically located hundreds or thousands of meters from the radar to insure that the target is in the far field. See *indoor range.* (Section 3.2.6)
oval of Cassini	closed curve in a plane, originally discovered by Cassini; in bistatic radar, the locus of target locations in a plane such that the received power is constant (Section 13.1.1)
ownship	aircraft carrying the radar that is being considered (Section 11.1)
paired echoes	in a SAR image, pattern characteristic of a vibrating target. A vibrating point target will appear in three locations: the main target image appears at the correct location, while a small fraction of the energy appears in each of two pixels displaced in crossrange—the paired echoes. (Section 7.5.2)
parabolic reflector	aperture antenna consisting of a paraboloid of revolution with the source of microwave radiation (*feed*) situated at the focal point. (Sometimes the paraboloid is cylindrical and the feed is a line source). (Section 2.3.2)
Parseval's theorem	theorem expressing the fact that, for a Fourier transform, with the proper multiplicative normalization

constant (independent of data content), the sum of V^2 (~ energy) in the time domain equals the sum of V^2 in the frequency domain. Applied to SAR/ISAR, it states that the sum of V^2 (~ energy) in the signal domain (*video phase history*) equals the sum of V^2 in the image domain (Section 8.1.1).

partially-adaptive STAP — a procedure that, using specified criteria, reduces the dimension of the $MN \times 1$ space-time snapshot and therefore of the interference covariance matrix, and consequently the computation time (Section 12.7).

passive array — phased array such that the microwave radiation is produced by a central transmitter and delivered via waveguide or through space (a *space-fed array*) to each element, which contains a *phase shifter* (Section 2.4).

peak power (P_{peak}) — average power transmitted during a pulse (Section 1.4).

peak-to-first-null value — for a *point-spread function (PSF)* or antenna *radiation pattern,* represented by a *sinc* or similar function, the interval of the independent variable between the peak and the first null (see *three-dB width*) (Section 2.3.2, 4.2.5).

period of a wave (T_p) — elapsed time between the passage of two consecutive crests past a particular point; $T_p = 1/f = 2\pi/\omega$ (Section 1.3).

periodogram — the expected value of the power image over an ensemble of data collections (Section 8.2.3).

permeability of free space (μ_0) — fundamental constant (MKS units) $= 4\pi \times 10^{-7}$ kg m coul^{-2} (Section 1.3).

permittivity of free space (ϵ_0) — fundamental constant (MKS units) $= 8.85 \times 10^{-12}$ kg^{-1} m^{-3} s^2 coul^2 (Section 1.3).

phase noise — noise due to the phase instability of the primary oscillator, which then propagates through the rest of the radar electronics (Section 10.2).

phase shifter — microwave component that shifts the phase of the output radiation relative to the input radiation; used in *array elements* of a *passive array* (Section 2.4).

phase — for a sinusoidal EM wave $[\mathbf{E}(\mathbf{r}, t) = \mathbf{E}_0 \cos(\omega t - \mathbf{k} \cdot \mathbf{r} + \phi_0)]$, the argument of the cosine; also refers to

ϕ_0, an additive constant in the argument, independent of space or time (Section 1.3)

phased array — antenna consisting of a planar array of small radiating *array elements,* the phases of which are controlled so as to steer the beam (transmit and/or receive) to a particular direction or directions (Section 2.4)

phase-locked loop (PLL) — electronic component that maintains phase coherence by means of a feedback loop (Section 2.2.1)

physical DPCA — *DPCA* procedure where the apertures of two side-looking antennas are aligned parallel to the aircraft flight direction. The PRF is adjusted so that, when pulse number $n + 1$ is transmitted, the second antenna is at the position that was occupied by the first antenna when pulse number n was transmitted. Stationary clutter is suppressed and moving targets are highlighted. See *electronic DPCA.* (Section 11.4.6)

physical optics (PO) — principle of optics based on Huygens' principle (Section 3.1.3)

physical optics approximation — approximation that essentially involves assuming the wavelength to be much less than any aperture dimension (Section 2.3.2)

Planck distribution — formula describing emission from a blackbody surface (Section 3.3)

plane EM wave — EM wave such that the electric vector is the same at all points on a plane normal to the propagation direction (this property remains as the wave propagates) (Section 1.3)

plan-position indicator (PPI) — "a display in which target echoes (blips) are shown in plan position, thus forming a map-like display, with radial distance from the center representing range and with the angle of the radius vector representing azimuth angle." [1] (Section 1.19)

point-spread function (PSF) — the SAR/ISAR image of a single point scatterer. The PSF is "spread out" relative to the point scatterer. [Also called "impulse-response" (IPR) function] (Section 5.4, 8.1.5)

polar format algorithm (PFA) — technique for processing collected spotlight-SAR or ISAR VPH data by interpolating to a square grid in

	k-space (or frequency space), thereby permitting a 2D FFT for efficient computation (Section 5.8)
polarimetric radar	radar capable of transmitting at either of two orthogonal polarizations (e.g., H and V, or R and L) and receiving at either of the transmitted polarizations; the choice of transmitted and received polarizations can be varied from pulse to pulse. Important application is polarimetric SAR. (Section 7.5.6)
polarization of an EM wave	property related to the direction (versus space and time) of the electric vector (Section 1.3)
polarization scattering matrix (PSM)	matrix summarizing the polarization relationship between the incident and scattered EM waves (polarimetric RCS) (Section 3.1.4.1)
polyphase modulation	phase modulation such that the phase is switched among several values, e.g., *quadriphase,* which uses 0°, 90°, 180°, and 270° (e.g., *Frank Code*) (Section 7.2.1)
positive frequency	frequency such that signal vector moves counter-clockwise in *I, Q* diagram (see *negative frequency*) (Section 4.2.1)
post-detection integration (PDI)	non-coherent integration of the voltage magnitudes resulting from different CPIs; see *detection,* definition (B) (Section 10.6)
power image	SAR/ISAR image where the square of the pixel voltage magnitude $V^2(x,y)$ is displayed (see *complex image, voltage image*) (Section 8.1.1)
power management	*LPI* technique based on reducing the transmitted power to a level just high enough to allow successful radar operation, but still to prevent detection of the radar by an enemy interceptor (Section 13.2)
power spectral density	expression describing the W/Hz of a signal as a function of frequency (Section 4.2.4)
Poynting vector (S)	vector representing instantaneous *flux density* (W/m^2) in an EM wave; given by $\|\mathbf{S}\| = \|\mathbf{E} \times \mathbf{H}\| = c\epsilon_o E^2 = c\mu_o H^2$ (Section 1.3)
prescreening	first step in typical SAR ATR algorithm, which selects *regions of interest* that may be targets (Section 9.3.1)
PRF jitter	simplified version of PRF switching, using only two PRFs; the radar switches between them to observe

	the change in the apparent target range, and thereby determines actual range (Section 10.5)
PRF switching	procedure for disambiguating range or velocity, in which the radar observes a particular target using one PRF; it then observes the same target using a second PRF, then perhaps a third, and so forth. The *Chinese Remainder Theorem* may be used to determine the actual range or velocity. (Section 10.5)
probability density function (PDF)	function $p_A(x)$ describing the probability $p_A(x)dx$ that a dependent variable (represented by A) will lie between x and $x + dx$; the integral of $p_A(x)$, over all possible values of x, is unity (Section 4.1.2)
projection-slice theorem	theorem used in *computer-aided tomography (CAT),* stating that the *Fourier transform* of a projection of the 2D target density function at an angle θ is equal to a radial slice at θ of the 2D Fourier transform of the target density function. By collecting the projections at many angles, an image of the target density may be obtained. (Section 6.9)
prominent point	relatively bright point-like scatterer on a target (or scene), necessary for certain radar image focusing algorithms (Section 6.12)
pulse	a short interval of transmitted energy followed by a longer interval of no transmission (Section 1.4)
pulse-chasing	in bistatic radar, term describing situation where an agile receive beam rapidly scans the volume covered by the transmitted radiation, essentially chasing the pulse as it propagates (Section 13.1.3)
pulse-compression	"a method for obtaining the resolution of a short pulse with the energy of a long pulse of width T by internally modulating the phase or frequency of a long pulse so as to increase its bandwidth $B >> 1/T$, and using a matched filter (also called a pulse compression filter) on reception to compress the pulse of width T to a width of approximately $1/B$." [1] (Section 4.2.6)
pulse-Doppler (PD) radar	(or "pulsed-Doppler radar") "a *Doppler radar* that uses pulsed transmissions" [2] (Section 10.0, 11.0)
pulse-repetition frequency (PRF)	"the number of pulses per unit of time, usually per second." [1] (Section 1.4)

pulse-repetition interval (PRI)	"time duration between successive pulses" [1], usually (but not always) constant over a time necessary to transmit many pulses (Section 1.4)
pulse-tagging	procedure by which successive pulses are made to be different by means of RF variation, phase modulation, frequency-modulation, pulse-width modulation, or amplitude modulation (the latter two are probably impractical). The technique has been used on RCS measurement ranges. (Section 7.5.7.4, 10.4)
pulsewidth	transmission time of a single pulse, typically expressed in nanoseconds or microseconds (Section 1.4)
quadrature (Q) signal	signal produced by mixing the IF echo signal with the signal from the COHO shifted in phase by 90° (Section 2.2.13, 4.2.1)
quadrature mixer	component that mixes the IF signal with two signals: (1) the signal from the COHO and (2) the signal from the COHO shifted in phase by 90°; *in-phase (I)* and *quadrature (Q)* signals are produced (Section 2.2.13)
quadriphase modulation	phase modulation among 0°, 90°, 180°, and 270° (e.g., *Frank code*) (Section 7.2.1)
quantization noise	effective noise resulting from roundoff error due to the least significant bit of the A/D (Section 2.2.14)
radar	"an electromagnetic system for the detection and location of objects that operates by transmitting electromagnetic signals, receiving echoes from objects (targets) within its volume of coverage, and extracting location and other information from the echo signal." [1]; originally an acronym for RAdio Detection and Ranging (Section 1.1)
radar cross-section (RCS, or σ)	"4π times the ratio of the power per unit solid angle scattered in a specified direction to the power per unit area in a plane wave incident on the scatterer from a specified direction" [1, 2] (Section 1.7, 3.1)
radar equation (or radar range equation)	"a mathematical expression that relates the range of a radar at which specific performance is obtained to the parameters characterizing the radar, target, and environment. *Note:* The parameters in the radar equation can include the transmitter power, antenna gain and effective area, frequency, radar cross section of

the target, range to the target, receiver noise figure, signal-to-noise ratio required for detection, losses in the radar system, and the effects of the propagation path." [1] (Section 1.11)

radar uncertainty relation
: statement that $\beta\alpha \geq \pi$; where β is the "effective bandwidth" and α is the "effective time duration" (Section 4.3.4)

radiated power (P_r)
: power radiated into the environment (e.g., free space) (Section 1.6)

radiation efficiency (η_r)
: ratio of radiated power to transmitted power (P_r/P_{trans}); reciprocal of one-way antenna loss; see *aperture efficiency* (Section 1.6)

radiation intensity [$\Phi(\theta,\phi)$]
: power per unit solid angle (watts/steradian) (Section 1.6)

radiation pattern (power) $|f(\theta,\phi)|^2$
: normalized pattern representing relative magnitude of the radiated power. The maximum value of $|f(\theta,\phi)|^2$ is unity. (Section 1.6)

radiation pattern (voltage) $f(\theta,\phi)$
: normalized complex (including phase) pattern representing relative magnitude of the radiated electric field; its magnitude is squared to produce the *power radiation pattern.* The maximum value of $|f(\theta,\phi)|$ is unity. (Section 1.6)

radio waves
: EM waves with frequency in the GHz range or lower (Section 1.3)

range
: "distance between a radar and a target." [1] (Section 1.8)

range (or radar range)
: facility, indoor or outdoor, at which objects are placed on rotating turntables or poles, and their radar properties, including images, are obtained for various frequencies and polarizations (Section 3.2.6, 6.2)

range gate
: for a monochromatic echo pulse of width τ, sampled at a particular instant: the interval of range in which a detected target may be located, equal to $c\tau/2$ (Section 1.18)

range migration
: term describing the motion of point scatterers through range cells during the data collection for a radar image ([9], p. 281). [10], Section 3.2.1, discusses the terms "motion through resolution cells (MTRC)," "range curvature," and "range walk," and states (p. 90)

	"Terminology regarding range curvature differs somewhat in the literature." (Section 7.1.6)
range migration algorithm (RMA)	algorithm, originally developed for seismic applications, that provides the most theoretically correct solution to the stripmap image problem. RMA involves substantial computational complexity; however, as processors become more sophisticated, this limitation is disappearing (Section 7.2.2)
range walk	"The migration of a point scatterer from range cell to range cell during the signal integration period." [1] (Section 7.1.6)
range window	downrange extent of a radar image (Section 8.1.6)
rank	of a matrix, number of independent rows (Section 8.2.5)
ray	an orthogonal to a wave front of an EM wave (Section 1.3)
Rayleigh criterion	resolution criterion stating simply that the image resolution is equal to the peak-to-first-null value of the PSF (Section 8.2)
Rayleigh PDF	*probability density function (PDF)* describing envelope-detected voltage values from noise only (see *Rician PDF*) (Section 4.1.3.2)
Rayleigh Scattering	scattering from a sphere, the radius a of which is much less than a wavelength. The RCS is much less than πa^2 and becomes progressively smaller ($\sim \lambda^{-4}$) for longer wavelengths (lower frequencies); this explains why lower frequencies penetrate rain better than higher ones. (Section 3.1.1.1, 14.1.2)
Rayleigh-Jeans region	for a Planck distribution, region where frequency << peak frequency, and emitted radiation (W m^{-2} Hz^{-1}) $\approx 2kT/\lambda^2$ (see *Wien region*) (Section 3.3)
real-aperture radar (RAR)	radar with a full aperture consisting of physical antenna structure, as distinct from a *synthetic aperture radar* (Section 5.2)
received power (P_{recd})	power collected at the radar receiver after scattering from some target(s) in the external environment (Section 1.6)

receiver	the portion of the radar electronics that accepts the echo signal and amplifies it for delivery to a final output device (Section 2.2.10)
receiver operating characteristic (ROC) curves	"plots of probability of detection versus probability of false alarm for various input signal-to-noise power ratios and detection threshold settings." [1] (Section 4.1.2)
receiver temperature (T_{rcvr})	the effective input temperature to the LNA if the antenna temperature and the temperature of the radar components were zero; T_{rcvr} accounts for noise generated within the LNA itself (Section 2.2.11)
rectangular function	function of x with value 1 if $\|x\| < 1/2$ and 0 if $\|x\| > 1/2$ (Section 4.2.5)
reference oscillator (RO)	an extremely stable oscillator that provides the basic reference frequency for the radar (Section 2.2.1)
region of interest (ROI)	portion of a SAR image that has been selected as relatively likely to contain a target of interest (Section 9.3.1)
residual video phase (RVP)	with respect to the formula for phase of a SAR echo, a term proportional to $(R_t - R_0)^2$, where R_t is target range and R_0 is range to the *motion compensation point* (Section 7.2.1.1)
resolution	the minimum distance between two point targets such that they can be individually resolved (observed) in an image. A "resolution criterion" must be stated, e.g., the *Rayleigh criterion.* "Resolution" also applies to range (delay) or velocity (Doppler frequency) individually. (Section 4.2.5, 8.2)
retroreflector	reflector with the property that an incident ray is reflected so as to emerge into the same direction from which it came ($\mathbf{k}_{\mathbf{reflected}} = -\mathbf{k}_{\mathbf{incident}}$) (Section 3.1.1)
Rician PDF	probability density function (PDF) describing envelope-detected voltage values from a target signal plus noise (see *Rayleigh PDF*) (Section 4.1.3.2)
rise time	of a pulse, ideally, the time between the beginning of transmission of a pulse and the time the power reaches essentially its full value; may be defined more precisely (e.g., time between 10% and 90% of full transmitted power; see *fall time*) (Section 1.4)

saturation — process by which strong signals *above* a certain level *L* are transformed to signals *at L,* resulting in loss of information about the original signal level (usually undesirable) (Section 2.2.9)

scanSAR — SAR mode where the beam observes a straight strip of terrain that is not parallel to the flight path. Clearly such a strip must be of finite length, since eventually the range becomes so great that the signal-to-noise ratio (SNR) is too low to produce clear imagery. (Section 7.1.1)

search radar — "a radar used primarily for the initial detection of targets in a particular volume of interest." [1] (Section 1.19)

search radar equation — radar equation modified to apply specifically to a volume-search radar; the SNR of a target detected in the search is, to first order, proportional to power-aperture product and independent of frequency band (Section 4.2.9)

sensitivity-time control (STC) — "programmed variation of the gain (sensitivity) of a radar receiver as a function of time within each pulse-repetition interval or observation time in order to prevent overloading of the receiver by strong echoes from targets or clutter at close ranges." [1] (Section 11.3.2)

short side — the edge of a triangular trihedral between a dihedral vertex and the trihedral vertex (Section 3.1.1.3)

side lobes — of an antenna radiation pattern the angular regions of relatively low radiated intensity (Section 1.6)

side-looking airborne radar (SLAR) — "a high-resolution (in both range and angle) airborne imaging radar, without synthetic-aperture radar (SAR) processing, directed sidelooking (perpendicular to the line of flight) using large, narrow-beamwidth antennas." [1] (Section 7.1)

sigma-delta STAP ($\Sigma\Delta$-STAP) — a two-aperture *STAP* technique, for which the returns from the two subapertures are summed and differenced (Section 12.8.4)

signal processing — processing of the radar echo signal, e.g., to form an image (Section 8.0)

signal-to-interference-plus-noise ratio (SINR)	term used in *STAP* to represent the ratio of the target signal to the noise-plus-interference (clutter, jamming) (Section 12.3)
signal-to-noise ratio (SNR)	"in radar, the ratio of the power corresponding to a specified target measured at some point in the receiver to the noise power at the same point in the absence of the received signal." [1] (Section 1.10, 4.2.9, 7.3.2)
$\mathrm{sinc}(x)$	$(\sin(x))/x$ (Section 2.3.2)
$\mathrm{sind}(N,x)$	as used in this book, $(\sin(Nx))/\sin(x)$; $\mathrm{sind}(N,0) = N$ (Section 4.2.7)
single-frequency imaging	theoretical *ISAR* or *spotlight-SAR* technique that relies on a synthetic aperture of large angle, without *pulse-compression,* to provide 2D resolution of a target (Section 5.6)
SINR improvement factor (I_{SINR})	in *STAP,* ratio of SINR (post-processing) to SINR (pre-processing) (Section 12.6.4)
SINR loss	in *STAP,* ratio of ideal SINR to observed SINR, often expressed in dB (i.e., 10 times the logarithm of this ratio) (Section 12.6.3)
slant plane	in SAR, plane determined by the *LOS* and its perpendicular in the ground plane (Section 7.1.4)
slant-plane image	*SAR* image representing range and crossrange in the slant plane; i.e., the image obtained from 2D FFTs without additional processing for projection to the ground plane (Section 7.1.4)
slow time	term describing time increments over many PRIs, e.g., within one spotlight SAR aperture time (t_A) (Section 7.2)
Snell's law	the principle relating the angle of refraction at a boundary between two media to the indices of refraction of the two media (Problem 2.7)
solid-state transmitter	a type of relatively low-power transmitter that uses only semiconductor components (rather than electron vacuum tubes such as the TWT) (Section 2.2.5)
sonar	system defined in a similar manner to radar, except that it uses sound waves rather than EM waves (Section 1.1)
space-fed array	passive phased array such that the radiation is delivered through space from a small antenna to the array elements (Section 2.4)

space-time adaptive processing (STAP)	"in airborne moving-target indication (MTI), a method of processing that compensates for the adverse effects of platform motion by adaptively placing antenna nulls in the directions of large clutter echoes and/or large noise or jamming sources. *Note:* It simultaneously employs the signals received from the multiple elements of an adaptive phased array antenna (spatial domain) and the signals from multiple pulse repetition periods (time domain) to provide adaptive processing in both the time and spatial domains." [1] (Chapter 12)
space-time snapshot	*STAP* term describing an $MN \times 1$ column vector used as an input to the *adaptive matched filter*; M is pulse number and N is antenna element (Section 12.1)
spatially-variant apodization (SVA)	procedure developed by H. Stankwitz et al. of Veridian ERIM-International, for forming a fine-resolution image, based on a convolution of the unweighted FFT complex image using only the nearest neighbors of each pixel. This simplicity means very little computational burden. (Section 8.2.7)
speckle	"a mottled effect in coherent radar images, such as those from synthetic-aperture radar (SAR) and laser radar, caused by random additive and subtractive interference of signals from individual scatterers within each resolution cell." [1] (Section 7.3.5)
spectral estimation	processing technique that implicitly makes an estimate of portions of the input signal not explicitly present, to obtain an improved output, i.e., narrowed mainlobe and/or reduced sidelobes for point scatterers (Section 8.2.1)
specular reflection	reflection from a perfectly smooth surface, such that the angle of reflection is equal to the angle of incidence (from the Latin "speculum," meaning "mirror") (see *diffuse scattering*) (Section 3.2.5)
speed of light	(or any EM wave) in vacuum (c) 299,792,458 meters/second; this expression is exact and forms the definition of the meter (Section 1.3)
spherical EM wave	EM wave such that the intensity is the same at all points on the surface of a sphere (this property remains as the wave propagates) (Section 2.3.1)

spotlight synthetic-aperture radar SAR	(or "spot" SAR) "a form of SAR in which very high resolution is obtained by steering the real antenna beam to dwell longer on a scene or target than allowed by a fixed antenna." [1] (Section 7.1.1)
spur	spurious peak in a phase-noise spectrum (Section 2.2.1)
square-law detector	envelope detector with output voltage proportional to the square of the signal input voltage (Section 4.1.1)
squint angle (θ_{sq})	in airborne radar, azimuth angle between ownship broadside direction and vector from ownship to target; broadside is at $\theta_{sq} = 0°$ (Section 7.1.1, 11.4.2)
squint-mode synthetic-aperture radar SAR	"a SAR in which the beam is pointed other than at right angles to the flight path of the airborne radar platform." [1] (Section 7.1.1)
stable local oscillator (STALO)	"a highly stable radio-frequency local oscillator used for heterodyning signals to produce an intermediate frequency (IF)." [1] The STALO oscillates at a frequency of $f_{LO} = f_{RF} - f_{IF}$, where f_{RF} is the carrier frequency and f_{IF} is the *intermediate frequency*. (Section 2.2.1)
stealth	technology used to reduce the radar cross section of targets to avoid their detection by an enemy radar (Section 1.7)
Stefan-Boltzmann law	expression stating that the total power per unit area emitted from a blackbody surface is equal to the *Stefan-Boltzmann constant* (σ) multiplied by the fourth power of the absolute temperature (K) (Section 3.3)
step-chirp	*waveform* such that the radar transmits identical groups of pulses, each consisting of N ($N >> 1$) monochromatic pulses; within a group, the frequency of a pulse is Δf greater than that of the previous pulse, and the radar transmits f_R/N groups/second; FJB with individual pulses monochromatic (Section 5.1, 7.2.1)
stereo	noncoherent comparison of two SAR images of the same scene, obtained from somewhat different locations, enabling estimation of object height. The technique is analogous to the method by which humans use two eyes to help estimate the distance of objects seen (Section 7.5.3.3)

straddling loss — term describing the fact that a target may not be in the center of a processing filter but rather at a noncentral location, where its received signal is less than optimum (Section 12.6.2)

stretch processing — SAR procedure involving dechirp when $2S/c < t_p$ (the usual case when dechirp is advantageous). The sampling time is "stretched" to $2S/c + t_p$ (S is the swath width, t_p the pulse width) (Section 7.2.1.2)

stripmap SAR (or "strip" SAR) — SAR mode such that the beam remains at a constant *squint angle* θ_{sq} to the perpendicular to the flight path (the latter is assumed to be a straight line) and continuously observes a *strip* of terrain parallel to the flight path; also called "search" SAR, since it is useful for imaging large areas at relatively coarse resolution (Section 7.1.1)

subaperture — a portion of a (usually flat) antenna aperture that collects the returned echo. In *monopulse,* complex echoes from two subapertures are typically combined to form sum and difference signals, which then are fed to separate receiver channels and separate A/Ds, for processing to determine precise angular direction of a target. In *STAP,* complex echoes from multiple subapertures are fed to separate receiver channels and processed to achieve suppression of clutter and/or jammer(s). (Section 5.0)

subpatch — relatively small portion of a SAR image that is processed separately, typically with the *polar format algorithm,* to avoid *range migration*; many subpatches are "stitched" together electronically to form the full image (may apply to either stripmap or spotlight processing) (Section 7.2.2)

sum beam — beam radiated when energy is transmitted from both *subapertures* of a *monopulse* antenna simultaneously, or receive beam (same shape) corresponding to addition of signals received simultaneously by the two subapertures (see *difference beam*) (Section 5.2)

super SVA — a bandwidth-extrapolation procedure for forming a fine-resolution image, developed by H. Stankwitz et al. of Veridian ERIM-International, which has been found experimentally to reduce the mainlobe width,

	and lower the sidelobes, of a point-target PSF. Super-SVA may also be used to interpolate over gaps in the bandwidth of a received signal. (Section 8.2.8)
superheterodyne	heterodyne process using a non-zero IF frequency (Section 2.2.4)
superresolution	a processing procedure that improves resolution beyond the limits of the DFT (Section 8.2)
surface clutter	radar returns from the Earth's surface (Section 3.2.7)
Swerling cases	four models for fluctuating targets, as defined by Peter Swerling (1954) (Section 4.1.4.1)
symmetric matrix	square matrix such that the transpose equals the matrix itself (Table 9.1)
synthetic-aperture radar	"a coherent radar system that generates a narrow cross range impulse response by signal processing (integrating) the amplitude and phase of the received signal over an angular rotation of the radar line of sight with respect to the object (target) illuminated. *Note:* Due to the change in line-of-sight direction, a synthetic aperture is produced by the signal processing that has the effect of an antenna with a much larger aperture (and hence a much greater angular resolution)." [1] (Section 5.2, Chapter 7)
system temperature (T_{sys})	temperature such that, when it is multiplied by Boltzmann's constant, yields noise power density (kT_s) (Section 1.9)
tail chase	term describing an aircraft that is "chasing" another aircraft ahead of it, which is moving in the same direction (Section 11.3.1)
tapered fully-adaptive STAP	*STAP* algorithm including a *tapering function,* producing slightly less *SINR* but lower sidelobes compared with optimum fully adaptive STAP (Section 12.6.1)
tapering function	(also *weighting function*) function describing the illumination function of an aperture antenna (more generally, a function multiplying a signal prior to a Fourier transform). The radiation pattern (more generally, the Fourier transform) using the tapered illumination function typically results in a broader mainlobe and lower sidelobes than the radiation pattern (Fourier

	transform) for the uniform illumination function. (Section 2.3.4)
target	"(A) Specifically, an object of radar search or tracking. (B) Broadly, any discrete object that scatters energy back to the radar." [1] (Section 1.7)
target steering vector	in *STAP*, the $MN \times 1$ vector representing the complex target signal received from the M pulses at the N subapertures (Section 12.2)
terrain delimitation	ATR step that compares a SAR image with a priori knowledge about the terrain, and eliminates all ROIs that occur in regions where targets of interest (e.g., tanks) could not plausibly be located (e.g., lakes or high mountains) (Section 9.4)
test data	data used to test an algorithm; it should be distinct from the *training data* (Section 9.4, 12.3)
test statistic	positive scalar compared with a threshold, to decide whether or not to declare the presence of a target (Section 4.1.2, 12.3)
theta-phi (θ-ϕ) coordinates	coordinate system generally preferred by theorists in describing a radiation pattern. The polar axis (z) of the coordinate system is perpendicular to the aperture plane. See *azimuth-elevation coordinates.* (Section 2.3.2)
three-dB width	for a *point-spread function (PSF)* or antenna *radiation pattern,* represented by a sinc or similar function, the interval of the independent variable between the half-power points (see *peak-to-first-null value*); known also as "half-power width," and in the optics community as "full-width half maximum (FWHM)" ([8], p. 37) (Section 2.3.2, 4.2.6)
three-dimensional (3D) ISAR	*ISAR* resulting in a 3D target image, from collection of data over intervals of both azimuth and elevation in target-centered coordinates (Section 5.7)
three-dimensional (3D) search	search performed in two angular dimensions (azimuth and elevation), the third dimension being range (same as *volume search*) (Section 1.19)
threshold	"a value of voltage or other measure that a signal must exceed in order to be detected or retained for further processing." [1] (Section 4.1.2)

thumbtack ambiguity function	an "ideal" ambiguity function, unequal to zero only near (0,0) (the integral must be unity) (Section 4.2.7.3)
time-bandwidth product	for a pulse, the product of the pulse width and signal bandwidth; ~1 for a monochromatic pulse, >>1 for a frequency-modulated pulse (Section 4.2.6)
time-delay shifter	microwave component that, given an input EM wave, produces an output EM wave that has a time delay relative to the input wave; the time delay is independent of frequency, thus distinguishing this component from a *phase shifter,* which produces a phase shift independent of frequency (Section 2.4.2)
timeline	a sequence of radar *modes* (Section 11.3.3)
total scattering cross section (meter^2)	the integral of *differential scattering cross section* over 4π steradians; if one multiplies the total scattering cross section times the incident power per unit area (w/meter^2), the result is the power scattered away from the original plane-wave beam (Section 14.1.2)
tracking	"the process of following a moving object or a variable input quantity. In radar, target tracking in angle, range, or Doppler frequency is accomplished by keeping a beam or angle cursor on the target angle, a range mark or gate on the delayed echo, or a narrowband filter on the signal frequency, respectively." [1] (Section 1.19)
trailing edge	of a pulse, the last part of a pulse to be radiated (Section 1.12).
training data	data used to develop an algorithm (see *test data*) (Section 9.4, 12.3)
training strategy	procedure used to decide what data to use for *training data* (Section 12.3)
transmit-receive (T/R) module	"an active TR electronic module, usually with integrated circuits, consisting of an antenna (or direct connection thereto), transmitter, receiver, duplexer, phase shifters, and power conditioner employed at the radiating elements of a phased array radar." [1] (Section 2.4)
transmitted power (P_{trans})	power accepted by the antenna from the radar hardware (watts) (Section 1.6)

transmitter	high-power amplifier that amplifies an input signal (typically milliwatts) to produce an output signal (typically watts, kilowatts, or megawatts), which is then sent to the antenna for radiation into the environment (Section 2.2.5)
transverse-electric (TE)	waveguide mode in which the electric field is purely transverse (no longitudinal component) (Section 2.2.6)
transverse-magnetic (TM)	waveguide mode in which the magnetic field is purely transverse (no longitudinal component) (Section 2.2.6)
traveling-wave tube (TWT)	a high-power amplifier used in transmitters, widely employed because it is both coherent and relatively wideband (Section 2.2.5)
trihedral corner reflector (or simply trihedral)	simple device used extensively for calibrating radars, consisting of three mutually perpendicular sheets of metal fastened together in a trihedral angle, forming a retroreflector. The individual sheets may be triangular (triangular trihedral), square (square trihedral), or in the shape of a quarter-circle (circular trihedral). (Section 3.1.1.3)
two-dimensional (2D) search	search performed in one angular dimension (usually azimuth), the two dimensions being angle and range (Section 1.19)
ultrasound	*sonar* system that utilizes relatively high *frequency* sound waves (MHz rather than kHz) (Section 1.1)
ultra-wideband SAR	SAR with relatively high *fractional bandwidth,* to obtain fine resolution at relatively low carrier frequency; typically needed for airborne *foliage-penetration* (Section 7.5.5)
unambiguous range interval	for a waveform characterized by a single PRF, f_R, a range interval of $c/(2f_R)$—the maximum interval in which target range may be measured unambiguously (Section 1.12)
unambiguous velocity interval	for a waveform characterized by a single PRF, f_R, a velocity interval of $f_R\lambda/2$—the maximum interval in which target velocity may be measured unambiguously (Section 1.16)

unfocused SAR (USAR)	SAR mode that involves a short synthetic aperture with a maximum two-way phase shift across the aperture of $\pi/2$; the returned pulses are processed coherently, and crossrange resolution is approximately $1/2\ (\lambda R)^{-1/2}$ (Section 7.1)
uniform-odd illumination	term referring to weighting function applied to a *monopulse difference beam*; each subaperture has uniform illumination, and the difference operation causes the phase of the weighting function to be opposite for one subaperture compared with the other (see *linear-odd*) (Section 5.2.2)
unitary matrix	square matrix such that the conjugate transpose equals the inverse (Table 9.1)
upconversion	mixing that results in a signal with higher carrier frequency (Section 2.2.4)
usable Doppler space fraction (UDSF)	in *STAP*, fraction of Doppler space such that SINR is acceptable (Section 12.6.5)
vertically polarized EM wave	EM wave with electric field vector essentially parallel to local gravitational field vector (Section 1.3)
video	"refers to the signal after envelope or phase detection, which in early radar was the displayed signal." [1] (Section 2.2.12)
video amplifier	amplifier following the mixing to baseband (video) (Section 2.2.14)
video integration	"a method of utilizing the redundancy of repetitive video signals to improve the output signal-to-noise ratio, by summing successive signals. Also called post-detection integration or noncoherent integration." [1] (Section 4.1.2)
video phase history (VPH)	the set of coherent digital voltage data collected by the radar, to be used for signal processing (may be I and Q data) (Section 2.2.14)
visible region	region of angle space, describing which *major lobes* appear in an *array pattern* (Section 2.4.1)
voltage image	SAR/ISAR image where only the pixel voltage magnitude $V(x,y)$ (not phase) is displayed (see *complex image, power image*) (Section 8.1.1)
volume clutter	radar returns from the Earth's atmosphere (Section 3.2.8)

volume search	3D search (Section 4.2.9)
wave front	a locus of constant phase of an EM wave (Section 1.3)
wave number	a vector with direction equal to that of the EM wave, and magnitude of $2\pi f/c$ (Section 1.3)
waveform	a sequence of similar pulses, usually at the same beam position; pulses may be modulated (Section 4.2, 11.3.3)
waveform generator	component that receives the analog waveform information, plus the IF from the COHO, and via a *mixer* produces a version of the transmitted waveform at IF and low power (also called *exciter*) (Section 2.2.2)
waveform information	information, typically provided by a control computer or resident software, containing the information about the signals the radar is to transmit, including the frequency, pulse width, PRF, start and stop times, and pulse characteristics (Section 2.2.2)
waveguide	a hollow metal pipe, with a cross section that is usually rectangular but may be circular or elliptical, through which microwave energy may pass in a controlled manner (Section 2.2.6)
wavelength	distance between adjacent crests of an EM wave (Section 1.3)
wavenumber space	2D or 3D coordinate system where the axes are the components of the wave number vector **k** (Section 5.8)
Weibull distribution	PDF of received signal voltage magnitude involving two parameters η and ν. For $\nu = 2$ the PDF is Rayleigh. The Weibull PDF is sometimes used to describe clutter. See *log-normal distribution.* (Section 4.1.7)
weighting function	*tapering function* (Section 2.3.4)
Wien region	for a Planck distribution, region where frequency >> peak frequency, and emitted radiation (w m^{-2} Hz^{-1}) ~ $\exp[-(hf/kT)]$ (see *Rayleigh-Jeans region*) (Section 3.3)
Wien's displacement law	expression stating that the wavelength corresponding to the peak of a blackbody frequency (or wavelength) distribution is inversely proportional to the absolute temperature (Section 3.3)

zero-padding — FT procedure for transforming from the time domain to produce interpolated (more finely sampled) data in the frequency domain (or vice versa). When M zeroes are added to the N time-domain samples, the frequency-domain function is sampled at $M + N$ locations over the interval N/T. (Section 8.1.3)

References

[1] *IEEE Standard Radar Definitions* (Std 686-1997, revision of IEEE Std 686-1990), New York: Institute of Electrical and Electronics Engineers, 1997.

[2] Kurpis, G. P., and C. J. Booth (Ed.), *The New IEEE Standard Dictionary of Electrical and Electronics Terms,* 5th Edition, New York: Institute of Electrical and Electronics Engineers, 1993.

[3] Barton, D. K., *Modern Radar System Analysis,* Norwood, MA: Artech House, 1988.

[4] Willis, N. J., *Bistatic Radar,* Norwood, MA: Artech House, 1991.

[5] Jakowatz, C. V., Jr., D. E. Wahl, P. H. Eichel, D. C. Ghiglia, and P. A. Thompson, *Spotlight-Mode SAR: A Signal-Processing Approach,* Boston: Kluwer Academic Publishers, 1996.

[6] Freedman, A., and N. Levanon, "Staggered Costas Signals," *IEEE Trans. Aerospace and Electronic Systems,* Vol. AES-22, No. 6, November 1986, pp. 695–701.

[7] Skolnik, M. I., *Introduction to Radar Systems,* 2nd Edition, New York: McGraw-Hill, 1980.

[8] Zissis, G. J. (ed.), "Sources of Radiation," Volume 1 of Accetta, J., and D. L. Shumaker, *The Infrared and Electro-Optical Systems Handbook,* Bellingham, WA: SPIE Optical Engineering Press, 1993.

[9] Levanon, N., *Radar Principles,* New York: Wiley-Interscience, 1988.

[10] Carrara, W. G., R. S. Goodman, and R. M. Majewski, *Spotlight Synthetic Aperture Radar,* Norwood, MA: Artech House, 1995.

About the Author

Dr. Roger J. Sullivan received his B.S. and Ph.D. degrees in physics from the Massachusetts Institute of Technology. At System Planning Corporation, Arlington, VA, he contributed to the development of a series of imaging instrumentation radars, which are operated on outdoor and indoor radar ranges. At the Environmental Research Institute of Michigan (now Veridian ERIM International), Ann Arbor, MI, he served as program manager for development of an X/C/L-band polarimetric synthetic aperture radar (SAR) carried by a U.S. Navy P-3 aircraft, and led many analyses concerning SAR performance and automatic target recognition. Currently, at the Institute for Defense Analyses, Alexandria, VA, he advises the U.S. government concerning SAR and ground-moving target indication (GMTI) radars on unmanned aerial vehicles (UAVs) and performs analyses concerning advanced radar concepts.

Index

Recent Titles in the Artech House Radar Library

David K. Barton, Series Editor

Airborne Pulsed Doppler Radar, Second Edition, Guy V. Morris and Linda Harkness, editors

Bayesian Multiple Target Tracking, Lawerence D. Stone, Carl A. Barlow, and Thomas L. Corwin

Design and Analysis of Modern Tracking Systems, Samuel Blackman and Robert Popoli

Digital Techniques for Wideband Receivers, James Tsui

Electronic Intelligence: The Analysis of Radar Signals, Second Edition, Richard G. Wiley

Electronic Warfare in the Information Age, D. Curtis Schleher

High-Resolution Radar, Second Edition, Donald R. Wehner

Introduction to Electronic Warfare, D. Curtis Schleher

Introduction to Multisensor Data Fusion: Multimedia Software and User's Guide, TECH REACH, Inc.

Microwave Radar: Imaging and Advanced Concepts, Roger J. Sullivan

Millimeter-Wave and Infared Multisensor Design and Signal Processing, Lawrence A. Klein

Modern Radar System Analysis, David K. Barton

Modern Radar System Analysis Software and User's Manual, David K. Barton and William F. Barton

Principles of High-Resolution Radar, August W. Rihaczek

Radar Cross Section, Second Edition, Eugene F. Knott, et al.

Radar Evalution Handbook, David K. Barton, et al.

Radar Meteorolgy, Henri Sauvageot

Radar Signal Processing and Adaptive Systems, Ramon Nitzberg

Radar Technology Encyclopedia, David K. Barton and Sergey A. Leonov, editors

For further information on these and other Artech House titles, including previously considered out-of-print books now available through our In-Print-Forever® (IPF®) program, contact:

Artech House
685 Canton Street
Norwood, MA 02062
Phone: 781-769-9750
Fax: 781-769-6334
e-mail: artech@artechhouse.com

Artech House
46 Gillingham Street
London SW1V 1AH UK
Phone: +44 (0)20 7596-8750
Fax: +44 (0)20 7630-0166
e-mail: artech-uk@artechhouse.com

Find us on the World Wide Web at:
www.artechhouse.com